Biomolecular Engineering: Structures and Functions

Biomolecular Engineering: Structures and Functions

Edited by
Francis Joyce

Larsen & Keller
www.larsen-keller.com

Biomolecular Engineering: Structures and Functions
Edited by Francis Joyce
ISBN: 978-1-63549-649-9 (Hardback)

Larsen & Keller
Published by Larsen and Keller Education,
5 Penn Plaza,
19th Floor,
New York, NY 10001, USA

Cataloging-in-Publication Data

Biomolecular engineering : structures and functions / edited by Francis Joyce.
p. cm.
Includes bibliographical references and index.
ISBN 978-1-63549-649-9
1. Biomolecules. 2. Genetic engineering. 3. Biochemical engineering. 4. Molecular genetics.
I. Joyce, Francis.
QH506 .B56 2018
572.8--dc23

Printed and bound in China.

For more information regarding Larsen and Keller Education and its products, please visit the publisher's website www.larsen-keller.com

Table of Contents

Preface

As a branch of engineering, biomolecular engineering is the process of manipulating the molecules by using the elements of engineering and making them more useful. It is also used to find solutions for the problems related to the fields of food production, biotechnology, energy, medicine, with the help of biological processes and chemical engineering and creating molecular level solutions. This book presents the complex subject of biomolecular engineering and its applications in the most comprehensible and easy to understand language. Some of the diverse topics covered in it address the varied branches that fall under this category. For all those who are interested in biomolecular engineering, this textbook can prove to be an essential guide.

A short introduction to every chapter is written below to provide an overview of the content of the book:

Chapter 1 - Cell is the fundamental structural and functional unit of any living organism. It can be classified into prokaryotic and eukaryotic. While prokaryotes are 0.5 micrometre in length and lack membrane-bound nucleus and organelles, eukaryotes can have a length of upto 40 micrometres and possess membrane-bound nucleus and organelles. This chapter will provide an integrated understanding of biomolecular engineering; **Chapter 2 -** Cellular transport can be defined as the movement of molecules or compounds across the cell membrane. This is done to maintain homeostasis, to take in or release compounds or to allow cells to regulate the characteristics of transport. Cellular transport can be categorized into passive or active transport. The section serves as a source to understand the major categories related to cellular transport; **Chapter 3 -** A carbohydrate is a molecule comprising carbon, oxygen and hydrogen atoms. It is also referred to as saccharides and can be classified as monosaccharide, disaccharide, oligosaccharide, and polysaccharide. This chapter discusses carbohydrates in a critical manner providing key analysis to the subject matter; **Chapter 4 -** Amino acids are the building blocks of proteins comprising amine, carboxyl and a side chain. Out of the many functions of proteins, a few are that it acts as a biocatalyst, helps in transportation of cells, provides mechanical support, and are antibodies to the immune system. The topics discussed in the section are of great importance to broaden the existing knowledge on biomolecular engineering; **Chapter 5 -** Among the four major biomolecules, one of them is lipids. It is insoluble in a water solution but soluble in organic solvents. It has many important biological functions. It acts as the energy source in animals, humans, birds, etc. It activates important mitochondrial enzymes. It is a major component of electron systems in mitochondria. The aspects elucidated in this chapter are of vital importance, and provide a better understanding of biomolecular engineering; **Chapter 6 -** Nucleic acids store DNA and RNA in them, which are the basic building-blocks of life. DNA is responsible for heredity. The structure of every protein is due to the information programmed in the nucleotide sequence of DNA. Nucleic acid regulates protein synthesis, and replicates itself during cell division. This section has been carefully written to provide an easy understanding of the varied facets of nucleic acids; **Chapter 7 -** According to central dogma, the hereditary information is stored in the DNA nucleus. The sequence in gene expression runs from DNA to RNA to protein. Therefore, the central dogma comprises of three important biomolecules, DNA, RNA, and protein. The chapter strategically encompasses and incorporates the major components and key concepts biomolecular engineering, providing a complete understanding.

Finally, I would like to thank my fellow scholars who gave constructive feedback and my family members who supported me at every step.

Editor

1

Structure and Function of Cell

Cell is the fundamental structural and functional unit of any living organism. It can be classified into prokaryotic and eukaryotic. While prokaryotes are 0.5 micrometre in length and lack membrane-bound nucleus and organelles, eukaryotes can have a length of upto 40 micrometres and possess membrane-bound nucleus and organelles. This chapter will provide an integrated understanding of biomolecular engineering.

Cell

- A Cell is the basic structural and functional unit of all living organisms.
- Types of cells:
  > Prokaryotic : Bacteria and Archaea (Sometimes Archaea are preferably classified altogether as a different class and not under Prokaryotes).

  > Eukaryotic : Unicellular (protists) and Multicellular (Fungi, plants and animals).

Cell (Biology)

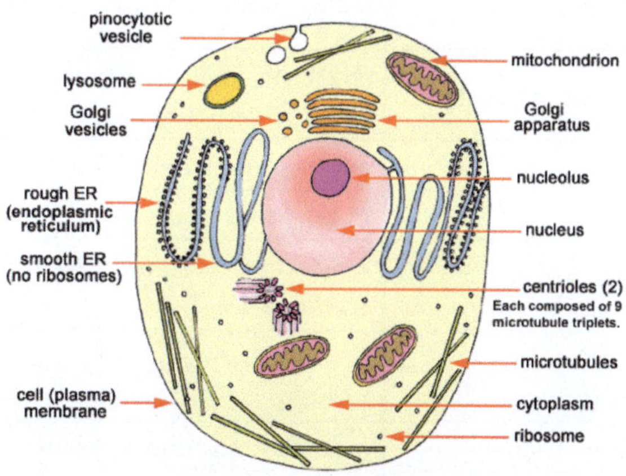

Structure of an animal cell

The cell (from Latin *cella*, meaning "small room") is the basic structural, functional, and biological unit of all known living organisms. A cell is the smallest unit of life that can replicate independently, and cells are often called the "building blocks of life". The study of cells is called cell biology.

Cells consist of cytoplasm enclosed within a membrane, which contains many biomolecules such as proteins and nucleic acids. Organisms can be classified as unicellular (consisting of a single cell; including bacteria) or multicellular (including plants and animals). While the number of cells in plants and animals varies from species to species, humans contain more than 10 trillion (10^{12}) cells. Most plant and animal cells are visible only under a microscope, with dimensions between 1 and 100 micrometres.

The cell was discovered by Robert Hooke in 1665, who named the biological unit for its resemblance to cells inhabited by Christian monks in a monastery. Cell theory, first developed in 1839 by Matthias Jakob Schleiden and Theodor Schwann, states that all organisms are composed of one or more cells, that cells are the fundamental unit of structure and function in all living organisms, that all cells come from preexisting cells, and that all cells contain the hereditary information necessary for regulating cell functions and for transmitting information to the next generation of cells. Cells emerged on Earth at least 3.5 billion years ago.

Anatomy

Comparison of features of prokaryotic and eukaryotic cells		
	Prokaryotes	**Eukaryotes**
Typical organisms	bacteria, archaea	protists, fungi, plants, animals
Typical size	~ 1–5 µm	~ 10–100 µm
Type of nucleus	nucleoid region; no true nucleus	true nucleus with double membrane
DNA	circular (usually)	linear molecules (chromosomes) with histone proteins
RNA/protein synthesis	coupled in the cytoplasm	RNA synthesis in the nucleus protein synthesis in the cytoplasm
Ribosomes	50S and 30S	60S and 40S
Cytoplasmic structure	very few structures	highly structured by endomembranes and a cytoskeleton
Cell movement	flagella made of flagellin	flagella and cilia containing microtubules; lamellipodia and filopodia containing actin
Mitochondria	none	one to several thousand
Chloroplasts	none	in algae and plants
Organization	usually single cells	single cells, colonies, higher multicellular organisms with specialized cells
Cell division	binary fission (simple division)	mitosis (fission or budding) meiosis
Chromosomes	single chromosome	more than one chromosome
Membranes	cell membrane	Cell membrane and membrane-bound organelles

Cells are of two types, eukaryotic, which contain a nucleus, and prokaryotic, which do not. Prokaryotes are single-celled organisms, while eukaryotes can be either single-celled or multicellular.

Prokaryotic Cells

Prokaryotic cells were the first form of life on Earth, characterised by having vital biological processes including cell signaling and being self-sustaining. They are simpler and smaller than eukaryotic cells, and lack membrane-bound organelles such as the nucleus. Prokaryotes include two of the domains of life, bacteria and archaea. The DNA of a prokaryotic cell consists of a single chromosome that is in direct contact with the cytoplasm. The nuclear region in the cytoplasm is called the nucleoid. Most prokaryotes are the smallest of all organisms ranging from 0.5 to 2.0 μm in diameter.

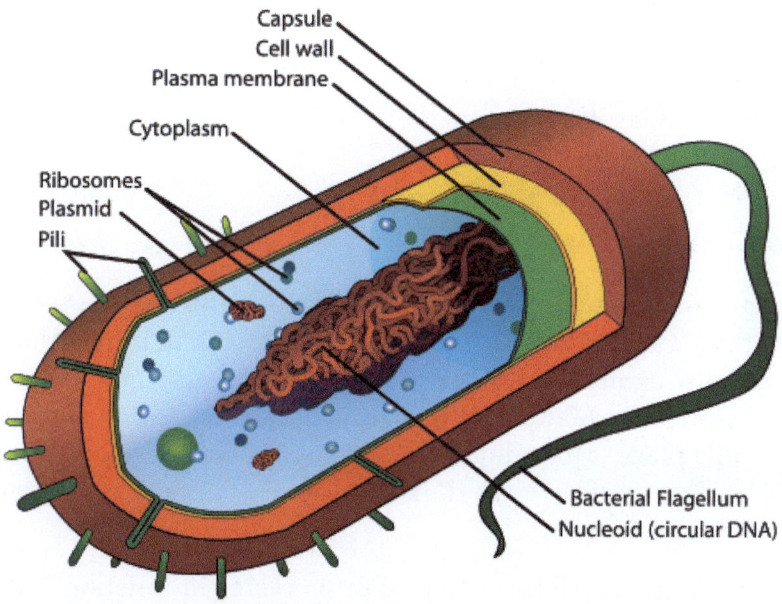

Structure of a typical prokaryotic cell

A prokaryotic cell has three architectural regions:

- Enclosing the cell is the cell envelope – generally consisting of a plasma membrane covered by a cell wall which, for some bacteria, may be further covered by a third layer called a capsule. Though most prokaryotes have both a cell membrane and a cell wall, there are exceptions such as *Mycoplasma* (bacteria) and *Thermoplasma* (archaea) which only possess the cell membrane layer. The envelope gives rigidity to the cell and separates the interior of the cell from its environment, serving as a protective filter. The cell wall consists of peptidoglycan in bacteria, and acts as an additional barrier against exterior forces. It also prevents the cell from expanding and bursting (cytolysis) from osmotic pressure due to a hypotonic environment. Some eukaryotic cells (plant cells and fungal cells) also have a cell wall.

- Inside the cell is the cytoplasmic region that contains the genome (DNA), ribosomes and various sorts of inclusions. The genetic material is freely found in the cytoplasm. Prokaryotes can carry extrachromosomal DNA elements called plasmids, which are usually circular. Linear bacterial plasmids have been identified in several species of spirochete bacteria, including members of the genus Borrelia notably *Borrelia burgdorferi*, which causes Lyme disease. Though not forming a *nucleus*, the DNA is condensed in a *nucleoid*. Plasmids encode additional genes, such as antibiotic resistance genes.

- On the outside, flagella and pili project from the cell's surface. These are structures (not present in all prokaryotes) made of proteins that facilitate movement and communication between cells.

Eukaryotic Cells

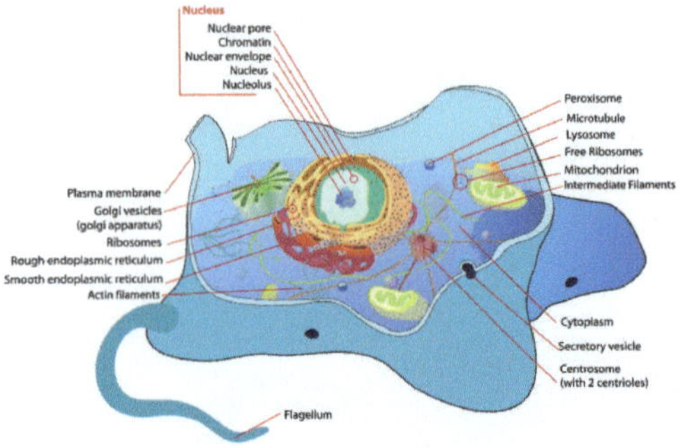

Structure of a typical animal cell

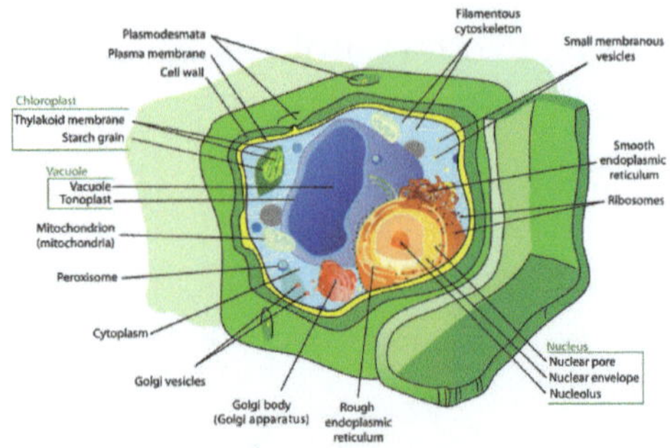

Structure of a typical plant cell

Plants, animals, fungi, slime moulds, protozoa, and algae are all eukaryotic. These cells are about fifteen times wider than a typical prokaryote and can be as much as a thousand times greater in volume. The main distinguishing feature of eukaryotes as compared to prokaryotes is compartmentalization: the presence of membrane-bound organelles (compartments) in which specific metabolic activities take place. Most important among these is a cell nucleus, an organelle that houses the cell's DNA. This nucleus gives the eukaryote its name, which means "true kernel (nucleus)". Other differences include:

- The plasma membrane resembles that of prokaryotes in function, with minor differences in the setup. Cell walls may or may not be present.
- The eukaryotic DNA is organized in one or more linear molecules, called chromosomes, which are associated with histone proteins. All chromosomal DNA is stored in the *cell nucleus*, separated from the cytoplasm by a membrane. Some eukaryotic organelles such as mitochondria also contain some DNA.
- Many eukaryotic cells are ciliated with *primary cilia*. Primary cilia play important roles in chemosensation, mechanosensation, and thermosensation. Cilia may thus be "viewed as a sensory cellular antennae that coordinates a large number of cellular signaling pathways, sometimes coupling the signaling to ciliary motility or alternatively to cell division and differentiation."
- Motile cells of eukaryotes can move using *motile cilia* or *flagella*. Motile cells are absent in conifers and flowering plants. Eukaryotic flagella are less complex than those of prokaryotes.

Subcellular Components

All cells, whether prokaryotic or eukaryotic, have a membrane that envelops the cell, regulates what moves in and out (selectively permeable), and maintains the electric potential of the cell. Inside the membrane, the cytoplasm takes up most of the cell's volume. All cells (except red blood cells which lack a cell nucleus and most organelles to accommodate maximum space for hemoglobin) possess DNA, the hereditary material of genes, and RNA, containing the information necessary to build various proteins such as enzymes, the cell's primary machinery. There are also other kinds of biomolecules in cells. This article lists these primary components of the cell, then briefly describes their function.

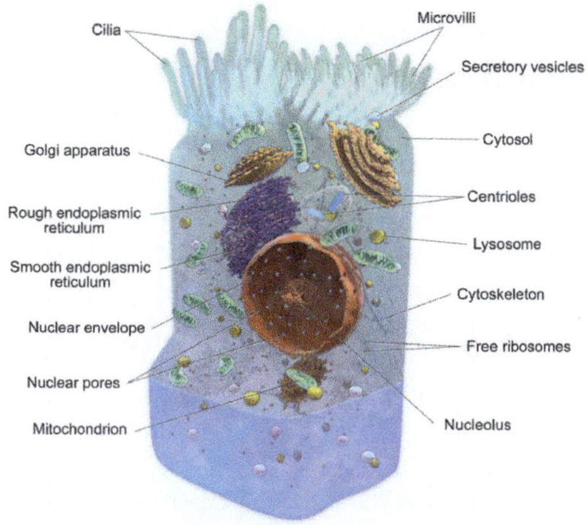

Anatomy of a Cell

Illustration depicting major structures inside a eukaryotic animal cell

Membrane

The cell membrane, or plasma membrane, is a biological membrane that surrounds the cytoplasm of a cell. In animals, the plasma membrane is the outer boundary of the cell, while in plants and prokaryotes it is usually covered by a cell wall. This membrane serves to separate and protect a cell from its surrounding environment and is made mostly from a double layer of phospholipids, which are amphiphilic (partly hydrophobic and partly hydrophilic). Hence, the layer is called a phospholipid bilayer, or sometimes a fluid mosaic membrane. Embedded within this membrane is a variety of protein molecules that act as channels and pumps that move different molecules into and out of the cell. The membrane is said to be 'semi-permeable', in that it can either let a substance (molecule or ion) pass through freely, pass through to a limited extent or not pass through at all. Cell surface membranes also contain receptor proteins that allow cells to detect external signaling molecules such as hormones.

Cytoskeleton

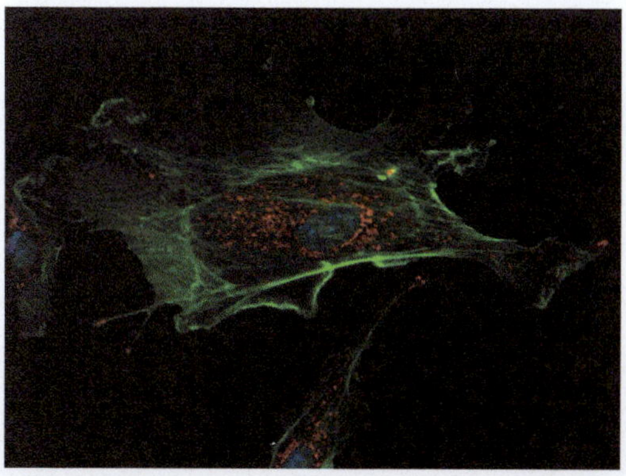

A fluorescent image of an endothelial cell. Nuclei are stained blue, mitochondria are stained red, and microfilaments are stained green.

The cytoskeleton acts to organize and maintain the cell's shape; anchors organelles in place; helps during endocytosis, the uptake of external materials by a cell, and cytokinesis, the separation of daughter cells after cell division; and moves parts of the cell in processes of growth and mobility. The eukaryotic cytoskeleton is composed of microfilaments, intermediate filaments and microtubules. There are a great number of proteins associated with them, each controlling a cell's structure by directing, bundling, and aligning filaments. The prokaryotic cytoskeleton is less well-studied but is involved in the maintenance of cell shape, polarity and cytokinesis. The subunit protein of microfilaments is a small, monomeric protein called actin. The subunit of microtubules is a dimeric molecule called tubulin. Intermediate filaments are heteropolymers whose subunits vary among the cell types in different tissues. But some of the subunit protein of intermediate filaments include vimentin, desmin, lamin (lamins A, B and C), keratin (multiple acidic and basic keratins), neurofilament proteins (NF - L, NF - M).

Genetic Material

Two different kinds of genetic material exist: deoxyribonucleic acid (DNA) and ribonucleic acid (RNA). Cells use DNA for their long-term information storage. The biological information contained in an organism is encoded in its DNA sequence. RNA is used for information transport (e.g., mRNA) and enzymatic functions (e.g., ribosomal RNA). Transfer RNA (tRNA) molecules are used to add amino acids during protein translation.

Prokaryotic genetic material is organized in a simple circular DNA molecule (the bacterial chromosome) in the nucleoid region of the cytoplasm. Eukaryotic genetic material is divided into different, linear molecules called chromosomes inside a discrete nucleus, usually with additional genetic material in some organelles like mitochondria and chloroplasts.

A human cell has genetic material contained in the cell nucleus (the nuclear genome) and in the mitochondria (the mitochondrial genome). In humans the nuclear genome is divided into 46 linear DNA molecules called chromosomes, including 22 homologous chromosome pairs and a pair of sex chromosomes. The mitochondrial genome is a circular DNA molecule distinct from the nuclear DNA. Although the mitochondrial DNA is very small compared to nuclear chromosomes, it codes for 13 proteins involved in mitochondrial energy production and specific tRNAs.

Foreign genetic material (most commonly DNA) can also be artificially introduced into the cell by a process called transfection. This can be transient, if the DNA is not inserted into the cell's genome, or stable, if it is. Certain viruses also insert their genetic material into the genome.

Organelles

Organelles are parts of the cell which are adapted and/or specialized for carrying out one or more vital functions, analogous to the organs of the human body (such as the heart, lung, and kidney, with each organ performing a different function). Both eukaryotic and prokaryotic cells have organelles, but prokaryotic organelles are generally simpler and are not membrane-bound.

There are several types of organelles in a cell. Some (such as the nucleus and golgi apparatus) are typically solitary, while others (such as mitochondria, chloroplasts, peroxisomes and lysosomes) can be numerous (hundreds to thousands). The cytosol is the gelatinous fluid that fills the cell and surrounds the organelles.

Eukaryotic

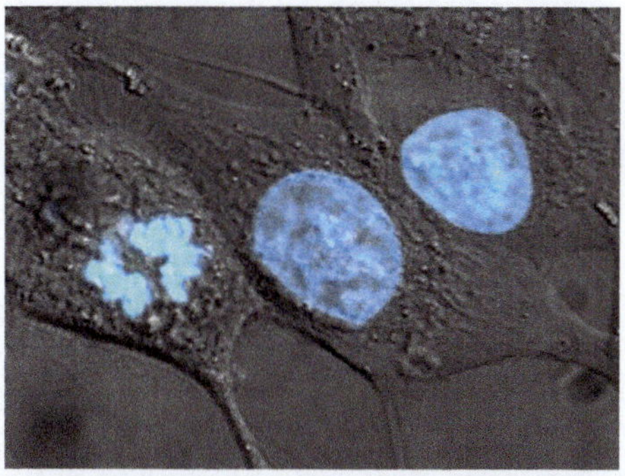

Human cancer cells with nuclei (specifically the DNA) stained blue. The central and rightmost cell are in interphase, so the entire nuclei are labeled. The cell on the left is going through mitosis and its DNA has condensed.

- Cell nucleus: A cell's information center, the cell nucleus is the most conspicuous organelle found in a eukaryotic cell. It houses the cell's chromosomes, and is the place where almost all DNA replication and RNA synthesis (transcription) occur. The nucleus is spherical and separated from the cytoplasm by a double membrane called the nuclear envelope. The nuclear envelope isolates and protects a cell's DNA from various molecules that could accidentally damage its structure or interfere with its processing. During processing, DNA is transcribed, or copied into a special RNA, called messenger RNA (mRNA). This mRNA is then transported out of the nucleus, where it is translated into a specific protein molecule. The nucleolus is a specialized region within the nucleus where ribosome subunits are assembled. In prokaryotes, DNA processing takes place in the cytoplasm.
- Mitochondria and Chloroplasts: generate energy for the cell. Mitochondria are self-replicating organelles that occur in various numbers, shapes, and sizes in the cytoplasm of all eukaryotic cells. Respiration occurs in the cell mitochondria, which generate the cell's energy by oxidative phosphorylation, using oxygen to release energy stored in cellular nutrients (typically pertaining to glucose) to generate ATP. Mitochondria multiply by binary fission, like prokaryotes. Chloroplasts can only be found in plants and algae, and they capture the sun's energy to make carbohydrates through photosynthesis.

Diagram of an endomembrane system

- Endoplasmic reticulum: The endoplasmic reticulum (ER) is a transport network for molecules targeted for certain modifications and specific destinations, as compared to molecules that float freely in the cytoplasm. The ER has two forms: the rough ER, which has ribosomes on its surface that secrete proteins into the ER, and the smooth ER, which lacks ribosomes. The smooth ER plays a role in calcium sequestration and release.
- Golgi apparatus: The primary function of the Golgi apparatus is to process and package the macromolecules such as proteins and lipids that are synthesized by the cell.
- Lysosomes and Peroxisomes: Lysosomes contain digestive enzymes (acid hydrolases). They digest excess or worn-out organelles, food particles, and engulfed viruses or bacteria. Peroxisomes have enzymes that rid the cell of toxic peroxides. The cell could not house these destructive enzymes if they were not contained in a membrane-bound system.
- Centrosome: the cytoskeleton organiser: The centrosome produces the microtubules of a cell – a key component of the cytoskeleton. It directs the transport through the ER and the Golgi apparatus. Centrosomes are composed of two centrioles, which separate during cell division and help in the formation of the mitotic spindle. A single centrosome is present in the animal cells. They are also found in some fungi and algae cells.
- Vacuoles: Vacuoles sequester waste products and in plant cells store water. They are often described as liquid filled space and are surrounded by a membrane. Some cells, most notably *Amoeba*, have contractile vacuoles, which can pump water out of the cell if there is too much water. The vacuoles of plant cells and fungal cells are usually larger than those of animal cells.

Eukaryotic and Prokaryotic

- Ribosomes: The ribosome is a large complex of RNA and protein molecules. They each consist of two subunits, and act as an assembly line where RNA from the nucleus is used to synthesise proteins from amino acids. Ribosomes can be found either floating freely or bound to a membrane (the rough endoplasmatic reticulum in eukaryotes, or the cell membrane in prokaryotes).

Structures Outside the Cell Membrane

Many cells also have structures which exist wholly or partially outside the cell membrane. These structures are notable because they are not protected from the external environment by the semipermeable cell membrane. In order to assemble these structures, their components must be carried across the cell membrane by export processes.

Cell Wall

Many types of prokaryotic and eukaryotic cells have a cell wall. The cell wall acts to protect the cell mechanically and chemically from its environment, and is an additional layer of protection to the cell membrane. Different types of cell have cell walls made up of different materials; plant cell walls are primarily made up of cellulose, fungi cell walls are made up of chitin and bacteria cell walls are made up of peptidoglycan.

Prokaryotic

Capsule

A gelatinous capsule is present in some bacteria outside the cell membrane and cell wall. The capsule may be polysaccharide as in pneumococci, meningococci or polypeptide as *Bacillus anthracis* or hyaluronic acid as in streptococci. Capsules are not marked by normal staining protocols and can be detected by India ink or methyl blue; which allows for higher contrast between the cells for observation.

Flagella

Flagella are organelles for cellular mobility. The bacterial flagellum stretches from cytoplasm through the cell membrane(s) and extrudes through the cell wall. They are long and thick thread-like appendages, protein in nature. A different type of flagellum is found in archaea and a different type is found in eukaryotes.

Fimbria

A fimbria also known as a pilus is a short, thin, hair-like filament found on the surface of bacteria. Fimbriae, or pili are formed of a protein called pilin (antigenic) and are responsible for attachment of bacteria to specific receptors of human cell (cell adhesion). There are special types of specific pili involved in bacterial conjugation.

Cellular Processes

Growth and Metabolism

Between successive cell divisions, cells grow through the functioning of cellular metabolism. Cell metabolism is the process by which individual cells process nutrient molecules. Metabolism has two distinct divisions: catabolism, in which the cell breaks down complex molecules to produce energy and reducing power, and anabolism, in which the cell uses energy and reducing power to construct complex molecules and perform other biological functions. Complex sugars consumed by the organism can be broken down into simpler sugar molecules called monosaccharides such as glucose. Once inside the cell, glucose is broken down to make adenosine triphosphate (ATP), a molecule that possesses readily available energy, through two different pathways.

Replication

Cell division involves a single cell (called a *mother cell*) dividing into two daughter cells. This leads to growth in multicellular organisms (the growth of tissue) and to procreation (vegetative reproduction) in unicellular organisms. Prokaryotic cells divide by binary fission, while eukaryotic cells usually undergo a process of nuclear division, called mitosis, followed by division of the cell, called cytokinesis. A diploid cell may also undergo meiosis to produce haploid cells, usually four. Haploid cells serve as gametes in multicellular organisms, fusing to form new diploid cells.

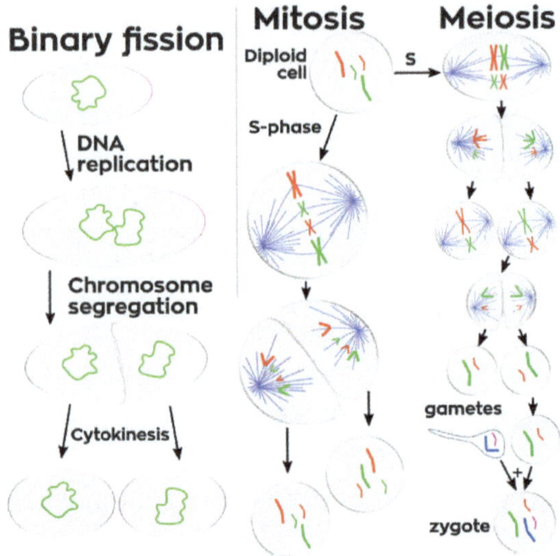

Bacteria divide by binary fission, while eukaryotes divide by mitosis or meiosis.

DNA replication, or the process of duplicating a cell's genome, always happens when a cell divides through mitosis or binary fission. This occurs during the S phase of the cell cycle.

In meiosis, the DNA is replicated only once, while the cell divides twice. DNA replication only occurs before meiosis I. DNA replication does not occur when the cells divide the second time, in meiosis II. Replication, like all cellular activities, requires specialized proteins for carrying out the job.

Protein Synthesis

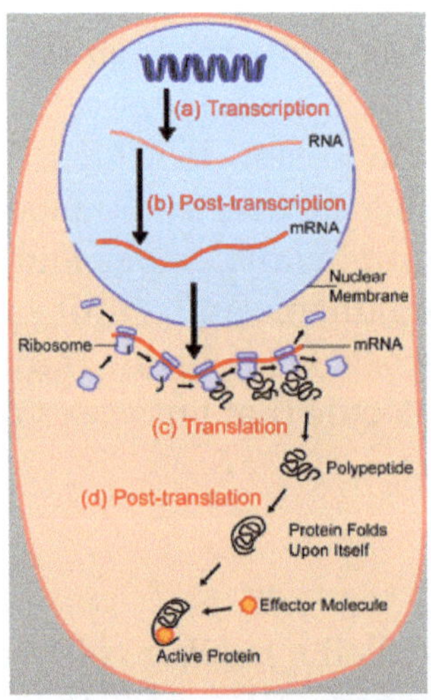

An overview of protein synthesis.
Within the nucleus of the cell (*light blue*), genes (DNA, *dark blue*) are transcribed into RNA. This RNA is then subject to post-transcriptional modification and control, resulting in a mature mRNA (*red*) that is then transported out of the nucleus and into the cytoplasm (*peach*), where it undergoes translation into a protein. mRNA is translated by ribosomes (*purple*) that match the three-base codons of the mRNA to the three-base anti-codons of the appropriate tRNA. Newly synthesized proteins (*black*) are often further modified, such as by binding to an effector molecule (*orange*), to become fully active.

Cells are capable of synthesizing new proteins, which are essential for the modulation and maintenance of cellular activities. This process involves the formation of new protein molecules from amino acid building blocks based on information encoded in DNA/RNA. Protein synthesis generally consists of two major steps: transcription and translation.

Transcription is the process where genetic information in DNA is used to produce a complementary RNA strand. This RNA strand is then processed to give messenger RNA (mRNA), which is free to migrate through the cell. mRNA molecules bind to protein-RNA complexes called ribosomes located in the cytosol, where they are translated into polypeptide sequences. The ribosome mediates the formation of a polypeptide sequence based on the mRNA sequence. The mRNA sequence directly relates to the polypeptide sequence by binding to transfer RNA (tRNA) adapter molecules in binding pockets within the ribosome. The new polypeptide then folds into a functional three-dimensional protein molecule.

Movement or Motility

Unicellular organisms can move in order to find food or escape predators. Common mechanisms of motion include flagella and cilia.

In multicellular organisms, cells can move during processes such as wound healing, the immune response and cancer metastasis. For example, in wound healing in animals, white blood cells move to the wound site to kill the microorganisms that cause infection. Cell motility involves many receptors, crosslinking, bundling, binding, adhesion, motor and other proteins. The process is divided into three steps – protrusion of the leading edge of the cell, adhesion of the leading edge and de-adhesion at the cell body and rear, and cytoskeletal contraction to pull the cell forward. Each step is driven by physical forces generated by unique segments of the cytoskeleton.

Cell Specialization

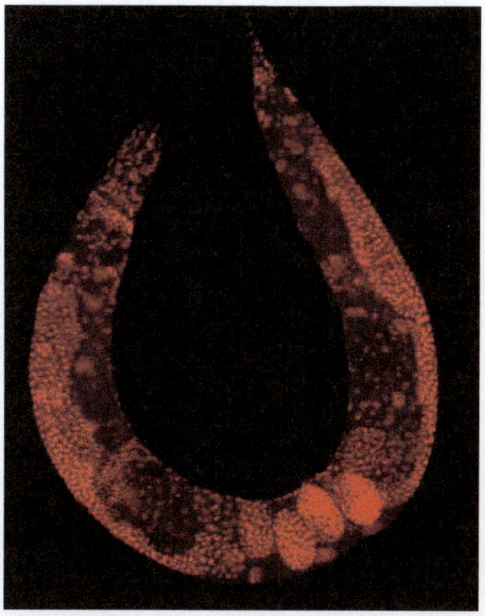

Staining of a *Caenorhabditis elegans* which highlights the nuclei of its cells.

Multicellular organisms are organisms that consist of more than one cell, in contrast to single-celled organisms.

In complex multicellular organisms, cells specialize into different cell types that are adapted to particular functions. In mammals, major cell types include skin cells, muscle cells, neurons, blood cells, fibroblasts, stem cells, and others. Cell types differ both in appearance and function, yet are genetically identical. Cells are able to be of the same genotype but of different cell type due to the differential expression of the genes they contain.

Most distinct cell types arise from a single totipotent cell, called a zygote, that differentiates into hundreds of different cell types during the course of development. Differentiation of cells is driven by different environmental cues (such as cell–cell interaction) and intrinsic differences (such as those caused by the uneven distribution of molecules during division).

Origin of Multicellularity

Multicellularity has evolved independently at least 25 times, including in some prokaryotes, like cyanobacteria, myxobacteria, actinomycetes, *Magnetoglobus multicellularis* or *Methanosarcina*. However, complex multicellular organisms evolved only in six eukaryotic groups: animals, fungi, brown algae, red algae, green algae, and plants. It evolved repeatedly for plants (Chloroplastida), once or twice for animals, once for brown algae, and perhaps several times for fungi, slime molds, and red algae. Multicellularity may have evolved from colonies of interdependent organisms, from cellularization, or from organisms in symbiotic relationships.

The first evidence of multicellularity is from cyanobacteria-like organisms that lived between 3 and 3.5 billion years ago. Other early fossils of multicellular organisms include the contested Grypania spiralis and the fossils of the black shales of the Palaeoproterozoic Francevillian Group Fossil B Formation in Gabon.

The evolution of multicellularity from unicellular ancestors has been replicated in the laboratory, in evolution experiments using predation as the selective pressure.

The origin of cells has to do with the origin of life, which began the history of life on Earth.

Origin of the First Cell

Stromatolites are left behind by cyanobacteria, also called blue-green algae. They are the oldest known fossils of life on Earth. This one-billion-year-old fossil is from Glacier National Park in the United States.

There are several theories about the origin of small molecules that led to life on the early Earth. They may have been carried to Earth on meteorites, created at deep-sea vents, or synthesized by lightning in a reducing atmosphere. There is little experimental data defining what the first self-replicating forms were. RNA is thought to be the earliest self-replicating molecule, as it is capable of both storing genetic information and catalyzing chemical reactions, but some other entity with the potential to self-replicate could have preceded RNA, such as clay or peptide nucleic acid.

Cells emerged at least 3.5 billion years ago. The current belief is that these cells were heterotrophs. The early cell membranes were probably more simple and permeable than modern ones, with only a single fatty acid chain per lipid. Lipids are known to spontaneously form bilayered vesicles in water, and could have preceded RNA, but the first cell membranes could also have been produced by catalytic RNA, or even have required structural proteins before they could form.

Origin of Eukaryotic Cells

The eukaryotic cell seems to have evolved from a symbiotic community of prokaryotic cells. DNA-bearing organelles like the mitochondria and the chloroplasts are descended from ancient symbiotic oxygen-breathing proteobacteria and cyanobacteria, respectively, which were endosymbiosed by an ancestral archaean prokaryote.

There is still considerable debate about whether organelles like the hydrogenosome predated the origin of mitochondria, or vice versa.

Prokaryotic vs. Eukaryotic Cells

	Prokaryotes	**Eukaryotes**
Size and external appearance	Small about 0.5 micrometers. Bacterial cells also contain flagellum, plasmid and capsule.	Up to 40 micrometers.
Organelles	Lacks membrane bound nucleus and membrane bound organelles.	Membrane bound organelles : nucleus, mitochondria & chloroplasts. GA (Golgi Apparatus), ER (Endoplasmic Reticulum) & lysosomes.
Cell wall structure	Rigid, formed from glycoproteins (mainly murein)	**Fungi:** rigid, formed from polysaccharide, chitin. **Plant:** rigid, formed from polysaccharide. E.g.: cellulose. Animals no cell wall.
Genetic material	Circular DNA in cytoplasm.	DNA in form of linear chromosomes (in nucleus).
Ribosome	70 S	80 S

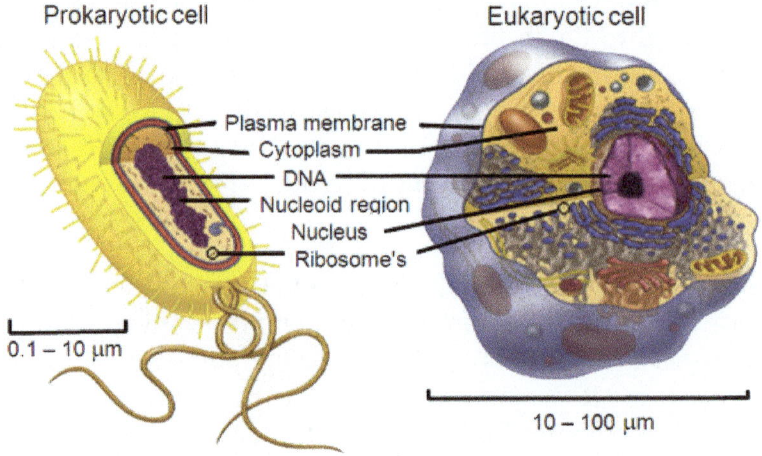

Prokaryotic Cell Walls

1. Bacterial Cell Wall:

Prokaryotes, especially bacteria are divided into Gram positive and Gram negative cells based on differential staining by crystal violet-iodine reagent (Gram's reagent).

Bacteria stained purple are Gram + : their cell walls are thick and composed of predominantly petidoglycan and teichoic acid polymers which lipids are conspicuousy absent. The teichoic acid polymers play key role in antigenicity of the cell.

Bacteria stained pink are Gram – : their cell walls have thin peptidoglycan (sandwiched between the cell membrane and outer envelope) and lipopolysaccharides with no teichoic acid.

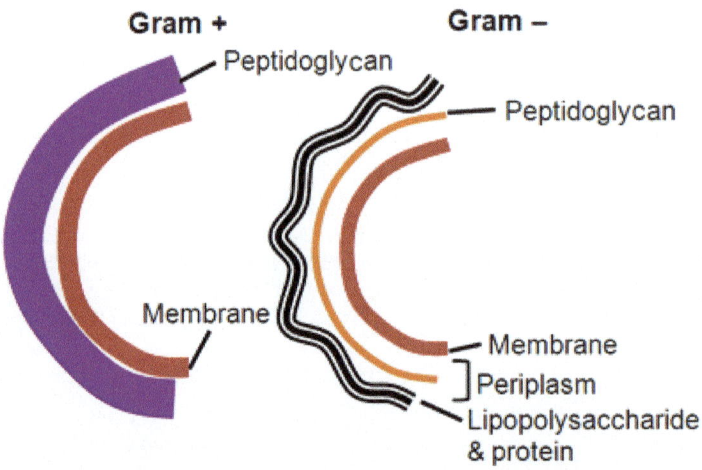

In Gram - positive bacteria, the purple crystal violet stain is trapped by the layer of peptidoglycan which forms the outer layer of the cell.

In Gram - negative bacteria, the outer membrane of lipopolysaccharides prevents the stain from reaching the peptidoglycan layer. The outer membrane is then permeabilized by acetone treatment, and the pink safranin counterstain is trapped by the peptidoglycan layer.

Eukaryote

A eukaryote is any organism whose cells have a nucleus and other organelles enclosed within membranes. Eukaryotes belong to the taxon Eukarya or Eukaryota. The defining feature that sets eukaryotic cells apart from prokaryotic cells (Bacteria and Archaea) is that they have membrane-bound organelles, especially the nucleus, which contains the genetic material and is enclosed by the nuclear envelope. The presence of a nucleus gives eukaryotes their name. Eukaryotic cells also contain other membrane-bound organelles such as mitochondria and the Golgi apparatus. In addition, plants and algae contain chloroplasts. Eukaryotic organisms may be unicellular or multicellular. Only eukaryotes form multicellular organisms consisting of many kinds of tissue made up of different cell types.

Eukaryotes can reproduce both asexually through mitosis and sexually through meiosis and gamete fusion. In mitosis, one cell divides to produce two genetically identical cells. In meiosis, DNA replication is followed by two rounds of cell division to produce four daughter cells each with half the number of chromosomes as the original parent cell (haploid cells). These act as sex cells (gametes – each gamete has just one complement of chromosomes, each a unique mix of the corresponding pair of parental chromosomes) resulting from genetic recombination during meiosis.

The domain Eukaryota appears to be monophyletic, and so makes up one of the three domains of life. The two other domains, Bacteria and Archaea, are prokaryotes and have none of the above features. Eukaryotes represent a tiny minority of all living things. However, due to their much larger size, eukaryotes' collective worldwide biomass is estimated at about equal to that of prokaryotes. Eukaryotes first developed approximately 1.6–2.1 billion years ago (during the Proterozoic eon).

History of the Eukaryote Concept

Konstantin Mereschkowski proposed a symbiotic origin for cells with nuclei.

In 1905 and 1910, the Russian biologist Konstantin Mereschkowsky (1855–1921) argued three things about the origin of nucleated cells. Firstly, plastids were reduced cyanobacteria in a symbiosis with a non-photosynthetic (heterotrophic) host. Secondly, the host had earlier in evolution formed by symbiosis between an amoeba-like host and a bacteria-like ("micrococcal") cell that formed the nucleus. Thirdly, plants inherited photosynthesis from cyanobacteria.

The split between the prokaryotes and eukaryotes was introduced in the 1960s. The concept of the eukaryote has been attributed to the French biologist Edouard Chatton (1883-1947). The terms prokaryote and eukaryote were more definitively reintroduced by the Canadian microbiologist Roger Stanier and the Dutch-American microbiologist C. B. van Niel in 1962. In his 1938 work *Titres et Travaux Scientifiques*, Chatton had proposed the two terms, calling the bacteria prokaryotes and organisms with nuclei in their cells eukaryotes. However he mentioned this in only one paragraph, and the idea was effectively ignored until Chatton's statement was rediscovered by Stanier and van Niel.

In 1967, Lynn Margulis provided microbiological evidence for endosymbiosis as the origin of chloroplasts and mitochondria in eukaryotic cells in her paper, *On the origin of mitosing cells*. In the 1970s, Carl Woese explored microbial phylogenetics, studying variations in 16S ribosomal RNA. This helped to uncover the origin of the eukaryotes and the symbiogenesis of two important eukaryote organelles, mitochondria and chloroplasts. In 1977, Woese and George Fox introduced a "third form of life", which they called the Archaebacteria; in 1990, Woese, Otto Kandler and Mark L. Wheeler renamed this the Archaea.

In 1979, G. W. Gould and G. J. Dring suggested that the eukaryotic cell's nucleus came from the ability of Gram-positive bacteria to form endospores. In 1987 and later papers, Thomas Cavalier-Smith proposed instead that the membranes of the nucleus and endoplasmic reticulum first formed by infolding a prokaryote's plasma membrane. In the 1990s, several other biologists proposed endosymbiotic origins for the nucleus, effectively reviving Mereschkowsky's theory.

Cell Features

Eukaryotic cells are typically much larger than those of prokaryotes. They have a variety of internal membrane-bound structures, called organelles, and a cytoskeleton composed of microtubules, microfilaments, and intermediate filaments, which play an important role in defining the cell's organization and shape. Eukaryotic DNA is divided into several linear bundles called chromosomes, which are separated by a microtubular spindle during nuclear division.

Internal Membrane

Eukaryote cells include a variety of membrane-bound structures, collectively referred to as the endomembrane system. Simple compartments, called vesicles or vacuoles, can form by budding off other membranes. Many cells ingest food and other materials through a process of endocytosis, where the outer membrane invaginates and then pinches off to form a vesicle. It is probable that most other membrane-bound organelles are ultimately derived from such vesicles. Alternatively some products produced by the cell can leave in a vesicle through exocytosis.

The nucleus is surrounded by a double membrane (commonly referred to as a nuclear membrane or nuclear envelope), with pores that allow material to move in and out. Various tube- and sheet-like extensions of the nuclear membrane form what is called the endoplasmic reticulum or ER, which is involved in protein transport and maturation. It includes the rough ER where ribosomes are attached to synthesize proteins, which enter the interior space or lumen. Subsequently, they generally enter vesicles, which bud off from the smooth ER. In most eukaryotes, these protein-carrying vesicles are released and further modified in stacks of flattened vesicles, called Golgi bodies or dictyosomes.

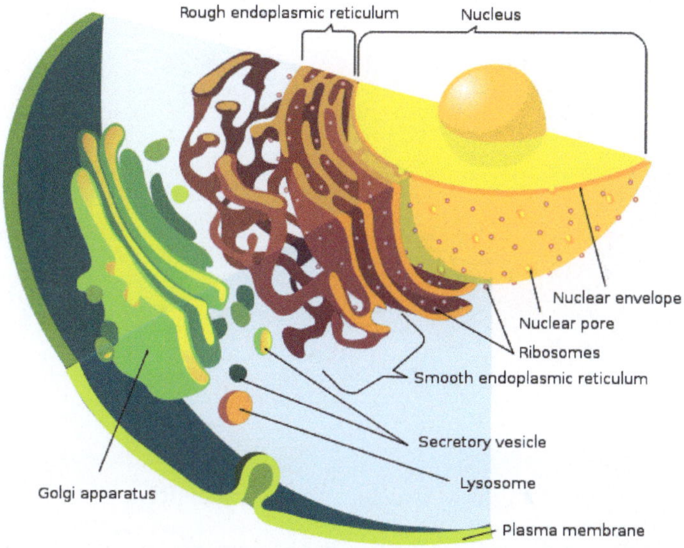

Detail of the endomembrane system and its components

Vesicles may be specialized for various purposes. For instance, lysosomes contain digestive enzymes that break down the contents of food vacuoles, and peroxisomes are used to break down peroxide, which is toxic otherwise. Many protozoa have contractile vacuoles, which collect and expel excess water, and extrusomes, which expel material used to deflect predators or capture prey. In higher plants, most of a cell's volume is taken up by a central vacuole, which primarily maintains its osmotic pressure.

A 3D rendering of an animal cell cut in half.

Mitochondria and Plastids

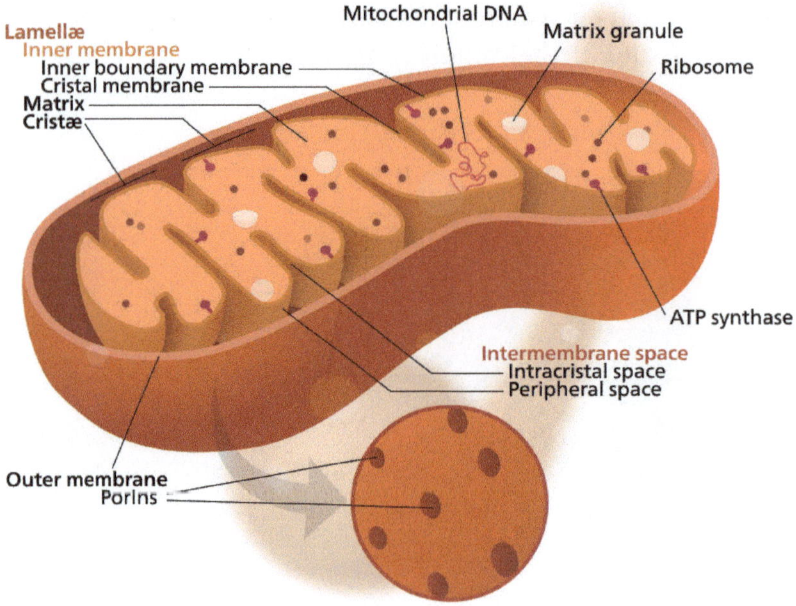

Simplified structure of a mitochondrion

Mitochondria are organelles found in nearly all eukaryotes that provide energy to the cell by converting ingested sugars into ATP. They are surrounded by two membranes (each a phospholipid bi-layer), the inner of which is folded into invaginations called cristae, where aerobic respiration takes place. Mitochondria contain their own DNA. They are now generally held to have developed from endosymbiotic prokaryotes, probably proteobacteria. Protozoa and microbes that lack mitochondria, such as the amoebozoan Pelomyxa and metamonads such as Giardia and Trichomonas, have usually been found to contain mitochondrion-derived organelles, such as hydrogenosomes and mitosomes, and thus probably lost the mitochondria secondarily.

In 2016, *Monocercomonoides*, a metamonad flagellate which resides in the intestines of the chinchilla, has been found to lack mitochondria entirely. *Monocercomonoides* obtains its energy by enzymatic action on nutrients absorbed from the environment. It has also acquired, by lateral gene transfer, a cytosolic sulphur mobilisation system which provides the clusters of iron and sulfur required for protein synthesis. The normal mitochondrial iron-sulphur cluster pathway is considered to have been lost secondarily.

Plants and various groups of algae also have plastids. Plastids have their own DNA and are developed from endosymbionts, in this case cyanobacteria. They usually take the form of chloroplasts, which like cyanobacteria contain chlorophyll and produce organic compounds (such as glucose) through photosynthesis. Others are involved in storing food. Although plastids probably had a single origin, not all plastid-containing groups are closely related. Instead, some eukaryotes have obtained them from others through secondary endosymbiosis or ingestion.

Endosymbiotic origins have also been proposed for the nucleus, and for eukaryotic flagella.

Cytoskeletal Structures

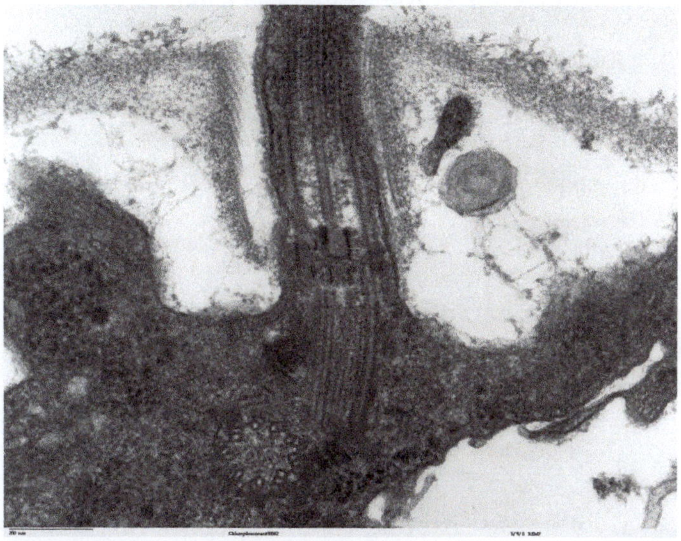

Longitudinal section through the flagellum of *Chlamydomonas reinhardtii*

Many eukaryotes have long slender motile cytoplasmic projections, called flagella, or similar structures called cilia. Flagella and cilia are sometimes referred to as undulipodia, and are variously involved in movement, feeding, and sensation. They are composed mainly of tubulin. These are entirely distinct from prokaryotic flagellae. They are supported by a bundle of microtubules arising from a basal body, also called a kinetosome or centriole, characteristically arranged as nine doublets surrounding two singlets. Flagella also may have hairs, or mastigonemes, and scales connecting membranes and internal rods. Their interior is continuous with the cell's cytoplasm.

Microfilamental structures composed of actin and actin binding proteins, e.g., α-actinin, fimbrin, filamin are present in submembraneous cortical layers and bundles, as well. Motor proteins of microtubules, e.g., dynein or kinesin and actin, e.g., myosins provide dynamic character of the network.

Centrioles are often present even in cells and groups that do not have flagella, but conifers and flowering plants have neither. They generally occur in groups of one or two, called kinetids, that give rise to various microtubular roots. These form a primary component of the cytoskeletal structure, and are often assembled over the course of several cell divisions, with one flagellum retained from the parent and the other derived from it. Centrioles may also be associated in the formation of a spindle during nuclear division.

The significance of cytoskeletal structures is underlined in the determination of shape of the cells, as well as their being essential components of migratory responses like chemotaxis and chemokinesis. Some protists have various other microtubule-supported organelles. These include the radiolaria and heliozoa, which produce axopodia used in flotation or to capture prey, and the haptophytes, which have a peculiar flagellum-like organelle called the haptonema.

Cell Wall

The cells of plants, fungi, and most chromalveolates have a cell wall, a layer outside the cell membrane, providing the cell with structural support, protection, and a filtering mechanism. The cell wall also prevents over-expansion when water enters the cell.

The major polysaccharides making up the primary cell wall of land plants are cellulose, hemicellulose, and pectin. The cellulose microfibrils are linked via hemicellulosic tethers to form the cellulose-hemicellulose network, which is embedded in the pectin matrix. The most common hemicellulose in the primary cell wall is xyloglucan.

Differences Among Eukaryotic Cells

There are many different types of eukaryotic cells, though animals and plants are the most familiar eukaryotes, and thus provide an excellent starting point for understanding eukaryotic structure. Fungi and many protists have some substantial differences, however.

Animal Cell

All animals consist of eukaryotic cells. Animal cells are distinct from those of other eukaryotes, most notably plants, as they lack cell walls and chloroplasts and have smaller vacuoles. Due to the lack of a cell wall, animal cells can adopt a variety of shapes. A phagocytic cell can even engulf other structures.

There are many other types of cell. For instance, there are approximately 210 distinct cell types in the adult human body.

Plant Cell

Plant cells are quite different from the cells of the other eukaryotic organisms. Their distinctive features are:

- A large central vacuole (enclosed by a membrane, the tonoplast), which maintains the cell's turgor and controls movement of molecules between the cytosol and sap
- A primary cell wall containing cellulose, hemicellulose and pectin, deposited by the protoplast on the outside of the cell membrane; this contrasts with the cell walls of fungi, which contain chitin, and the cell envelopes of prokaryotes, in which peptidoglycans are the main structural molecules
- The plasmodesmata, linking pores in the cell wall that allow each plant cell to communicate with other adjacent cells; this is different from the functionally analogous system of gap junctions between animal cells.
- Plastids, especially chloroplasts that contain chlorophyll, the pigment that gives plants their green color and allows them to perform photosynthesis
- Bryophytes and seedless vascular plants lack flagellae and centrioles except in the sperm cells. Sperm of cycads and *Ginkgo* are large, complex cells that swim with hundreds to thousands of flagellae.

- Conifers (Pinophyta) and flowering plants (Angiospermae) lack the flagellae and centrioles that are present in animal cells.

Fungal Cell

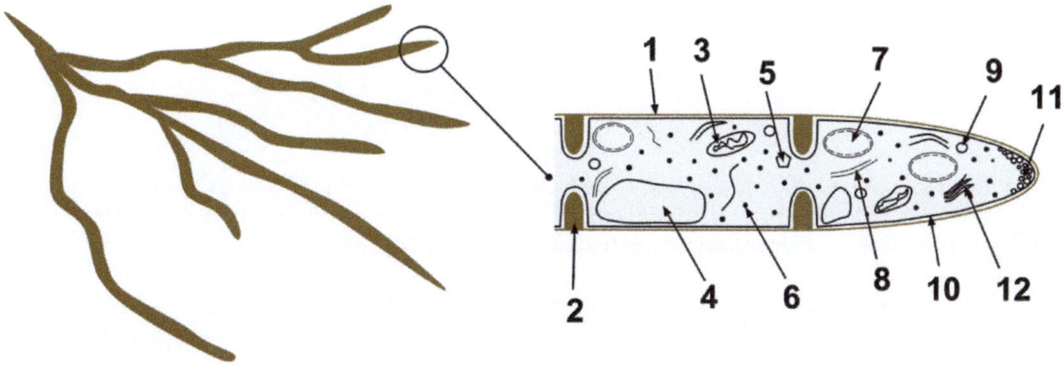

Fungal Hyphae Cells : 1- Hyphal wall 2- Septum 3- Mitochondrion 4- Vacuole 5- Ergosterol crystal 6- Ribosome 7- Nucleus 8- Endoplasmic reticulum 9- Lipid body 10- Plasma membrane 11- Spitzenkörper 12- Golgi apparatus

Fungal cells are most similar to animal cells, with the following exceptions:

- A cell wall that contains chitin
- Less definition between cells; the hyphae of higher fungi have porous partitions called septa, which allow the passage of cytoplasm, organelles, and, sometimes, nuclei. Primitive fungi have few or no septa, so each organism is essentially a giant multinucleate supercell; these fungi are described as coenocytic.
- Only the most primitive fungi, chytrids, have flagella.

Other Eukaryotic Cells

Eukaryotes are a very diverse group, and their cell structures are equally diverse. Many have cell walls; many do not. Many have chloroplasts, derived from primary, secondary, or even tertiary endosymbiosis; and many do not. Some groups have unique structures, such as the cyanelles of the glaucophytes, the haptonema of the haptophytes, or the ejectisomes of the cryptomonads. Other structures, such as pseudopods, are found in various eukaryote groups in different forms, such as the lobose amoebozoans or the reticulose foraminiferans.

Reproduction

Cell division generally takes place asexually by mitosis, a process that allows each daughter nucleus to receive one copy of each chromosome. In most eukaryotes, there is also a process of sexual reproduction, typically involving an alternation between haploid generations, wherein only one copy of each chromosome is present, and diploid generations, wherein two copies of each chromosome are present, occurring through meiosis. There is considerable variation in this pattern.

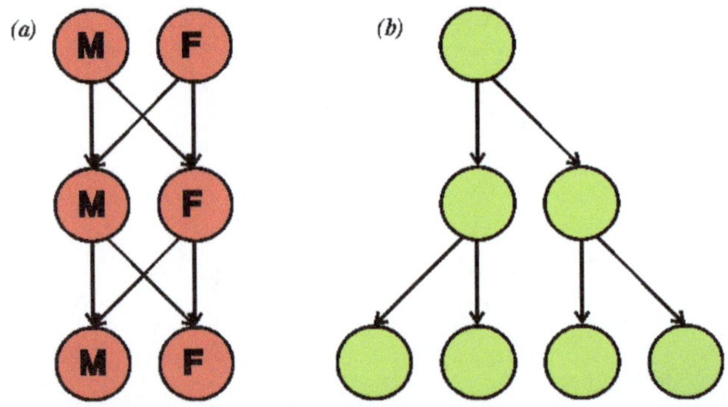

This diagram illustrates the *twofold cost of sex*. If each individual were to contribute to the same number of offspring (two), *(a)* the sexual population remains the same size each generation, where the *(b)* asexual population doubles in size each generation.

Eukaryotes have a smaller surface area to volume ratio than prokaryotes, and thus have lower metabolic rates and longer generation times. In some multicellular organisms, cells specialized for metabolism will have enlarged surface areas, such as intestinal vili.

The evolution of sexual reproduction may be a primordial and fundamental characteristic of eukaryotes. Based on a phylogenetic analysis, Dacks and Roger proposed that facultative sex was present in the common ancestor of all eukaryotes. A core set of genes that function in meiosis is present in both *Trichomonas vaginalis* and *Giardia intestinalis*, two organisms previously thought to be asexual. Since these two species are descendants of lineages that diverged early from the eukaryotic evolutionary tree, it was inferred that core meiotic genes, and hence sex, were likely present in a common ancestor of all eukaryotes. Other studies on eukaryotic species once thought to be asexual have revealed evidence for a sexual cycle. For instance, parasitic protozoa of the genus Leishmania have recently been shown to have a sexual cycle. Also, evidence now indicates that amoeba, that were previously regarded as asexual, are anciently sexual and that the majority of present-day asexual groups likely arose recently and independently.

Classification

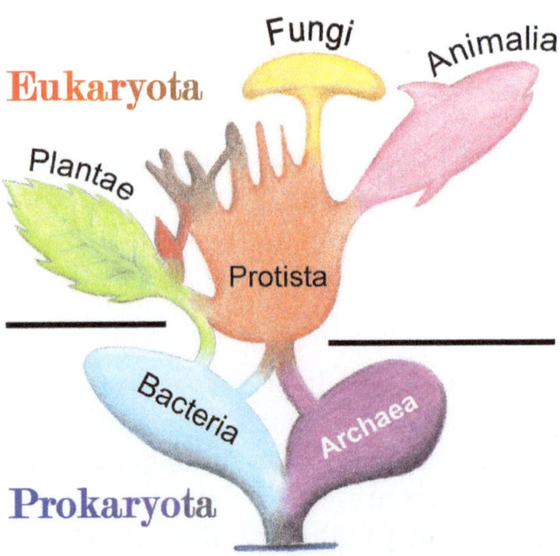

Phylogenetic and symbiogenetic tree of living organisms, showing a view of the origins of eukaryotes & prokaryotes

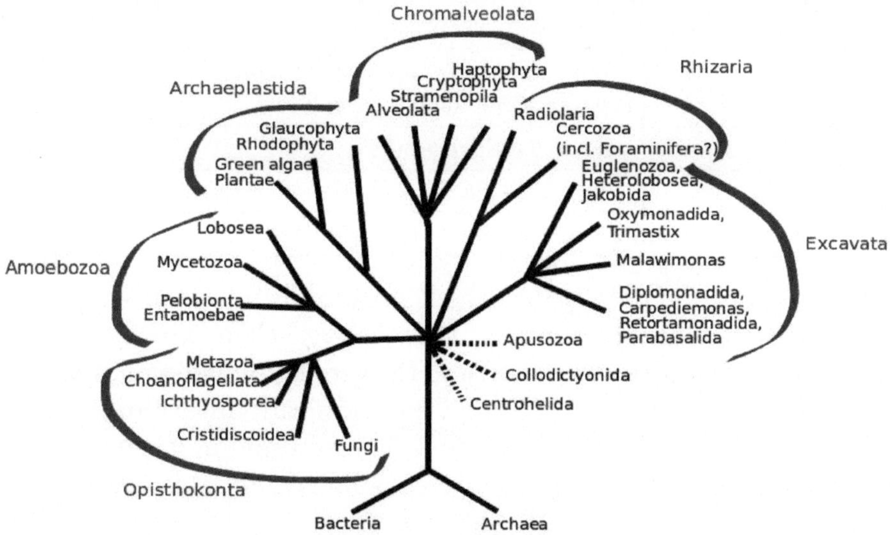

One hypothesis of eukaryotic relationships. The Opisthokonta group includes both animals (Metazoa) and fungi. Plants (Plantae) are placed in Archaeplastida.

In antiquity, the two clades of animals and plants were recognized. They were given the taxonomic rank of Kingdom by Linnaeus. Though he included the fungi with plants with some reservations, it was later realized that they are quite distinct and warrant a separate kingdom, the composition of which was not entirely clear until the 1980s. The various single-cell eukaryotes were originally placed with plants or animals when they became known. In 1830, the German biologist Georg A. Goldfuss coined the word *protozoa* to refer to organisms such as ciliates, and this group was expanded until it encompassed all single-celled eukaryotes, and given their own kingdom, the Protista, by Ernst Haeckel in 1866. The eukaryotes thus came to be composed of four kingdoms:

- Kingdom Protista
- Kingdom Plantae
- Kingdom Fungi
- Kingdom Animalia

The protists were understood to be "primitive forms", and thus an evolutionary grade, united by their primitive unicellular nature. The disentanglement of the deep splits in the tree of life only really started with DNA sequencing, leading to a system of domains rather than kingdoms as top level rank being put forward by Carl Woese, uniting all the eukaryote kingdoms under the eukaryote domain. At the same time, work on the protist tree intensified, and is still actively going on today. Several alternative classifications have been forwarded, though there is no consensus in the field.

A classification produced in 2005 for the International Society of Protistologists, which reflected the consensus of the time, divided the eukaryotes into six supposedly monophyletic 'supergroups'. However, in the same year (2005), doubts were expressed as to whether some of these supergroups were monophyletic, particularly the Chromalveolata, and a review in 2006 noted the lack of evidence for several of the supposed six supergroups. A revised classification in 2012 recognizes five supergroups.

Archaeplastida (or Primoplantae)	Land plants, green algae, red algae, and glaucophytes
SAR supergroup	Stramenopiles (brown algae, diatoms, etc.), Alveolata, and Rhizaria (Foraminifera, Radiolaria, and various other amoeboid protozoa).
Excavata	Various flagellate protozoa
Amoebozoa	Most lobose amoeboids and slime molds
Opisthokonta	Animals, fungi, choanoflagellates, etc.

There are also smaller groups of eukaryotes whose position is uncertain or seems to fall outside the major groups. In particular, Haptophyta, Cryptophyta, Centrohelida, Telonemia, Picozoa, Apusomonadida, Ancyromonadida, Breviatea, and the genus *Collodictyon*. Overall, it seems that, although progress has been made, there are still very significant uncertainties in the evolutionary history and classification of eukaryotes. As Roger & Simpson said in 2009 "with the current pace of change in our understanding of the eukaryote tree of life, we should proceed with caution."

In an article published in *Nature Microbiology* the authors, "reinforced once again that the life we see around us – plants, animals, humans and other so-called eukaryotes – represent a tiny percentage of the world's biodiversity." They classified eukaryote "based on the inheritance of their information systems as opposed to lipid or other cellular structures." Jillian F. Banfield of the University of California, Berkeley and fellow scientists used a super computer to generate a diagram of a new tree of life based on DNA from 3000 species including 2,072 known species and 1,011 newly reported microbial organisms, whose DNA they had gathered from diverse environments. As the capacity to sequence DNA became easier, Banfield and team were able to do metagenomic sequencing—"sequencing whole communities of organisms at once and picking out the individual groups based on their genes alone."

Phylogeny

The rRNA trees constructed during the 1980s and 1990s left most eukaryotes in an unresolved "crown" group (not technically a true crown), which was usually divided by the form of the mitochondrial cristae. The few groups that lack mitochondria branched separately, and so the absence was believed to be primitive; but this is now considered an artifact of long-branch attraction, and they are known to have lost them secondarily.

As of 2011, there is widespread agreement that the Rhizaria belong with the Stramenopiles and the Alveolata, in a clade dubbed the SAR supergroup, so that Rhizaria is not one of the main eukaryote groups; also that the Amoebozoa and Opisthokonta are each monophyletic and form a clade, often called the unikonts. Beyond this, there does not appear to be a consensus.

It has been estimated that there may be 75 distinct lineages of eukaryotes. Most of these lineages are protists.

The known eukaryote genome sizes vary from 8.2 megabases (Mb) in *Babesia bovis* to 112,000–220,050 Mb in the dinoflagellate *Prorocentrum micans*, suggesting that the genome of the ancestral eukaryote has undergone considerable variation during its evolution. The last common ancestor of all eukaryotes is believed to have been a phagotrophic protist with a nucleus, at least one centriole and cilium, facultatively aerobic mitochondria, sex (meiosis and syngamy), a dormant cyst with a cell wall of chitin and/or cellulose and peroxisomes. Later endosymbiosis led to the spread of plastids in some lineages.

Five Supergroups

A global tree of eukaryotes from a consensus of phylogenetic evidence (in particular, phylogenomics), rare genomic signatures, and morphological characteristics is presented in Adl *et al.* 2012 and Burki 2014.

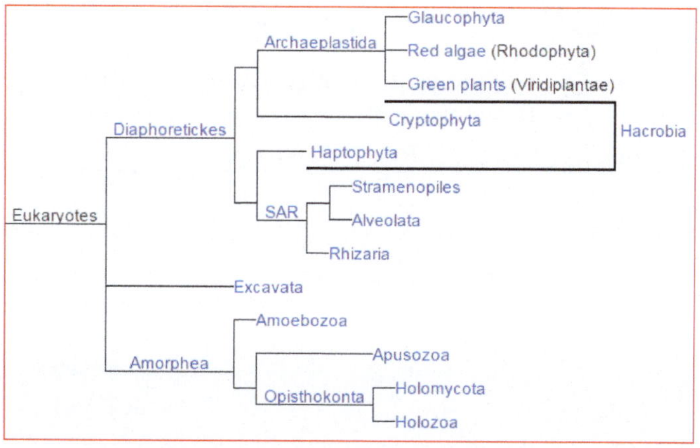

In some analyses, the Hacrobia group (Haptophyta + Cryptophyta) is placed next to Archaeplastida, but in other ones it is nested inside the Archaeplastida. However, several recent studies have concluded that Haptophyta and Cryptophyta do not form a monophyletic group. The former could be a sister group to the SAR group, the latter cluster with the Archaeplastida (plants in the broad sense). As of February 2012, it remains unclear whether the Hacrobia forms a monophyletic group.

The division of the eukaryotes into two primary clades, bikonts (Archaeplastida + SAR + Excavata) and unikonts (Amoebozoa + Opisthokonta), derived from an ancestral biflagellar organism and an ancestral uniflagellar organism, respectively, had been suggested earlier. A 2012 study produced a somewhat similar division, although noting that the terms "unikonts" and "bikonts" were not used in the original sense.

Cavalier-Smith's Tree

Cavalier-Smith 2010, 2013, and 2014 places the eukaryotic tree's root between Excavata (with ventral feeding groove supported by a microtubular root) and the grooveless Euglenozoa:

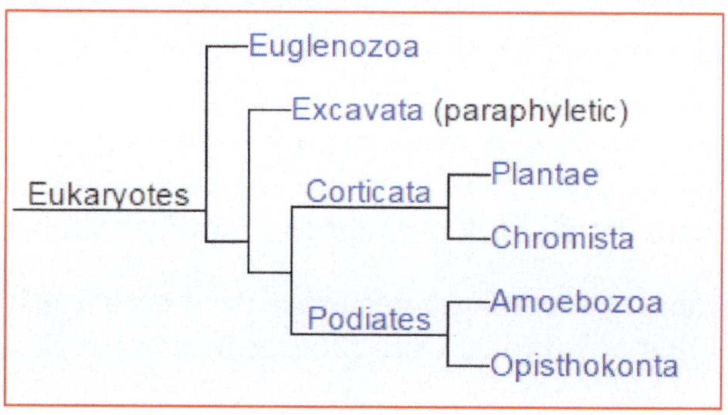

Origin of Eukaryotes

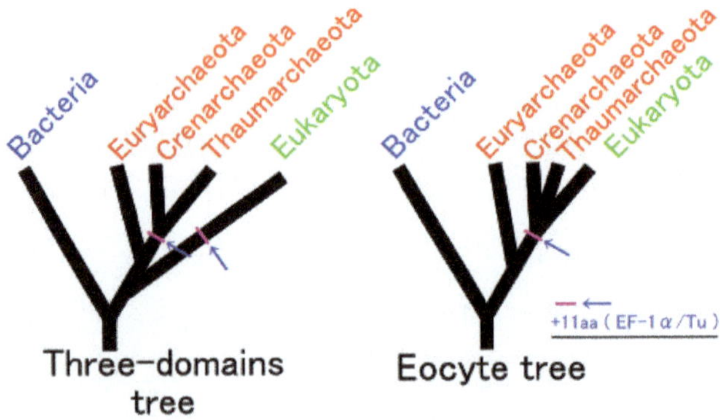

The three-domains tree and the Eocyte hypothesis.

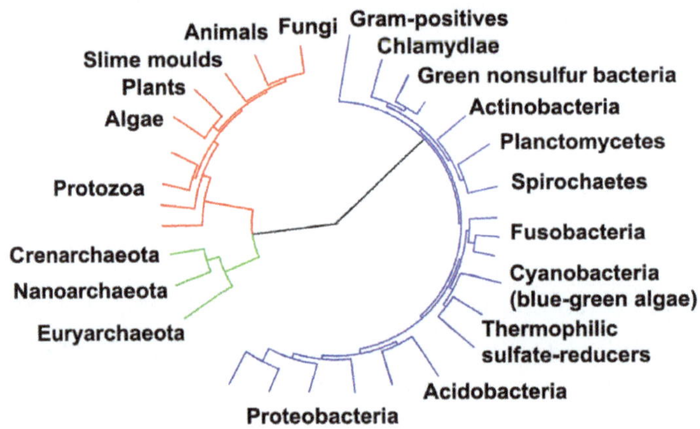

Phylogenetic tree showing the relationship between the eukaryotes and other forms of life. Eukaryotes are colored red, archaea green and bacteria blue.

Fossils

The origin of the eukaryotic cell is considered a milestone in the evolution of life, since eukaryotes include all complex cells and almost all multicellular organisms. The timing of this series of events is hard to determine; Knoll (2006) suggests they developed approximately 1.6–2.1 billion years ago. Some acritarchs are known from at least 1.65 billion years ago, and the possible alga *Grypania* has been found as far back as 2.1 billion years ago. The *Geosiphon*-like fossil fungus *Diskagma* has been found in paleosols 2.2 billion years old

Organized living structures have been found in the black shales of the Palaeoproterozoic Francevillian B Formation in Gabon, dated at 2.1 billion years old. Eukaryotic life could have evolved at that time. Fossils that are clearly related to modern groups start appearing an estimated 1.2 billion years ago, in the form of a red alga, though recent work suggests the existence of fossilized filamentous algae in the Vindhya basin dating back perhaps to 1.6 to 1.7 billion years ago.

Biomarkers suggest that at least stem eukaryotes arose even earlier. The presence of steranes in Australian shales indicates that eukaryotes were present in these rocks dated at 2.7 billion years old.

Relationship to Archaea

Eukaryotes are more closely related to Archaea than Bacteria, at least in terms of nuclear DNA and genetic machinery, and one controversial idea is to place them with Archaea in the clade Neomura. However, in other respects, such as membrane composition, eukaryotes are similar to Bacteria. Three main explanations for this have been proposed:

- Eukaryotes resulted from the complete fusion of two or more cells, wherein the cytoplasm formed from a eubacterium, and the nucleus from an archaeon, from a virus, or from a pre-cell.
- Eukaryotes developed from Archaea, and acquired their eubacterial characteristics from the proto-mitochondrion.
- Eukaryotes and Archaea developed separately from a modified eubacterium.

The chronocyte hypothesis for the origin of the eukaryotic cell postulates that a primitive eukaryotic cell was formed by the endosymbiosis of both archaea and bacteria by a third type of cell, termed a chronocyte.

A possible cladogram for the positioning of eukaryotes within Archaea, based on phylogenomic analyses of the Asgard archaea:

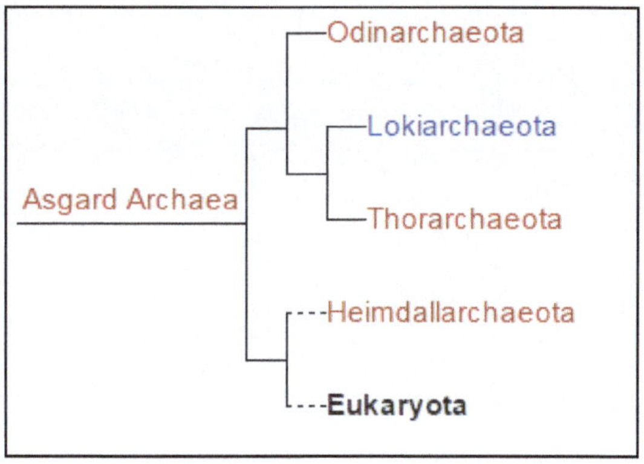

Endomembrane System and Mitochondria

The origins of the endomembrane system and mitochondria are also unclear. The phagotrophic hypothesis proposes that eukaryotic-type membranes lacking a cell wall originated first, with the development of endocytosis, whereas mitochondria were acquired by ingestion as endosymbionts. The syntrophic hypothesis proposes that the proto-eukaryote relied on the proto-mitochondrion for food, and so ultimately grew to surround it. Here the membranes originated after the engulfment of the mitochondrion, in part thanks to mitochondrial genes (the hydrogen hypothesis is one particular version).

In a study using genomes to construct supertrees, Pisani *et al.* (2007) suggest that, along with evidence that there was never a mitochondrion-less eukaryote, eukaryotes evolved from a syntrophy between an archaea closely related to Thermoplasmatales and an α-proteobacterium, likely a symbiosis driven by sulfur or hydrogen. The mitochondrion and its genome is a remnant of the α-proteobacterial endosymbiont.

Hypotheses For the Origin of Eukaryotes

Different hypotheses have been proposed as to how eukaryotic cells came into existence. These hypotheses can be classified into two distinct classes – autogenous models and chimeric models.

Autogenous Models

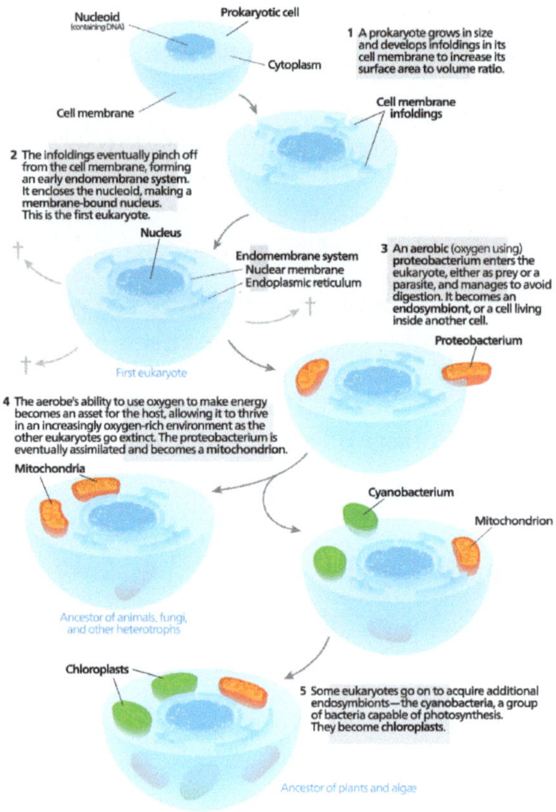

An autogenous model for the origin of eukaryotes.

Autogenous models propose that a proto-eukaryotic cell containing a nucleus existed first, and later acquired mitochondria. According to this model, a large prokaryote developed invaginations in its plasma membrane in order to obtain enough surface area to service its cytoplasmic volume. As the invaginations differentiated in function, some became separate compartments—giving rise to the endomembrane system, including the endoplasmic reticulum, golgi apparatus, nuclear membrane, and single membrane structures such as lysosomes. Mitochondria are proposed to come from the endosymbiosis of an aerobic proteobacterium, and it's assumed that all the eukaryotic lineages that did not acquire mitochondria became extinct. Chloroplasts came about from another endosymbiotic event involving cyanobacteria. Since all eukaryotes have mitochondria, but not all have chloroplasts, mitochondria are thought to have come first. This is the serial endosymbiosis theory.

Some models propose that the origins of double layered organelles, such as mitochondria and chloroplasts, in the proto-eukaryotic cell is due to the compartmentalization of DNA vesicles that were formed from the secondary invaginations or more detailed infoldings of cellular membrane.

Chimeric Models

Chimeric models claim that two prokaryotic cells existed initially – an archaeon and a bacterium. These cells underwent a merging process, either by a physical fusion or by endosymbiosis, thereby leading to the formation of a eukaryotic cell. Within these chimeric models, some studies further claim that mitochondria originated from a bacterial ancestor while others emphasize the role of endosymbiotic processes behind the origin of mitochondria.

Based on the process of mutualistic symbiosis, the hypotheses can be categorized as – the serial endosymbiotic theory (SET), the hydrogen hypothesis (mostly a process of symbiosis where hydrogen transfer takes place among different species), and the syntrophy hypothesis.

According to serial endosymbiotic theory (championed by Dr. Lynn Margulis), a union between a motile anaerobic bacterium (like *Spirochaeta*) and a thermoacidophilic crenarchaeon (like *Thermoplasma* which is sulfidogenic in nature) gave rise to the present day eukaryotes. This union established a motile organism capable of living in the already existing acidic and sulfurous waters. Oxygen is known to cause toxicity to organisms that lack the required metabolic machinery. Thus, the archaeon provided the bacterium with a highly beneficial reduced environment (sulfur and sulfate were reduced to sulfide). In microaerophilic conditions, oxygen was reduced to water thereby creating a mutual benefit platform. The bacterium on the other hand, contributed the necessary fermentation products and electron acceptors along with its motility feature to the archaeon thereby gaining a swimming motility for the organism. From a consortium of bacterial and archaeal DNA originated the nuclear genome of eukaryotic cells. Spirochetes gave rise to the motile features of eukaryotic cells. Endosymbiotic unifications of the ancestors of alpha-proteobacteria and cyanobacteria, led to the origin of mitochondria and plastids respectively. For example, *Thiodendron* has been known to have originated via an ectosymbiotic process based on a similar syntrophy of sulfur existing between the two types of bacteria – *Desulphobacter* and *Spirochaeta*. However, such an association based on motile symbiosis have never been observed practically. Also there is no evidence of archaeans and spirochetes adapting to intense acid-based environments.

In the hydrogen hypothesis, the symbiotic linkage of an anaerobic and autotrophic methanogenic archaeon (host) with an alpha-proteobacterium (the symbiont) gave rise to the eukaryotes. The host utilized hydrogen (H_2) and carbon dioxide (CO_2) to produce methane while the symbiont, capable of aerobic respiration, expelled H_2 and CO_2 as byproducts of anaerobic fermentation process. The host's methanogenic environment worked as a sink for H_2, which resulted in heightened bacterial fermentation. Endosymbiotic gene transfer (EGT) acted as a catalyst for the host to acquire the symbionts' carbohydrate metabolism and turn heterotrophic in nature. Subsequently, the host's methane forming capability was lost. Thus, the origins of the heterotrophic organelle (symbiont) are identical to the origins of the eukaryotic lineage. In this hypothesis, the presence of H_2 represents the selective force that forged eukaryotes out of prokaryotes.

The syntrophy hypothesis was developed in contrast to the hydrogen hypothesis and proposes the existence of two symbiotic events. According to this theory, eukaryogenesis (i.e. origin of eukaryotic cells) occurred based on metabolic symbiosis (syntrophy) between a methanogenic archaeon and a delta-proteobacterium. This syntrophic symbiosis was initially facilitated by H_2 transfer between different species under anaerobic environments. In earlier stages, an alpha-proteobacterium became a member of this integration, and later developed into the mitochondrion. Gene transfer from a delta-proteobacterium to an archaeon led to the methanogenic archaeon developing into a nucleus. The archaeon constituted the genetic apparatus while the delta-proteobacterium contributed towards the cytoplasmic features. This theory incorporates two selective forces that were needed to be considered during the time of nucleus evolution – (a) presence of metabolic partitioning in order to avoid the harmful effects of the co-existence of anabolic and catabolic cellular pathways, and (b) prevention of abnormal biosynthesis of proteins that occur due to a vast spread of introns in the archaeal genes after acquiring the mitochondrion and the loss of methanogenesis.

Thus, the origin of eukaryotes by endosymbiotic processes has been broadly recognized and accepted so far. Mitochondria and plastids have been known to originate from a bacterial ancestor during parallel adaptation to anaerobiosis. However, there still remains a greater need in assessing the question of how much eukaryotic complexity is being originated via an implementation of these symbiogenetic theories.

Cell Wall

A cell wall is a structural layer surrounding some types of cells, situated outside the cell membrane. It can be tough, flexible, and sometimes rigid. It provides the cell with both structural support and protection, and also acts as a filtering mechanism. Cell walls are present in most prokaryotes (except mycoplasma bacteria), in algae, plants and fungi but rarely in other eukaryotes including animals. A major function is to act as pressure vessels, preventing over-expansion of the cell when water enters.

The composition of cell walls varies between species and may depend on cell type and developmental stage. The primary cell wall of land plants is composed of the polysaccharides cellulose, hemicellulose and pectin. Often, other polymers such as lignin, suberin or cutin are anchored to or embedded in plant cell walls. Algae possess walls made of glycoproteins and polysaccharides such as carrageenan and agar that are absent from land plants. In bacteria, the cell wall is composed of peptidoglycan. The cell walls of archaea have various compositions, and may be formed of glycoprotein S-layers, pseudopeptidoglycan, or polysaccharides. Fungi possess cell walls made of the glucosamine polymer chitin. Unusually, diatoms have a cell wall composed of biogenic silica.

History

A plant cell wall was first observed and named (simply as a "wall") by Robert Hooke in 1665. However, "the dead excrusion product of the living protoplast" was forgotten, for almost three centuries, being the subject of scientific interest mainly as a resource for industrial processing or in relation to animal or human health.

In 1804, Karl Rudolphi and J.H.F. Link proved that cells had independent cell walls. Before, it had been thought that cells shared walls and that fluid passed between them this way.

The mode of formation of the cell wall was controversial in the 19th century. Hugo von Mohl (1853, 1858) advocated the idea that the cell wall grows by apposition. Carl Nägeli (1858, 1862, 1863) believed that the growth of the wall in thickness and in area was due to a process termed intussusception. Each theory was improved in the following decades: the apposition (or lamination) theory by Eduard Strasburger (1882, 1889), and the intussusception theory by Julius Wiesner (1886).

In 1930, Ernst Münch coined the term *apoplast* in order to separate the "living" symplast from the "dead" plant region, the latter of which included the cell wall.

By the 1980s, some authors suggested replacing the term "cell wall", particularly as it was used for plants, with the more precise term "extracellular matrix", as used for animal cells, but others preferred the older term.

Properties

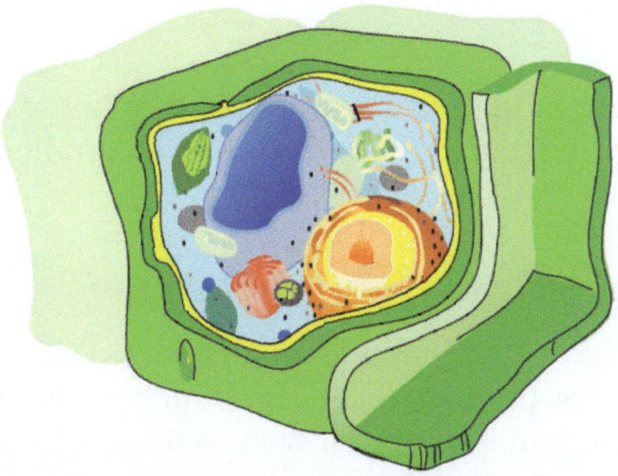

Diagram of the plant cell, with the cell wall in green.

Cell walls serve similar purposes in those organisms that possess them. They may give cells rigidity and strength, offering protection against mechanical stress. In multicellular organisms, they permit the organism to build and hold a definite shape (morphogenesis). Cell walls also limit the entry of large molecules that may be toxic to the cell. They further permit the creation of stable osmotic environments by preventing osmotic lysis and helping to retain water. Their composition, properties, and form may change during the cell cycle and depend on growth conditions.

Rigidity of Cell Walls

In most cells, the cell wall is flexible, meaning that it will bend rather than holding a fixed shape, but has considerable tensile strength. The apparent rigidity of primary plant tissues is enabled by cell walls, but is not due to the walls' stiffness. Hydraulic turgor pressure creates this rigidity, along with the wall structure. The flexibility of the cell walls is seen when plants wilt, so that the stems and leaves begin to droop, or in seaweeds that bend in water currents. As John Howland explains:

Think of the cell wall as a wicker basket in which a balloon has been inflated so that it exerts pressure from the inside. Such a basket is very rigid and resistant to mechanical damage. Thus does the prokaryote cell (and eukaryotic cell that possesses a cell wall) gain strength from a flexible plasma membrane pressing against a rigid cell wall.

The apparent rigidity of the cell wall thus results from inflation of the cell contained within. This inflation is a result of the passive uptake of water.

In plants, a secondary cell wall is a thicker additional layer of cellulose which increases wall rigidity. Additional layers may be formed by lignin in xylem cell walls, or suberin in cork cell walls. These compounds are rigid and waterproof, making the secondary wall stiff. Both wood and bark cells of trees have secondary walls. Other parts of plants such as the leaf stalk may acquire similar reinforcement to resist the strain of physical forces.

Permeability

The primary cell wall of most plant cells is freely permeable to small molecules including small proteins, with size exclusion estimated to be 30-60 kDa. The pH is an important factor governing the transport of molecules through cell walls.

Evolution

Cell walls evolved independently in many groups, even in the photosynthetic eukaryotes. In these lineages, the cell wall is closely related to the evolution of multicellularity, terrestrialization and vascularization.

Plant Cell Walls

The walls of plant cells must have sufficient tensile strength to withstand internal osmotic pressures of several times atmospheric pressure that result from the difference in solute concentration between the cell interior and external solutions. Plant cell walls vary from 0.1 to several μm in thickness.

Layers

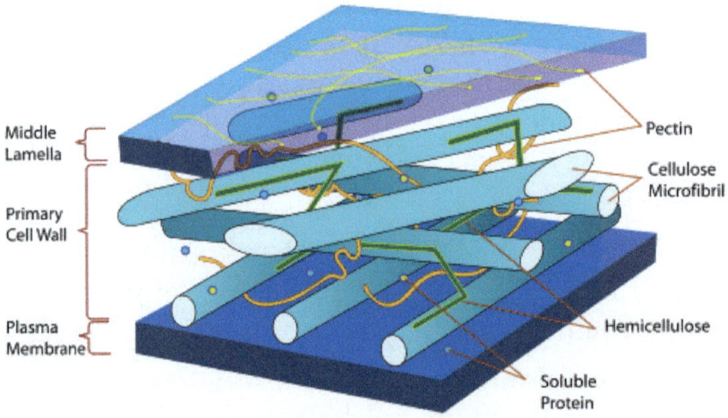

Molecular structure of the primary cell wall in plants.

Up to three strata or layers may be found in plant cell walls:

- The primary cell wall, generally a thin, flexible and extensible layer formed while the cell is growing.
- The secondary cell wall, a thick layer formed inside the primary cell wall after the cell is fully grown. It is not found in all cell types. Some cells, such as the conducting cells in xylem, possess a secondary wall containing lignin, which strengthens and waterproofs the wall.

- The middle lamella, a layer rich in pectins. This outermost layer forms the interface between adjacent plant cells and glues them together.

Composition

In the primary (growing) plant cell wall, the major carbohydrates are cellulose, hemicellulose and pectin. The cellulose microfibrils are linked via hemicellulosic tethers to form the cellulose-hemicellulose network, which is embedded in the pectin matrix. The most common hemicellulose in the primary cell wall is xyloglucan. In grass cell walls, xyloglucan and pectin are reduced in abundance and partially replaced by glucuronarabinoxylan, another type of hemicellulose. Primary cell walls characteristically extend (grow) by a mechanism called acid growth, which involves turgor-driven movement of the strong cellulose microfibrils within the weaker hemicellulose/pectin matrix, catalyzed by expansin proteins. The outer part of the primary cell wall of the plant epidermis is usually impregnated with cutin and wax, forming a permeability barrier known as the plant cuticle.

Secondary cell walls contain a wide range of additional compounds that modify their mechanical properties and permeability. The major polymers that make up wood (largely secondary cell walls) include:

- cellulose, 35-50%
- xylan, 20-35%, a type of hemicellulose
- lignin, 10-25%, a complex phenolic polymer that penetrates the spaces in the cell wall between cellulose, hemicellulose and pectin components, driving out water and strengthening the wall.

Additionally, structural proteins (1-5%) are found in most plant cell walls; they are classified as hydroxyproline-rich glycoproteins (HRGP), arabinogalactan proteins (AGP), glycine-rich proteins (GRPs), and proline-rich proteins (PRPs). Each class of glycoprotein is defined by a characteristic, highly repetitive protein sequence. Most are glycosylated, contain hydroxyproline (Hyp) and become cross-linked in the cell wall. These proteins are often concentrated in specialized cells and in cell corners. Cell walls of the epidermis may contain cutin. The Casparian strip in the endodermis roots and cork cells of plant bark contain suberin. Both cutin and suberin are polyesters that function as permeability barriers to the movement of water. The relative composition of carbohydrates, secondary compounds and proteins varies between plants and between the cell type and age. Plant cells walls also contain numerous enzymes, such as hydrolases, esterases, peroxidases, and transglycosylases, that cut, trim and cross-link wall polymers.

Secondary walls - especially in grasses - may also contain microscopic silica crystals, which may strengthen the wall and protect it from herbivores.

Cell walls in some plant tissues also function as storage deposits for carbohydrates that can be broken down and resorbed to supply the metabolic and growth needs of the plant. For example, endosperm cell walls in the seeds of cereal grasses, nasturtium and other species, are rich in glucans and other polysaccharides that are readily digested by enzymes during seed germination to form simple sugars that nourish the growing embryo.

Formation

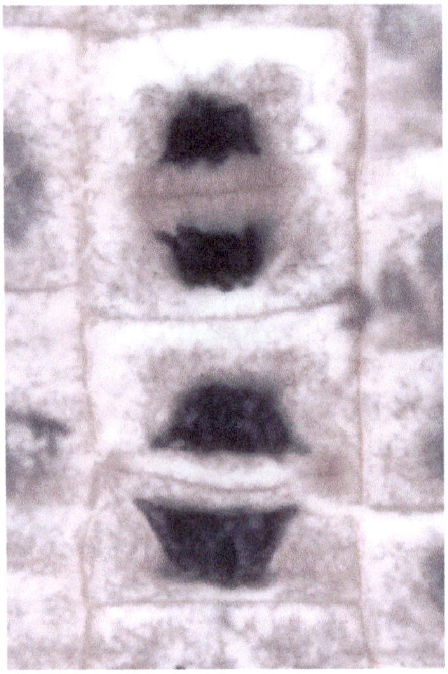

Photomicrograph of onion root cells, showing the centrifugal development of new cell walls (phragmoplast).

The middle lamella is laid down first, formed from the cell plate during cytokinesis, and the primary cell wall is then deposited inside the middle lamella. The actual structure of the cell wall is not clearly defined and several models exist - the covalently linked cross model, the tether model, the diffuse layer model and the stratified layer model. However, the primary cell wall, can be defined as composed of cellulose microfibrils aligned at all angles. Cellulose microfibrils are produced at the plasma membrane by the cellulose synthase complex, which is proposed to be made of a hexameric rosette that contains three cellulose synthase catalytic subunits for each of the six units. Microfibrils are held together by hydrogen bonds to provide a high tensile strength. The cells are held together and share the gelatinous membrane called the *middle lamella*, which contains magnesium and calcium pectates (salts of pectic acid). Cells interact though plasmodesmata, which are inter-connecting channels of cytoplasm that connect to the protoplasts of adjacent cells across the cell wall.

In some plants and cell types, after a maximum size or point in development has been reached, a *secondary wall* is constructed between the plasma membrane and primary wall. Unlike the primary wall, the cellulose microfibrils are aligned parallel in layers, the orientation changing slightly with each additional layer so that the structure becomes helicoidal. Cells with secondary cell walls can be rigid, as in the gritty sclereid cells in pear and quince fruit. Cell to cell communication is possible through pits in the secondary cell wall that allow plasmodesmata to connect cells through the secondary cell walls.

Fungal Cell Walls

There are several groups of organisms that have been called "fungi". Some of these groups (Oomycete and Myxogastria) have been transferred out of the Kingdom Fungi, in part because of fundamental biochemical differences in the composition of the cell wall. Most true fungi have a cell wall consisting largely of chitin and other polysaccharides. True fungi do not have cellulose in their cell walls.

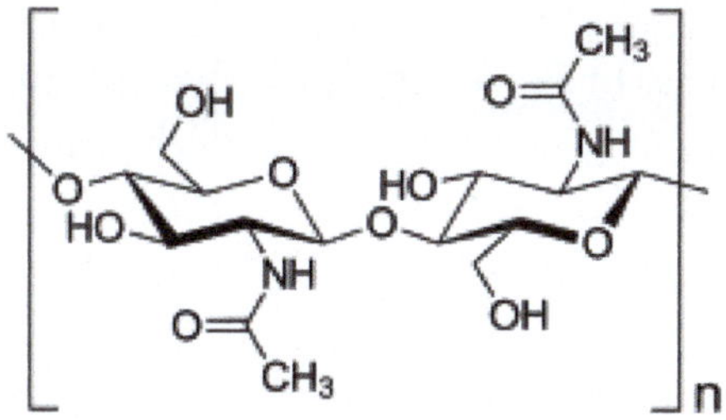

Chemical structure of a unit from a chitin polymer chain.

True Fungi

In fungi, the cell wall is the outer-most layer, external to the plasma membrane. The fungal cell wall is a matrix of three main components:

- chitin: polymers consisting mainly of unbranched chains of β-(1,4)-linked-N-Acetylglucosamine in the Ascomycota and Basidiomycota, or poly-β-(1,4)-linked-N-Acetylglucosamine (chitosan) in the Zygomycota. Both chitin and chitosan are synthesized and extruded at the plasma membrane.
- glucans: glucose polymers that function to cross-link chitin or chitosan polymers. β-glucans are glucose molecules linked via β-(1,3)- or β-(1,6)- bonds and provide rigidity to the cell wall while α-glucans are defined by α-(1,3)- and/or α-(1,4) bonds and function as part of the matrix.
- proteins: enzymes necessary for cell wall synthesis and lysis in addition to structural proteins are all present in the cell wall. Most of the structural proteins found in the cell wall are glycosylated and contain mannose, thus these proteins are called mannoproteins or mannans.

Other Eukaryotic Cell Walls

Algae

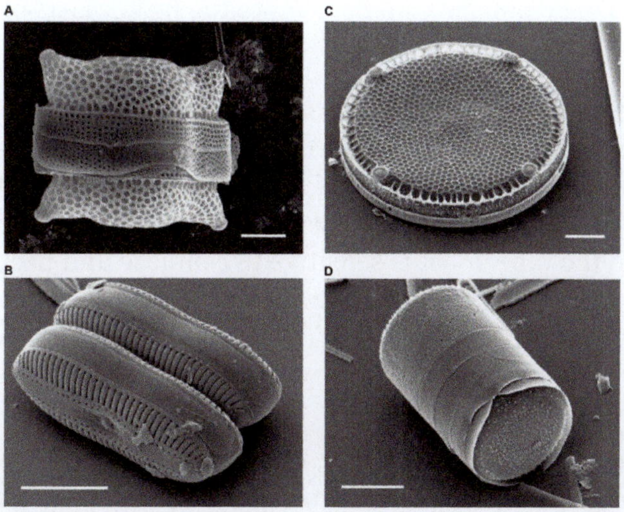

Scanning electron micrographs of diatoms showing the external appearance of the cell wall

Like plants, algae have cell walls. Algal cell walls contain either polysaccharides (such as cellulose (a glucan)) or a variety of glycoproteins (Volvocales) or both. The inclusion of additional polysaccharides in algal cells walls is used as a feature for algal taxonomy.

- Mannans: They form microfibrils in the cell walls of a number of marine green algae including those from the genera, *Codium*, *Dasycladus*, and *Acetabularia* as well as in the walls of some red algae, like *Porphyra* and *Bangia*.
- Xylans:
- Alginic acid: It is a common polysaccharide in the cell walls of brown algae.
- Sulfonated polysaccharides: They occur in the cell walls of most algae; those common in red algae include agarose, carrageenan, porphyran, furcelleran and funoran.

Other compounds that may accumulate in algal cell walls include sporopollenin and calcium ions.

The group of algae known as the diatoms synthesize their cell walls (also known as frustules or valves) from silicic acid (specifically orthosilicic acid, H_4SiO_4). The acid is polymerised intra-cellularly, then the wall is extruded to protect the cell. Significantly, relative to the organic cell walls produced by other groups, silica frustules require less energy to synthesize (approximately 8%), potentially a major saving on the overall cell energy budget and possibly an explanation for higher growth rates in diatoms.

In brown algae, phlorotannins may be a constituent of the cell walls.

Water Molds

The group Oomycetes, also known as water molds, are saprotrophic plant pathogens like fungi. Until recently they were widely believed to be fungi, but structural and molecular evidence has led to their reclassification as heterokonts, related to autotrophic brown algae and diatoms. Unlike fungi, oomycetes typically possess cell walls of cellulose and glucans rather than chitin, although some genera (such as *Achlya* and *Saprolegnia*) do have chitin in their walls. The fraction of cellulose in the walls is no more than 4 to 20%, far less than the fraction of glucans. Oomycete cell walls also contain the amino acid hydroxyproline, which is not found in fungal cell walls.

Slime Molds

The dictyostelids are another group formerly classified among the fungi. They are slime molds that feed as unicellular amoebae, but aggregate into a reproductive stalk and sporangium under certain conditions. Cells of the reproductive stalk, as well as the spores formed at the apex, possess a cellulose wall. The spore wall has been shown to possess three layers, the middle of which is composed primarily of cellulose, and the innermost is sensitive to cellulase and pronase.

Prokaryotic Cell Walls

Bacterial Cell Walls

Around the outside of the cell membrane is the bacterial cell wall. Bacterial cell walls are made of peptidoglycan (also called murein), which is made from polysaccharide chains cross-linked by unusual peptides containing D-amino acids. Bacterial cell walls are different from the cell walls of plants and fungi which are made of cellulose and chitin, respectively. The cell wall of bacteria is also distinct from that of Archaea, which do not contain peptidoglycan. The cell wall is essential to the survival of many bacteria, although L-form bacteria can be produced in the laboratory that lack a cell wall. The antibiotic penicillin is able to kill bacteria by preventing the cross-linking of peptidoglycan and this causes the cell wall to weaken and lyse. The lysozyme enzyme can also damage bacterial cell walls.

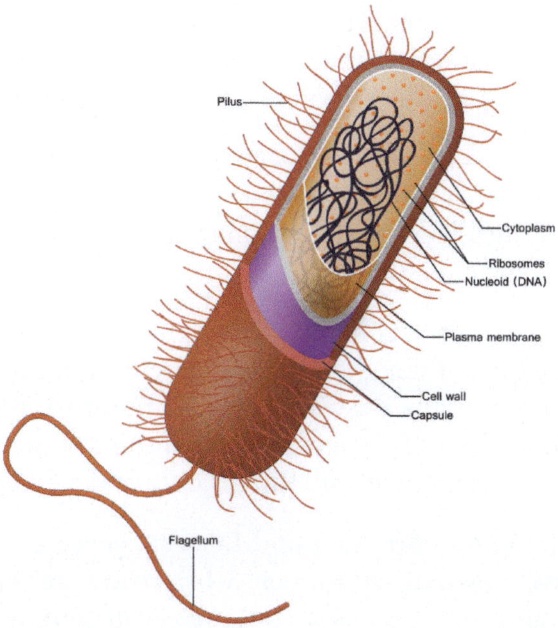

Diagram of a typical Gram-positive bacterium. The cell envelope comprises a plasma membrane, seen here in light brown, and a thick peptidoglycan-containing cell wall (the purple layer). No outer lipid membrane is present, as would be the case in Gram-negative bacteria. The red layer, known as the capsule, is distinct from the cell envelope.

There are broadly speaking two different types of cell wall in bacteria, called Gram-positive and Gram-negative. The names originate from the reaction of cells to the Gram stain, a test long-employed for the classification of bacterial species.

Gram-positive bacteria possess a thick cell wall containing many layers of peptidoglycan and teichoic acids. In contrast, Gram-negative bacteria have a relatively thin cell wall consisting of a few layers of peptidoglycan surrounded by a second lipid membrane containing lipopolysaccharides and lipoproteins. Most bacteria have the Gram-negative cell wall and only the Firmicutes and Actinobacteria (previously known as the low G+C and high G+C Gram-positive bacteria, respectively) have the alternative Gram-positive arrangement. These differences in structure can produce differences in antibiotic susceptibility, for instance vancomycin can kill only Gram-positive bacteria and is ineffective against Gram-negative pathogens, such as *Haemophilus influenzae* or *Pseudomonas aeruginosa*.

Archaeal Cell Walls

Although not truly unique, the cell walls of Archaea are unusual. Whereas peptidoglycan is a standard component of all bacterial cell walls, all archaeal cell walls lack peptidoglycan, with the exception of one group of methanogens. In that group, the peptidoglycan is a modified form very different from the kind found in bacteria. There are four types of cell wall currently known among the Archaea.

One type of archaeal cell wall is that composed of pseudopeptidoglycan (also called pseudomurein). This type of wall is found in some methanogens, such as *Methanobacterium* and *Methanothermus*. While the overall structure of archaeal *pseudo*peptidoglycan superficially resembles that of bacterial peptidoglycan, there are a number of significant chemical differences. Like the peptidoglycan found in bacterial cell walls, pseudopeptidoglycan consists of polymer chains of glycan cross-linked by short peptide connections. However, unlike peptidoglycan, the sugar N-acetylmuramic acid is replaced by N-acetyltalosaminuronic acid, and the two sugars are bonded with a β,1-3 glycosidic linkage instead of β,1-4. Additionally, the cross-linking peptides are L-amino acids rather than D-amino acids as they are in bacteria.

A second type of archaeal cell wall is found in *Methanosarcina* and *Halococcus*. This type of cell wall is composed entirely of a thick layer of polysaccharides, which may be sulfated in the case of *Halococcus*. Structure in this type of wall is complex and not fully investigated.

A third type of wall among the Archaea consists of glycoprotein, and occurs in the hyperthermophiles, *Halobacterium*, and some methanogens. In *Halobacterium*, the proteins in the wall have a high content of acidic amino acids, giving the wall an overall negative charge. The result is an unstable structure that is stabilized by the presence of large quantities of positive sodium ions that neutralize the charge. Consequently, *Halobacterium* thrives only under conditions with high salinity.

In other Archaea, such as *Methanomicrobium* and *Desulfurococcus*, the wall may be composed only of surface-layer proteins, known as an *S-layer*. S-layers are common in bacteria, where they serve as either the sole cell-wall component or an outer layer in conjunction with polysaccharides. Most Archaea are Gram-negative, though at least one Gram-positive member is known.

Other Cell Coverings

Many protists and bacteria produce other cell surface structures apart from cell walls, external (extracellular matrix) or internal. Many algae have a sheath or envelope of mucilage outside the cell made of exopolysaccharides. Diatoms build a frustule from silica extracted from the surrounding water; radiolarians, foraminiferans, testate amoebae and silicoflagellates also produce a skeleton from minerals, called test in some groups. Many green algae, such as *Halimeda* and the Dasycladales, and some red algae, the Corallinales, encase their cells in a secreted skeleton of calcium carbonate. In each case, the wall is rigid and essentially inorganic. It is the non-living component of cell. Some golden algae, ciliates and choanoflagellates produces a shell-like protective outer covering called lorica. Some dinoflagellates have a theca of cellulose plates, and coccolithophorids have coccoliths.

An extracellular matrix is also present in metazoans. Its composition varies between cells, but collagens are the most abundant protein in the ECM.

Cell Organelles, Their Molecular Composition, Structure and Function

Nucleus :

The nucleus is the largest organelle in the cell. It is surrounded by double layered cell membrane, which regulates passage of products of nuclear biosynthesis from the nucleus to surrounding cytoplasm.

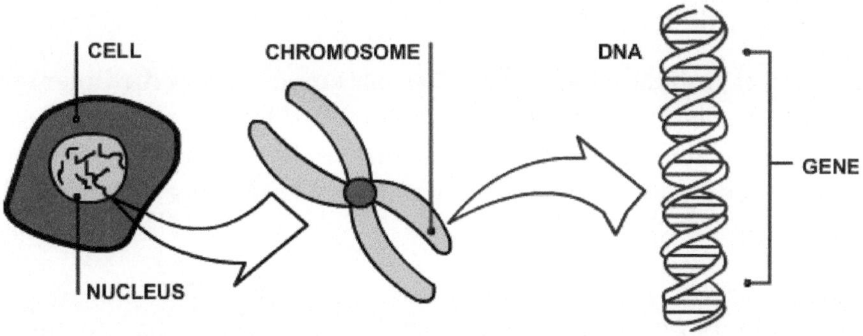

The nucleus contain chromosome which appear as dense network called chromatin. The chromosomes are comprised of DNA (Deoxyribonucleic Acid) and histone / non – histone proteins. The DNA is the genetic material of the cell which contains all the information required for cells to live, perform their functions and reproduce.

Inside the nucleus is another organelle called the nucleolus, responsible for making rRNA. The nucleolus is also dense but devoid of any membrane. Nucleolus contains enzymes responsible for synthesis of RNA (Ribonucleic acid)

Mitochondria :

Mitochondria are cell membrane-enclosed organelles distributed through the cytosol of most eukaryotic cells. These organelles are the sites of cellular respiration, a catabolic oxygen-requiring process that uses energy extracted from organic macromolecules to produce ATP. It has its own DNA.

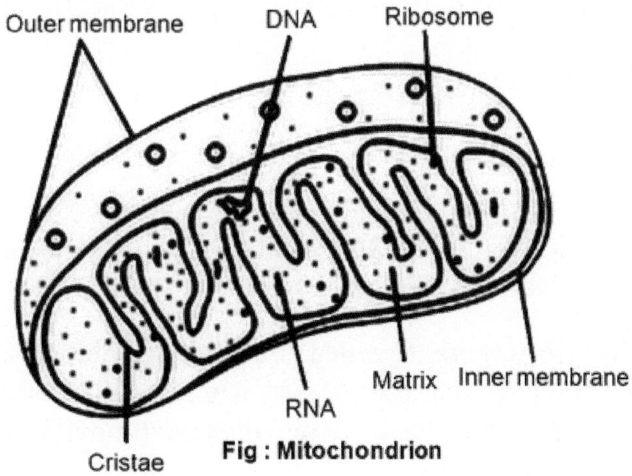

Fig : Mitochondrion

Mitochondria have :

- An outer smooth membrane that encloses the entire structure.
- An inner membrane that encloses a fluid-filled matrix between the two is the intermembrane space.

- The inner membrane is elaborately folded with shelflike cristae projecting into the matrix.
- A large number of enzymes belonging to Kreb cycle and β - oxidation are associated with the inner mitochondrial membrane.
- A large number of globular shaped particles are embedded on the inner surface of the cristae membranes. Such globular particulates are called Racker /F1 particles. They act as ATP synthetases. The cristae membranes also contain a large number of electron transporting enzymes and other oxido reductases deeply buried in the core of the lipid bilayers.

Chloroplast

- Chloroplasts are plant organelles which are a class of plant cell structures called plastids. Plastids may be of varying kinds :

  -> amyloplasts : store starch in roots & tubers.

  -> chromoplasts : store pigments for fruits & flowers.

  -> chloroplasts : store chlorophyll & perform photosynthesis. Present in leaves, other green structures of plants & in eukaryotic algae.

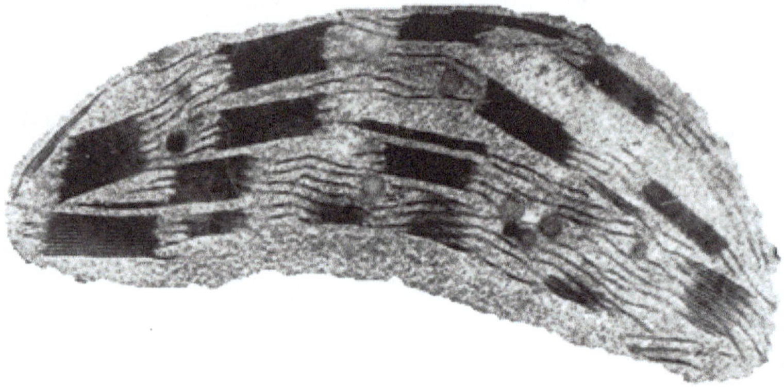

- It is the main site of photosynthesis inside the plant cell wherein light energy of the sun is converted into chemical energy.
- The chemical energy that is produced by chloroplasts is finally used to make carbohydrates like starch, that get stored in the plant.
- Chloroplasts contain tiny pigments called chlorophylls (chlorophyll a and b). Chlorophylls are responsible for trapping the light energy from the sun. Other pigments found within the chloroplast include carotenes, xanthophylls, phaetophytin a and b (orange, yellow and brown pigments respectively)

Structure

- Membrane bi-layer :

  -> outer membrane.

-> inner membrane.

- They contain structure called thylakoids :

-> Thylakoids are membranous sacs.

-> The thylakoid membrane of the chloroplast is impregnated with photosynthetic pigments (i.e., chlorophyll a and b, carotenoids).

-> Grana are stacks of thylakoids.

-> Surrounding the thylakoid is a fluid filled matrix called stroma which contains enzymes to enable ractions occuring here.

-> Starch grains are found in chloroplasts.

-> Also contain DNA, ribosomes & enzymes.

Function

- Site of photosynthesis.
- Generate ATP & synthesize sugars :

-> Transform solar energy into chemical energy.

-> Produce sugars from CO_2 & H_2O

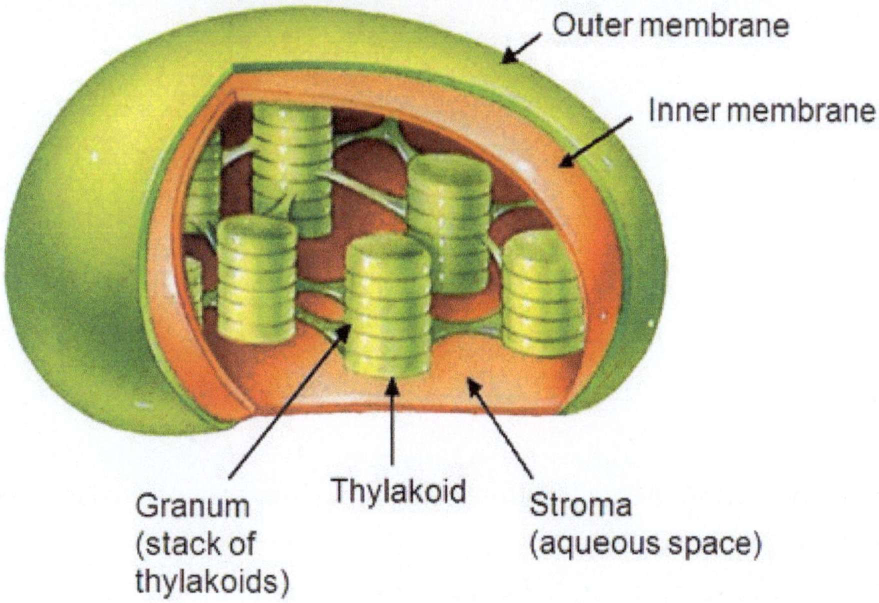

Ribosomes

Organelles that help in the synthesis of proteins. Ribosomes are made up of subunits necessary for protein synthesis in the cell. The prokaryotes contain a large 50S and 23S subunit while most eukaryotes contain 60S and 26-28S subunit. The smaller subunits for prokaryotes have characteristic 16S RNA while for eukaryotes have 18S rRNA. When the two units are docked together with by a messenger RNA, protein synthesis takes place. In eukaryotic cells, some ribosomes are found in the cytoplasm, but most are attached to the endoplasmic reticulum. While attached to the ER, ribosomes make proteins. The newly synthesized proteins are secreted into vesicular system and then transferred to the Golgi apparatus for further use.

Endoplasmic Reticulum(Er)

It is a network of membranous channels throughout the cytoplasm of the cell. It is absent in prokaryotic cells however present in all eukaryotic cells. The ER connects the cell membrane, coats the nucleus and extends to the Golgi apparatus.

There are two types of ER :

- The smooth ER : which have ribosomes are attached on to them.
- The rough ER : which have no ribosomes attached.

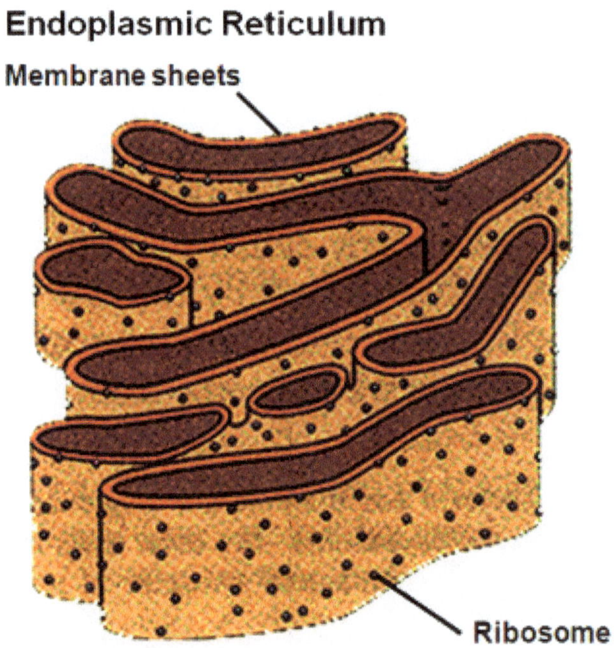

Functions

- The ribosomes present in rough endoplasmic reticulum are involved in the protein synthesis of the cell.
- The function of the smooth endoplasmic reticulum is to synthesize lipids in the cell. The smooth ER is also helps in the detoxification of harmful substances in the cell.

Golgi Bodies

These are also called DICTYOSOMES. It is made up of a stack of flattened out sacs. The golgi complex is responsible for post translational modifications, sorting and correctly sending the newly synthesised proteins receieved from the ER. The post translationally modified proteins leave the apparatus are delivered to target site such as plasma membrane and lysosome etc

Vescicles and Vacuoles

Vescicles are transient structures that helps store and transport products produced by the cell to the exterior or from the exterior to interior. The vesicles are the transport and delivery vehicles. Some vesicles deliver materials to parts of the cell and others transport materials outside the cell in a process called exocytosis. Vacuoles are storage organelles in cells. Vacuoles are larger than vesicles. Either structure may store water, waste products, food, and other cellular materials. In plant cells, the vacuole may take up most of the cell's volume. The membrane surrounding the plant cell vacuole is called the tonoplast.

Lysosomes

Lysosomes function as the cell's recycling compartment. It is a membrane bound large organelle consisting a matrix rich in hydrolytic enzymes, acid nucleases, acid glycosidases for digestion of proteins, RNA and DNA and polysaccharides respectively.

Lysosomes receive cellular and endocytosed proteins and lipids that need to be digested. The metabolites that result are transported either by vesicles or directly across the membrane.

References

- European Bioinformatics Institute, Karyn's Genomes: Borrelia burgdorferi, part of 2can on the EBI-EMBL database. Retrieved 5 August 2012
- Griffiths G (December 2007). "Cell evolution and the problem of membrane topology". Nature Reviews. Molecular Cell Biology. 8 (12): 1018–24. doi:10.1038/nrm2287. PMID 17971839
- Peter Hamilton Raven; George Brooks Johnson (2002). Biology. McGraw-Hill Education. p. 68. ISBN 978-0-07-112261-0. Retrieved 7 July 2013
- Cavalier-Smith, T (2006). "Protist phylogeny and the high-level classification of Protozoa". European Journal of Protistology. 39 (4): 338–348. doi:10.1078/0932-4739-00002
- Campbell, Neil A.; Brad Williamson; Robin J. Heyden (2006). Biology: Exploring Life. Boston, Massachusetts: Pearson Prentice Hall. ISBN 0-13-250882-6
- Michie K, Löwe J (2006). "Dynamic filaments of the bacterial cytoskeleton". Annu Rev Biochem. 75: 467–92. doi:10.1146/annurev.biochem.75.103004.142452. PMID 16756499
- Adl, Sina M.; et al. (September 2012). "The revised classification of eukaryotes" (PDF). Journal of Eukaryotic Microbiology. 59 (5): 429–514. doi:10.1111/j.1550-7408.2012.00644.x
- Lane, Nick (2016). The Vital Question: Why is Life the Way it is? (paperback ed.). Profile Books. pp. 157–191. ISBN 978-1-781-25037-2
- Martin W (December 2005). "Archaebacteria (Archaea) and the origin of the eukaryotic nucleus". Curr. Opin. Microbiol. 8 (6): 630–7. doi:10.1016/j.mib.2005.10.004. PMID 16242992
- Takemura M (May 2001). "Poxviruses and the origin of the eukaryotic nucleus". J. Mol. Evol. 52 (5): 419–25. doi:10.1007/s002390010171. PMID 11443345
- Lynn Margulis, Heather I. McKhann & Lorraine Olendzenski (ed.), Illustrated Glossary of Protoctista, Jones and Bartlett Publishers, Boston, 1993, p.xviii. ISBN 0-86720-081-2

- Bell PJ (September 2001). "Viral eukaryogenesis: was the ancestor of the nucleus a complex DNA virus?". J. Mol. Evol. 53 (3): 251–6. doi:10.1007/s002390010215. PMID 11523012
- Sendbusch, Peter V. (2003-07-31). "Cell Walls of Algae Archived November 28, 2005, at the Wayback Machine.". Botany Online. Retrieved on 2007-10-29
- Howland, John L. (2000). The Surprising Archaea: Discovering Another Domain of Life. Oxford: Oxford University Press. pp. 69–71. ISBN 0-19-511183-4
- Kalenderblatt Dezember 2013 – Mathematisch-Naturwissenschaftliche Fakultät – Universität Rostock. Math-nat.uni-rostock.de (2013-11-28). Retrieved on 2015-10-15
- Pisani D, Cotton JA, McInerney JO (2007). "Supertrees disentangle the chimerical origin of eukaryotic genomes". Mol Biol Evol. 24 (8): 1752–60. doi:10.1093/molbev/msm095. PMID 17504772
- Campbell, Neil A.; Reece, Jane B.; Urry, Lisa A.; Cain, Michael L.; Wasserman, Steven A.; Minorsky, Peter V.; Jackson, Robert B. (2008). Biology (8th ed.). p. 118. ISBN 978-0-8053-6844-4
- Martin W, Müller M (March 1998). "The hydrogen hypothesis for the first eukaryote". Nature. 392 (6671): 37–41. Bibcode:1998Natur.392...37M. doi:10.1038/32096. PMID 9510246
- Lopez-Garcia, P.; Moreira, D. (2006). "Selective forces for the origin of the eukaryotic nucleus". BioEssays. 28 (5): 525–533. doi:10.1002/bies.20413. PMID 16615090
- Buchanan; Gruissem, Jones (2000). Biochemistry & molecular biology of plants (1st ed.). American society of plant physiology. ISBN 0-943088-39-9

2

Cellular Transport: An Integrated Study

Cellular transport can be defined as the movement of molecules or compounds across the cell membrane. This is done to maintain homeostasis, to take in or release compounds or to allow cells to regulate the characteristics of transport. Cellular transport can be categorized into passive or active transport. The section serves as a source to understand the major categories related to cellular transport.

Cellular Transport

The movement of compounds or molecules across the membrane of the cell is referred to as cellular transport. The purpose of cell transport is :

-> to maintain homeostasis.

-> to take in or release compounds as per the biological needs of the cell.

-> allows the cells to regulate the characteristics of transport across the membrane.

Cell transport may be divided into two categories: those that require energy (passive transport) and those that do not (active transport). Various kinds of molecules can move in and out of the cell during cellular transport. Molecules and ions that are transported across the plasma membrane are glucose, Na^{+}, Ca^{2+}. In eukaryotic cells, transport of proteins, mRNA, Ca^{2+}, ATP across membrane-bounded intracellular compartments such as the nucleus, endoplasmic reticulum, and mitochondria.The lipid bilayer of the cell membrane is freely permeable to water molecules as well as small, uncharged, molecules like oxygen and carbon dioxide.

Types of Cellular Transport

- Passive Transport :

  1. Diffusion.

  2. Facilitated Diffusion.

  3. Osmosis.

- Active Transport :

  1. Endocytosis.

  2. Exocytosis.

Active Transport :

- Active transport is the process in which cell needs to utilize energy for transport of molecules.
- This process actively moves molecules to where they are needed.
- Movement of molecules takes place against the concentration gradient, i.e from an area of low concentration to an area of high concentration.
- Three types of active transport mechanisms occur :

  1. Protein pumps.

  2. Endocytosis.

  3. Exocytosis.
- Protein Pumps :– The mode of active transport uses transport proteins as pumps which normally uses the chemical energy of ATP. Protein changes shape to move molecules. For example, metal ions, such as Na^+, K^+, Mg^{2+}, or Ca^{2+}, require ion pumps or ion channels to cross membranes and distribute through the body. The pump for transport of sodium and potassium is called sodium-potassium pump or Na^+/K^+ ATPase. This protein pumps Na^+ ions from the inside to the outside of the nerve cell and pumps K^+ ions in the opposite direction using energy from ATP. Similarly, in the mitochondrion, hydrogen ion pumps pump hydrogen ions into the intermembrane space of the organelle during the energy generation process.
- Endocytosis :- also called cell eating is the process by which a cell takes in molecules and other objects that are too large to pass through the cellular membrane. The molecule to be taken in causes the cell membrane to bulge inward, forming a vesicle. Cell membrane in-folds around food particle, forms food vacuole and digests food. The useful products of digestion are released inside the cell and waste products are released out. In some cases endocytosis is mediated by receptors. This occurs when the material to be transported binds to certain specific molecules in the membrane.

Phagocytosis is the type of endocytosis where an entire cell is engulfed.

Pinocytosis is another type of endocytosis, when the external fluid is engulfed.

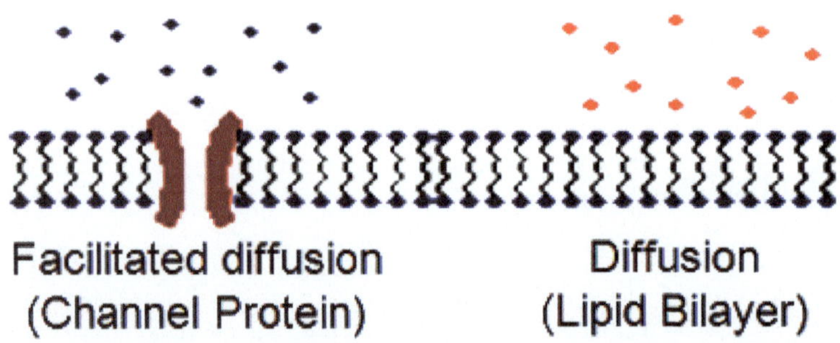

- Exocytosis :- the process by which a cell expels molecules and other objects out of the cell. These are molecules that are too large to be able to cross the cellular membrane. In this case, the cell forces material out of cell. The membrane of the vescicle surrounding the material fuses with cell membrane. Dissolved fluid contents in these vesicles are discharged into the extracellular fluid (secretion). The vesicles turn inside out upon fusion with the plasma membrane and the molecules is then forced out of the cellular membrane. Energy is required since the cell changes shape. For example, hormones or wastes released from cell by exocytosis.

Passive Cellular Transport

Passive Transport

- In the passive process of transport, cell requires no energy for transport.
- The movement of molecules is random.
- Molecules move across a concentration gradient i.e from an area of high concentration to an area of low concentration.
- Three types of Passive Transport mechanisms exist :

  1. Diffusion .

  2. Facilitative Diffusion – diffusion with the help of transport proteins.

  3. Osmosis – diffusion of water.

Diffusion

Diffusion is the net movement of molecules (or ions) from a region of their high concentration to a region of their lower concentration.

- It being a passive process does not require energy. Molecules have kinetic energy, which makes them move about randomly.
- As a result of diffusion, molecules reach an equilibrium state (concentrations on both side becomes same), where after there is no net movement of molecules from either side.
- Molecules that diffuse through cell membranes.

  Oxygen :– Non-polar so diffuses very quickly.

  Carbon dioxide :– Polar but very small so diffuses quickly.

  Water :– Polar but also very small so diffuses quickly.

More examples of diffusion will be welcome.

Facilitated Diffusion

- Large polar molecules such as glucose and amino acids, cannot diffuse across the phospholipid bilayer. Also ions such as Na^+ or Cl^- cannot pass.
- These molecules pass through protein channels instead. Diffusion through these channels is called facilitated diffusion.
- Movement of molecules is still passive just like ordinary diffusion, the only difference is, the molecules go through a protein channel instead of passing between the phospholipids.

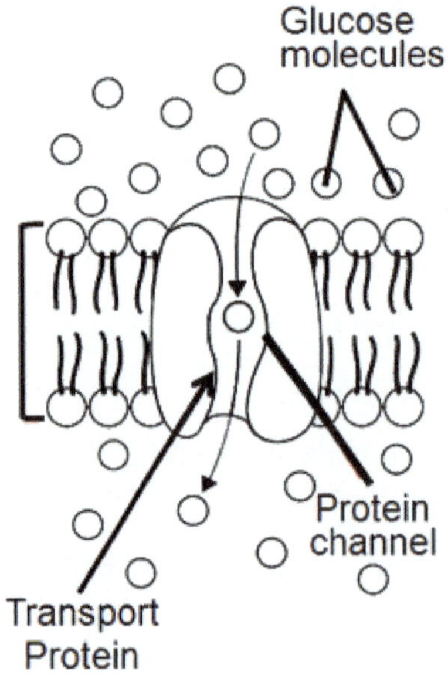

Molecules will randomly move through the opening like pore, by diffusion. This requires no energy, it is a PASSIVE process. Molecules move from an area of high concentration to an area of low conc. Diffusion of specific particles across the cell membrane occurs through transport proteins found in the membrane. Transport Proteins are specific – they "select" only certain molecules to cross the membrane. Transports larger or charged molecules.

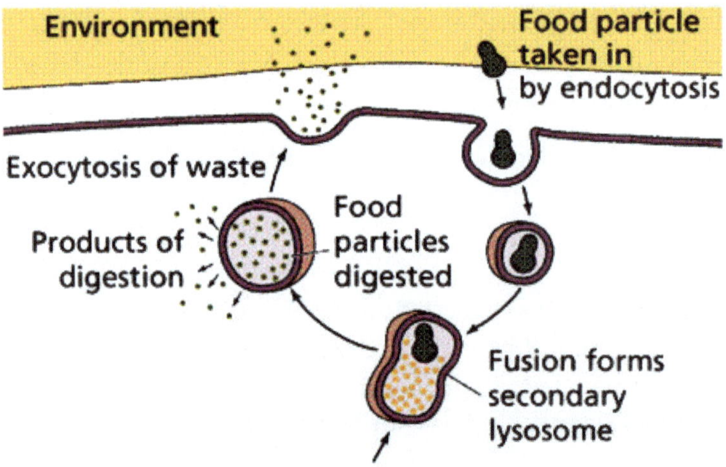

Osmosis

The diffusion of water from an area of high concentration of water molecules (high water potential) to an area of low concentration of water (low water potential) across a partially permeable membrane.

Endocytosis

Endocytosis is a form of active transport in which a cell transports molecules (such as proteins) into the cell (*endo- + cytosis*) by engulfing them in an energy-using process. Endocytosis and its counterpart, exocytosis, are used by all cells because most chemical substances important to them are large polar molecules that cannot pass through the hydrophobic plasma or cell membrane by passive means.

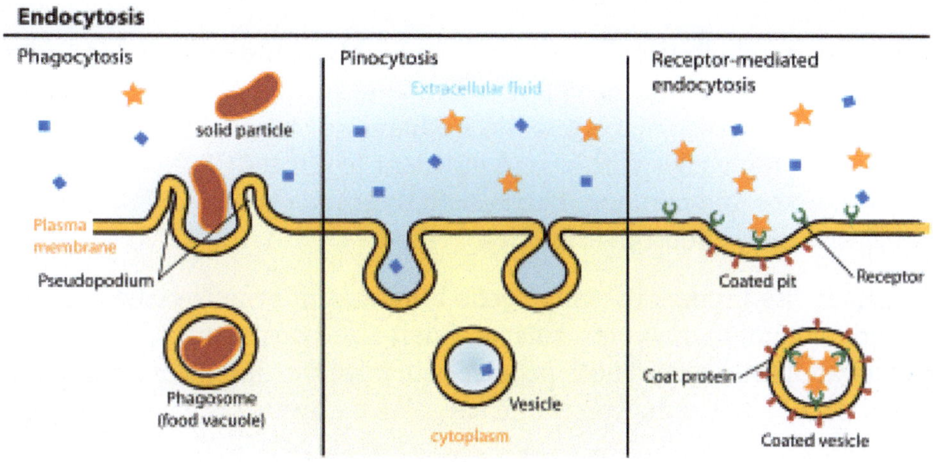

The different types of endocytosis

Endocytosis includes pinocytosis (cell drinking) and phagocytosis (cell eating).

History

The term was proposed by De Duve in 1963. Phagocytosis was discovered by Élie Metchnikoff.

Endocytosis Pathways

Endocytosis pathways can be subdivided into four categories: namely, receptor-mediated endocytosis, caveolae, macropinocytosis, and phagocytosis.

- Clathrin-mediated endocytosis is mediated by the production of small (approx. 100 nm in diameter) vesicles that have a morphologically characteristic coat made up of the cytosolic protein clathrin. Clathrin-coated vesicles (CCVs) are found in virtually all cells and form domains of the plasma membrane termed clathrin-coated pits. Coated pits can concentrate large extracellular molecules that have different receptors responsible for the receptor-mediated endocytosis of ligands, e.g. low density lipoprotein, transferrin, growth factors, antibodies and many others.
- Caveolae are the most common reported non-clathrin-coated plasma membrane buds, which exist on the surface of many, but not all cell types. They consist of the cholesterol-binding protein caveolin (Vip21) with a bilayer enriched in cholesterol and glycolipids. Caveolae are small (approx. 50 nm in diameter) flask-shape pits in the membrane that resemble the shape of a cave (hence the name caveolae). They can constitute up to a third of the plasma membrane area of the cells of some tissues, being especially abundant in smooth muscle, type I pneumocytes, fibroblasts, adipocytes, and endothelial cells. Uptake of extracellular molecules is also believed to be specifically mediated via receptors in caveolae.

- Potocytosis is a form of receptor-mediated endocytosis that uses caveolae vesicles to bring molecules of various sizes into the cell. Unlike most endocytosis that uses caveolae to deliver contents of vesicles to lysosomes or other organelles, material endocytosed via potocytosis is released into the cytosol.
- Macropinocytosis, which usually occurs from highly ruffled regions of the plasma membrane, is the invagination of the cell membrane to form a pocket, which then pinches off into the cell to form a vesicle (0.5–5 μm in diameter) filled with a large volume of extracellular fluid and molecules within it (equivalent to ~100 CCVs). The filling of the pocket occurs in a non-specific manner. The vesicle then travels into the cytosol and fuses with other vesicles such as endosomes and lysosomes.
- Phagocytosis is the process by which cells bind and internalize particulate matter larger than around 0.75 μm in diameter, such as small-sized dust particles, cell debris, micro-organisms and apoptotic cells. These processes involve the uptake of larger membrane areas than clathrin-mediated endocytosis and caveolae pathway.

More recent experiments have suggested that these morphological descriptions of endocytic events may be inadequate, and a more appropriate method of classification may be based upon the clathrin-dependence of particular pathways, with multiple subtypes of clathrin-dependent and clathrin-independent endocytosis. Mechanistic insight into non-phagocytic, clathrin-independent endocytosis has been lacking, but a recent study has shown how Graf1 regulates a highly prevalent clathrin-independent endocytic pathway known as the CLIC/GEEC pathway.

Principal Components of Endocytic Pathway

The endocytic pathway of mammalian cells consists of distinct membrane compartments, which internalize molecules from the plasma membrane and recycle them back to the surface (as in early endosomes and recycling endosomes), or sort them to degradation (as in late endosomes and lysosomes). The principal components of the endocytic pathway are:

- Early endosomes are the first compartment of the endocytic pathway. Early endosomes are often located in the periphery of the cell, and receive most types of vesicles coming from the cell surface. They have a characteristic tubulo-vesicular structure (vesicles up to 1 μm in diameter with connected tubules of approx. 50 nm diameter) and a mildly acid pH. They are principally sorting organelles where many endocytosed ligands dissociate from their receptors in the acid pH of the compartment, and from which many of the receptors recycle to the cell surface (via tubules). It is also the site of sorting into transcytotic pathway to later compartments (like late endosomes or lysosomes) via transvesicular compartments (like multivesicular bodies (MVB) or endosomal carrier vesicles (ECVs)).
- Late endosomes receive endocytosed material en route to lysosomes, usually from early endosomes in the endocytic pathway, from trans-Golgi network (TGN) in the biosynthetic pathway, and from phagosomes in the phagocytic pathway. Late endosomes often contain proteins characteristic of lysosomes, including lysosomal membrane glycoproteins and acid hydrolases. They are acidic (approx. pH 5.5), and are part of the trafficking pathway of mannose-6-phosphate receptors. Late endosomes are thought to mediate a final set of sorting events prior to delivery of material to lysosomes.

- Lysosomes are the last compartment of the endocytic pathway. Their chief function is to break down cellular waste products, fats, carbohydrates, proteins, and other macromolecules into simple compounds. These are then returned to the cytoplasm as new cell-building materials. To accomplish this, lysosomes use some 40 different types of hydrolytic enzymes, all of which are manufactured in the endoplasmic reticulum, modified in the Golgi apparatus and function in a acidic environment. The approximate pH of a lysosome is 4.8 and by electron microscopy (EM) usually appear as large vacuoles (1-2 μm in diameter) containing electron dense material. They have a high content of lysosomal membrane proteins and active lysosomal hydrolases, but no mannose-6-phosphate receptor. They are generally regarded as the principal hydrolytic compartment of the cell.

It was recently found that an eisosome serves as a portal of endocytosis in yeast.

Clathrin-mediated Endocytosis

The major route for endocytosis in most cells, and the best-understood, is that mediated by the molecule clathrin. This large protein assists in the formation of a coated pit on the inner surface of the plasma membrane of the cell. This pit then buds into the cell to form a coated vesicle in the cytoplasm of the cell. In so doing, it brings into the cell not only a small area of the surface of the cell but also a small volume of fluid from outside the cell.

Coats function to deform the donor membrane to produce a vesicle, and they also function in the selection of the vesicle cargo. Coat complexes that have been well characterized so far include coat protein-I (COP-I), COP-II, and clathrin. Clathrin coats are involved in two crucial transport steps: (i) receptor-mediated and fluid-phase endocytosis from the plasma membrane to early endosome and (ii) transport from the TGN to endosomes. In endocytosis, the clathrin coat is assembled on the cytoplasmic face of the plasma membrane, forming pits that invaginate to pinch off (scission) and become free CCVs. In cultured cells, the assembly of a CCV takes ~ 1min, and several hundred to a thousand or more can form every minute. The main scaffold component of clathrin coat is the 190-kD protein called clathrin heavy chain (CHC), which is associated with a 25- kD protein called clathrin light chain (CLC), forming three-legged trimers called triskelions.

Vesicles selectively concentrate and exclude certain proteins during formation and are not representative of the membrane as a whole. AP2 adaptors are multisubunit complexes that perform this function at the plasma membrane. The best-understood receptors that are found concentrated in coated vesicles of mammalian cells are the LDL receptor (which removes LDL from circulating blood), the transferrin receptor (which brings ferric ions bound by transferrin into the cell) and certain hormone receptors (such as that for EGF).

At any one moment, about 25% of the plasma membrane of a fibroblast is made up of coated pits. As a coated pit has a life of about a minute before it buds into the cell, a fibroblast takes up its surface by this route about once every 16 minutes. Coated vesicles formed from the plasma membrane have a diameter of about 36 nm and a lifetime measured in a few seconds. Once the coat has been shed, the remaining vesicle fuses with endosomes and proceeds down the endocytic pathway. The actual budding-in process, whereby a pit is converted to a vesicle, is carried out by clathrin assisted by a set of cytoplasmic proteins, which includes dynamin and adaptors such as adaptin.

Coated pits and vesicles were first seen in thin sections of tissue in the electron microscope by Matt Lions and Parker George. The importance of them for the clearance of LDL from blood was discovered by Richard G. Anderson, Michael S. Brown and Joseph L. Goldstein in 1977. Coated vesicles were first purified by Barbara Pearse, who discovered the clathrin coat molecule in 1976.

Exocytosis

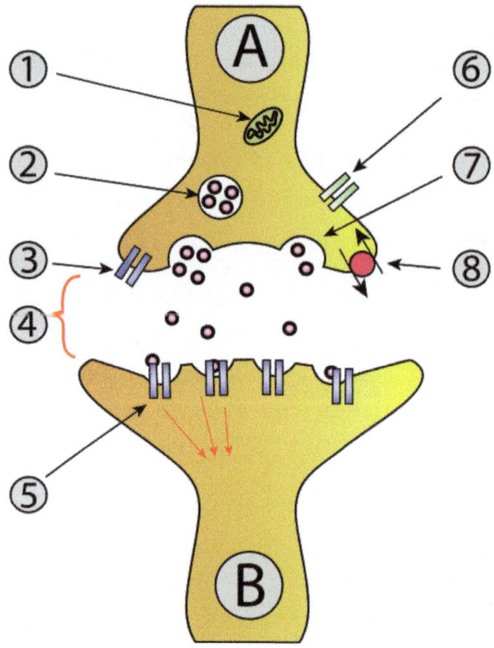

Neuron A (transmitting) to neuron B (receiving)
1. Mitochondrion 2. Synaptic vesicle with neurotransmitters 3. Autoreceptor 4. Synapse with neurotransmitter released (serotonin) 5. Postsynaptic receptors activated by neurotransmitter (induction of a postsynaptic potential) 6. Calcium channel 7. Exocytosis of a vesicle 8. Recaptured neurotransmitter

Exocytosis is a form of active transport in which a cell transports molecules (such as proteins) out of the cell (*exo- + cytosis*) by expelling them in an energy-using process. Exocytosis and its counterpart, endocytosis, are used by all cells because most chemical substances important to them are large polar molecules that cannot pass through the hydrophobic portion of the cell membrane by passive means.

In exocytosis, membrane-bound secretory vesicles are carried to the cell membrane, and their contents (water-soluble molecules such as proteins) are secreted into the extracellular environment. This secretion is possible because the vesicle transiently fuses with the outer cell membrane.

Exocytosis is also a mechanism by which cells are able to insert membrane proteins (such as ion channels and cell surface receptors), lipids, and other components into the cell membrane. Vesicles containing these membrane components fully fuse with and become part of the outer cell membrane.

History

The term was proposed by De Duve in 1963.

Types

In eukaryotes there are two types of exocytosis: 1) Ca^{2+} triggered non-constitutive (i.e., regulated exocytosis) and 2) non-Ca^{2+} triggered constitutive (i.e., non-regulated). *Ca^{2+} triggered non-constitutive* exocytosis requires an external signal, a specific sorting signal on the vesicles, a clathrin coat, as well as an increase in intracellular calcium. Exocytosis in neuronal chemical synapses is Ca^{2+} triggered and serves interneuronal signalling. *Constitutive exocytosis* is performed by all cells and serves the release of components of the extracellular matrix or delivery of newly synthesized membrane proteins that are incorporated in the plasma membrane after the fusion of the transport vesicle.

Vesicular exocytosis in prokaryote gram negative bacteria is a third mechanism and latest finding in exocytosis. The periplasm is pinched off as bacterial outer membrane vesicles (OMVs) for translocating microbial biochemical signals into eukaryotic host cells or other microbes located nearby, accomplishing control of the secreting microbe on its environment - including invasion of host, endotoxemia, competing with other microbes for nutrition, etc. This finding of membrane vesicle trafficking occurring at the host-pathogen interface also dispels the myth that exocytosis is purely a eukaryotic cell phenomenon.

Steps

Five steps are involved in exocytosis:

Vesicle Trafficking

Certain vesicle-trafficking steps require the transportation of a vesicle over a moderately small distance. For example, vesicles that transport proteins from the Golgi apparatus to the cell surface area, will be likely to use motor proteins and a cytoskeletal track to get closer to their target. Before tethering would have been appropriate, many of the proteins used for the active transport would have been instead set for passive transport, because the Golgi apparatus does not require ATP to transport proteins. Both the actin- and the microtubule-base are implicated in these processes, along with several motor proteins. Once the vesicles reach their targets, they come into contact with tethering factors that can restrain them.

Vesicle Tethering

It is useful to distinguish between the initial, loose *tethering* of vesicles to their objective from the more stable, *packing* interactions. Tethering involves links over distances of more than about half the diameter of a vesicle from a given membrane surface (>25 nm). Tethering interactions are likely to be involved in concentrating synaptic vesicles at the synapse.

Tethered vesicles are also involved in regular cell's transcription processes.

Vesicle Docking

Secretory vesicles transiently dock at the cell plasma membrane, preceding the formation of a tight t-/v-SNARE complex, leading to priming and the establishment of continuity between the opposing bilayers.

Vesicle Priming

In neuronal exocytosis, the term *priming* has been used to include all of the molecular rearrangements and ATP-dependent protein and lipid modifications that take place after initial docking of a synaptic vesicle but before exocytosis, such that the influx of calcium ions is all that is needed to trigger nearly instantaneous neurotransmitter release. In other cell types, whose secretion is constitutive (i.e. continuous, calcium ion independent, non-triggered) there is no priming.

Vesicle Fusion

Transient vesicle fusion is driven by SNARE proteins, resulting in release of vesicle contents into the extracellular space (or in case of neurons in the synaptic cleft).

The merging of the donor and the acceptor membranes accomplishes three tasks:

- The surface of the plasma membrane increases (by the surface of the fused vesicle). This is important for the regulation of cell size, e.g., during cell growth.
- The substances within the vesicle are released into the exterior. These might be waste products or toxins, or signaling molecules like hormones or neurotransmitters during synaptic transmission.
- Proteins embedded in the vesicle membrane are now part of the plasma membrane. The side of the protein that was facing the *inside* of the vesicle now faces the *outside* of the cell. This mechanism is important for the regulation of transmembrane and transporters.

Vesicle Retrieval

Retrieval of synaptic vesicles occurs by endocytosis. Some synaptic vesicles are recycled without a full fusion into the membrane (kiss-and-run fusion), while others require a complete reformation of synaptic vesicles from the membrane by a specialized complex of proteins (clathrin). Non-constitutive exocytosis and subsequent endocytosis are highly energy expending processes, and thus, are dependent on mitochondria.

Examination of cells following secretion using electron microscopy demonstrate increased presence of partially empty vesicles following secretion. This suggested that during the secretory process, only a portion of the vesicular content is able to exit the cell. This could only be possible if the vesicle were to temporarily establish continuity with the cell plasma membrane, expel a portion of its contents, then detach, reseal, and withdraw into the cytosol (endocytose). In this way, the secretory vesicle could be reused for subsequent rounds of exo-endocytosis, until completely empty of its contents.

References

- Rappoport JZ (June 2008). "Focusing on clathrin-mediated endocytosis". The Biochemical Journal. 412 (3): 415–23. doi:10.1042/BJ20080474. PMID 18498251
- Marsh, M.; McMahon, HT (July 1999). "The structural era of endocytosis". Science. 285 (5425): 215–20. doi:10.1126/science.285.5425.215. PMID 10398591. Retrieved 2009-06-19
- Luzio JP, Rous BA, Bright NA, Pryor PR, Mullock BM, Piper RC (May 1, 2000). "Lysosome-endosome fusion and lysosome biogenesis". Journal of Cell Science. 113 (9): 1515–24. PMID 10751143. Retrieved 2009-06-19

- Gaidarov I, Santini F, Warren RA, Keen JH (May 1999). "Spatial control of coated-pit dynamics in living cells". Nature Cell Biology. 1 (1): 1–7. doi:10.1038/8971. PMID 10559856
- Parton RG, Simons K (March 2007). "The multiple faces of caveolae". Nature Reviews Molecular Cell Biology. 8 (3): 185–94. doi:10.1038/nrm2122. PMID 17318224
- Mineo, Chieko; Anderson, Richard G. (2001). "Potocytosis". Histochemistry and Cell Biology. 116 (2): 109–118. doi:10.1007/s004180100289. PMID 11685539
- Mukherjee S, Ghosh RN, Maxfield FR (July 1997). "Endocytosis". Physiological Reviews. 77 (3): 759–803. PMID 9234965. Retrieved 2009-06-19
- Benmerah A, Lamaze C (August 2007). "Clathrin-coated pits: vive la différence?". Traffic (Copenhagen, Denmark). 8 (8): 970–82. doi:10.1111/j.1600-0854.2007.00585.x. PMID 17547704
- Robinson MS (March 1997). "Coats and vesicle budding". Trends in Cell Biology. 7 (3): 99–102. doi:10.1016/S0962-8924(96)10048-9. PMID 17708916
- Glick BS, Malhotra V (December 1998). "The curious status of the Golgi apparatus". Cell. 95 (7): 883–9. doi:10.1016/S0092-8674(00)81713-4. PMID 9875843
- Mellman I (1996). "Endocytosis and molecular sorting". Annual Review of Cell and Developmental Biology. 12: 575–625. doi:10.1146/annurev.cellbio.12.1.575. PMID 8970738
- Gruenberg J, Maxfield FR (August 1995). "Membrane transport in the endocytic pathway". Current Opinion in Cell Biology. 7 (4): 552–63. doi:10.1016/0955-0674(95)80013-1. PMID 7495576

3

Carbohydrates and its Types

A carbohydrate is a molecule comprising carbon, oxygen and hydrogen atoms. It is also referred to as saccharides and can be classified as monosaccharide, disaccharide, oligosaccharide, and polysaccharide. This chapter discusses carbohydrates in a critical manner providing key analysis to the subject matter.

Carbohydrates

Carbohydrates are defined as polyhydroxy aldehyde or ketone with empirical formula $(CH_2O)_n$., the simplest being glyceraldehyde or dihydroxy acetone. Carbohydrates include sugars, starches, cellulose and many other compounds found in living organisms.

What are Saccharides?

- Saccharide is a term derived from the Latin for sugar (origin = "sweet sand").The term carbohydrate is most common in biochemistry where it is a synonym of saccharide.
- Carbohydrates are often classified according to the number of saccharide units they contain. They are divided into four chemical groupings: monosaccharides, disaccharides, oligosaccharides and ploysaccharides.
- In their basic form, carbohydrates are simple sugars or monosaccharides. These simple sugars can combine with each other to form more complex carbohydrates. The combination of two simple sugars is a disaccharide. Carbohydrates comprising of 2-10 monosaccharide units are called oligosaccharides, and those with a larger number are called polysaccharides.

Basic Structure

- The saccharides encountered in living sytems are generally built of monosaccharides with general formula $(CH_2O)_n$ where n is three or more.
- A typical monosaccharide has the structure H -$(CHOH)_x$ (C = O) - $(CHOH)_y$ - H, containing a ketonic functional group or H - (C = O) - $(CHOH)_x$ - H, containing an aldehydic functional group. It has hydroxyl groups on each carbon atom, excluding the functional group carbon atom.
- Glucose, fructose , ribose, deoxyribose and glyceraldehyde are most commonly occurring monosacchrides in the living systems.
- It is important to note that there are many chemicals that may have the same formula but are not considered to be monosaccharides (e.g., formaldehyde CH_2O and inositol $(CH_2O)_6$.

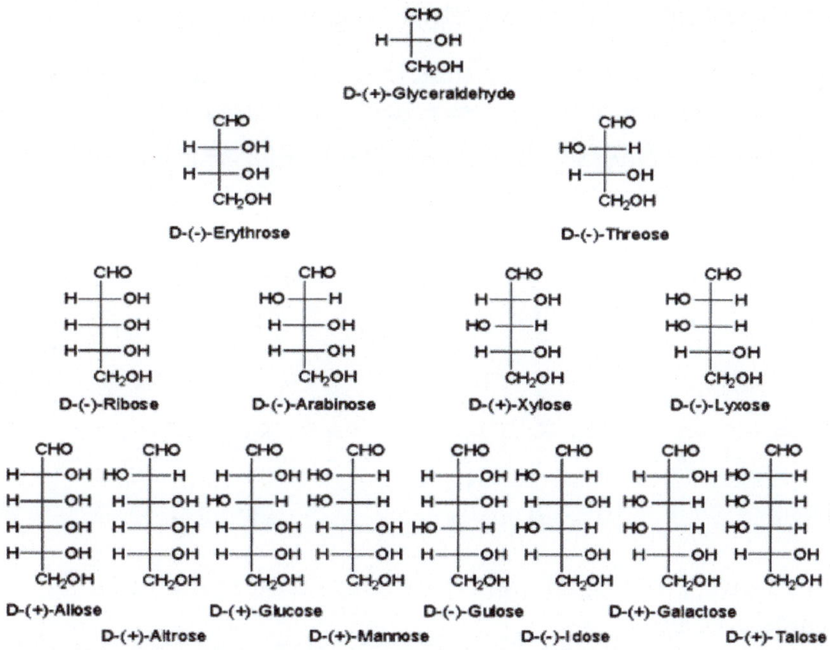

Study of Carbohydrates Necessitates the Concept of Isomerism

- Two broad categories for isomeric forms are :

  (i) Structural isomers.

  (ii) Stereo isomers.

- The structural isomers are defined as isomers having same molecule formula but different structures.
- The stereo isomers have same molecular and structural formula but differ in configuration i.e. arrangement of atoms in space.
- Stereo isomers are further sub grouped into optical isomers and geometrical isomers.
- Optical isomerism are more relevant in carbohydrate chemistry.
- Optical isomerism stems from the presence of chiral centre (asymmetric carbon atom). Chiral center refers to the carbon atom having four different groups attached to it. This leads to two possibilities by which atoms can be arranged as shown in figure below :

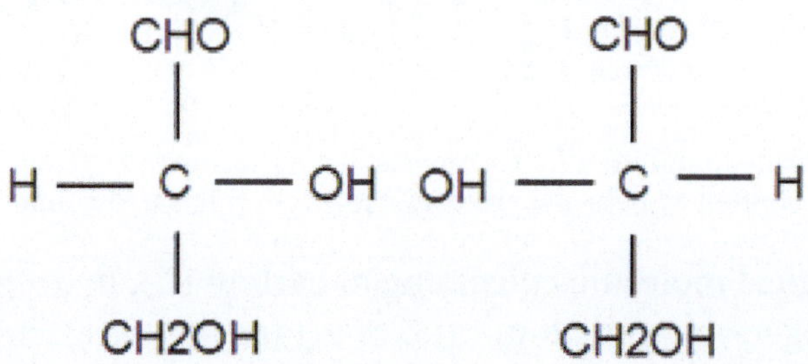

- In case of carbohydrates, D-glyceraldehyde is used as reference compound.
- D- represents the hydroxyl group on right hand side, whereas L- represents hydroxyl group on the left hand side. These two forms reflect mirror image of each others and calledc enantiomers. The stereoisomers which are not enantiomers are termed as distereoisomers.

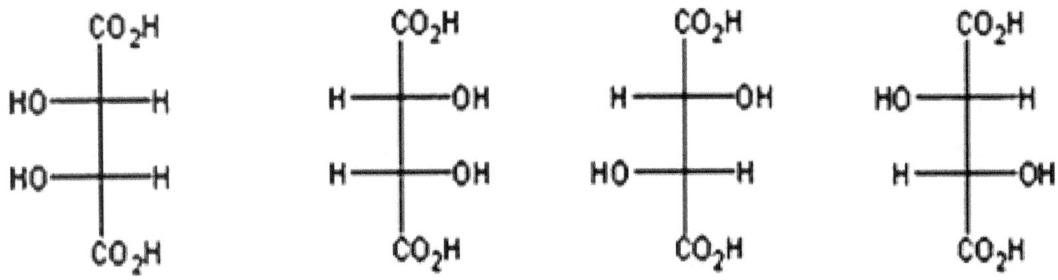

- In the figure above, I and II, III and IV are enantiomers whereas I, III, IV and II,III, IV are related as diastereomers.
- Enantiomers have same physical properties like melting point, boiling point, solubility in various solvents but they rotate plane polarized light in opposite directions. Those which rotate plane polarized light in clock wise directionare called dextrorotatory (represented by +) and those which rotate in anticlock wise direction are called levorotatory (represented by -). Thus D - Glucose can exist as both dextrorotatory (+) and Levorotatory (-).
- Van't Hoff formula of 2n works gives the numbers of possible optical isomers, where n is the number of chiral carbon. A triose will have two optical isomers and a tetrose will have four.
- D - Glucose and D - Mannose have different configuration only at C - 2 carbon. Such carbohydrates which differ in configuration only at one carbon atom are designated as epimers of each other.

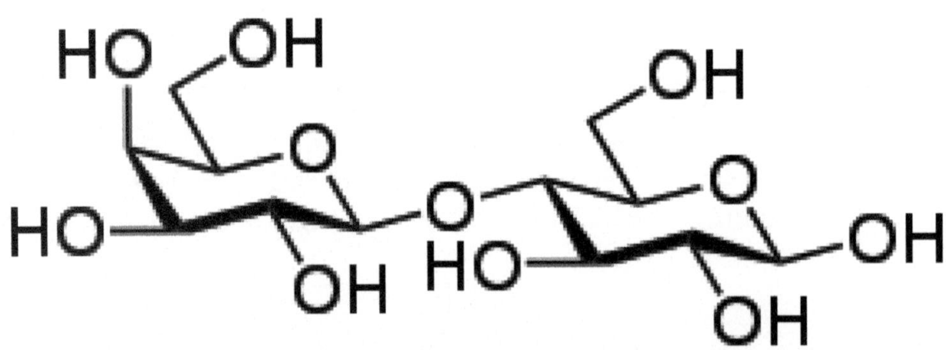

Lactose is a disaccharide found in milk. It consists of a molecule of D-galactose and a molecule of D-glucose bonded by beta-1-4 glycosidic linkage. It has a formula of $C_{12}H_{22}O_{11}$.

A carbohydrate is a biological molecule consisting of carbon (C), hydrogen (H) and oxygen (O) atoms, usually with a hydrogen–oxygen atom ratio of 2:1 (as in water); in other words, with the empirical formula $C_m(H_2O)_n$ (where m could be different from n). This formula holds true for monosaccharides. Some exceptions exist; for example, deoxyribose, a sugar component of DNA, has the empirical formula $C_5H_{10}O_4$. Carbohydrates are technically hydrates of carbon; structurally it is more accurate to view them as polyhydroxy aldehydes and ketones.

The term is most common in biochemistry, where it is a synonym of 'saccharide', a group that includes sugars, starch, and cellulose. The saccharides are divided into four chemical groups: monosaccharides, disaccharides, oligosaccharides, and polysaccharides. Monosaccharides and disaccharides, the smallest (lower molecular weight) carbohydrates, are commonly referred to as sugars. The word *saccharide* comes from the Greek word (*sákkharon*), meaning "sugar". While the scientific nomenclature of carbohydrates is complex, the names of the monosaccharides and disaccharides very often end in the suffix -ose. For example, grape sugar is the monosaccharide glucose, cane sugar is the disaccharide sucrose, and milk sugar is the disaccharide lactose.

Carbohydrates perform numerous roles in living organisms. Polysaccharides serve for the storage of energy (e.g. starch and glycogen) and as structural components (e.g. cellulose in plants and chitin in arthropods). The 5-carbon monosaccharide ribose is an important component of coenzymes (e.g. ATP, FAD and NAD) and the backbone of the genetic molecule known as RNA. The related deoxyribose is a component of DNA. Saccharides and their derivatives include many other important biomolecules that play key roles in the immune system, fertilization, preventing pathogenesis, blood clotting, and development.

In food science and in many informal contexts, the term carbohydrate often means any food that is particularly rich in the complex carbohydrate starch (such as cereals, bread and pasta) or simple carbohydrates, such as sugar (found in candy, jams, and desserts).

Often in lists of nutritional information, such as the USDA National Nutrient Database, the term "carbohydrate" (or "carbohydrate by difference") is used for everything other than water, protein, fat, ash, and ethanol. This will include chemical compounds such as acetic or lactic acid, which are not normally considered carbohydrates. It also includes "dietary fiber" which is a carbohydrate but which does not contribute much in the way of food energy (calories), even though it is often included in the calculation of total food energy just as though it were a sugar.

Carbohydrates are found in wide variety of foods. The important sources are cereals (wheat, maize, rice), potatoes, sugarcane, fruits, table sugar(sucrose), bread, milk, etc. Starch and sugar are the important carbohydrates in our diet. Starch is abundant in potatoes, maize, rice and other cereals. Sugar appears in our diet mainly as sucrose(table sugar) which is added to drinks and many prepared foods such as jam, biscuits and cakes. Glucose and fructose are found naturally in many fruits and some vegetables. Glycogen is carbohydrate found in the liver and muscles (as animal source). Cellulose in the cell wall of all plant tissue is a carbohydrate. It is important in our diet as fibre which helps to maintain a healthy digestive system.

Structure

Formerly the name "carbohydrate" was used in chemistry for any compound with the formula $C_m(H_2O)_n$. Following this definition, some chemists considered formaldehyde (CH_2O) to be the simplest carbohydrate, while others claimed that title for glycolaldehyde. Today, the term is generally understood in the biochemistry sense, which excludes compounds with only one or two carbons and includes many biological carbohydrates which deviate from this formula. For example, while the above representative formulas would seem to capture the commonly known carbohydrates, ubiquitous and abundant carbohydrates often deviate from this. For example, carbohydrates often display chemical groups such as: *N*-acetyl (e.g. chitin), sulphate (e.g. glycosaminoglycans), carboxylic acid (e.g. sialic acid) and deoxy modifications (e.g. fucose and sialic acid).

Natural saccharides are generally built of simple carbohydrates called monosaccharides with general formula $(CH_2O)_n$ where n is three or more. A typical monosaccharide has the structure $H-(CHOH)_x(C=O)-(CHOH)_y-H$, that is, an aldehyde or ketone with many hydroxyl groups added, usually one on each carbon atom that is not part of the aldehyde or ketone functional group. Examples of monosaccharides are glucose, fructose, and glyceraldehydes. However, some biological substances commonly called "monosaccharides" do not conform to this formula (e.g. uronic acids and deoxy-sugars such as fucose) and there are many chemicals that do conform to this formula but are not considered to be monosaccharides (e.g. formaldehyde CH_2O and inositol $(CH_2O)_6$).

The open-chain form of a monosaccharide often coexists with a closed ring form where the aldehyde/ketone carbonyl group carbon (C=O) and hydroxyl group (–OH) react forming a hemiacetal with a new C–O–C bridge.

Monosaccharides can be linked together into what are called polysaccharides (or oligosaccharides) in a large variety of ways. Many carbohydrates contain one or more modified monosaccharide units that have had one or more groups replaced or removed. For example, deoxyribose, a component of DNA, is a modified version of ribose; chitin is composed of repeating units of N-acetyl glucosamine, a nitrogen-containing form of glucose.

Division

Carbohydrates are polyhydroxy aldehydes, ketones, alcohols, acids, their simple derivatives and their polymers having linkages of the acetal type. They may be classified according to their degree of polymerization and may be divided initially into three principal groups, namely sugars, oligosaccharides and polysaccharides

The major dietary carbohydrates		
Class (DP*)	**Subgroup**	**Components**
Sugars (1–2)	Monosaccharides	Glucose, galactose, fructose, xylose
	Disaccharides	Sucrose, lactose, maltose, trehalose
	Polyols	Sorbitol, mannitol
Oligosaccharides (3–9)	Malto-oligosaccharides	Maltodextrins
	Other oligosaccharides	Raffinose, stachyose, fructo-oligosaccharides
Polysaccharides (>9)	Starch	Amylose, amylopectin, modified starches
	Non-starch polysaccharides	Cellulose, hemicellulose, pectins, hydrocolloids

DP * = Degree of polymerization

Monosaccharides

Monosaccharides are the simplest carbohydrates in that they cannot be hydrolyzed to smaller carbohydrates. They are aldehydes or ketones with two or more hydroxyl groups. The general chemical formula of an unmodified monosaccharide is $(C \bullet H_2O)_n$, literally a "carbon hydrate". Monosaccharides are important fuel molecules as well as building blocks for nucleic acids. The smallest monosaccharides, for which n=3, are dihydroxyacetone and D- and L-glyceraldehydes.

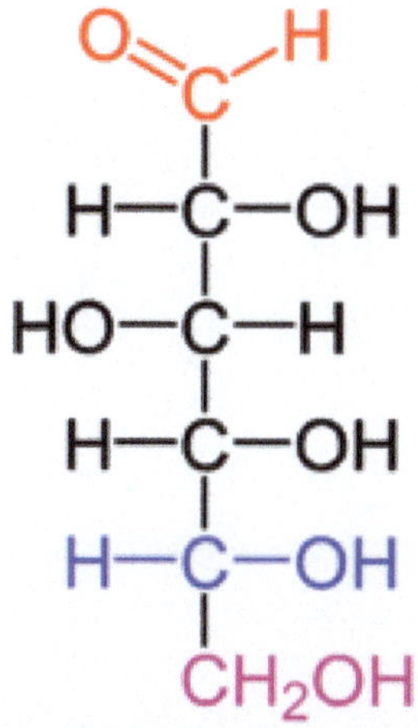

D-glucose is an aldohexose with the formula $(C \cdot H_2O)_6$. The red atoms highlight the aldehyde group and the blue atoms highlight the asymmetric center furthest from the aldehyde; because this -OH is on the right of the Fischer projection, this is a D sugar.

Classification of Monosaccharides

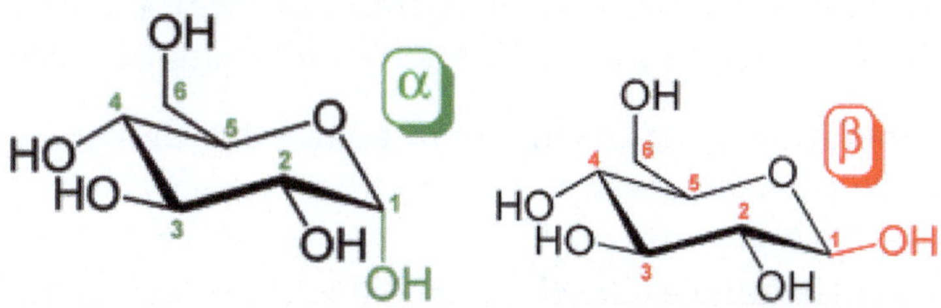

The α and β anomers of glucose. Note the position of the hydroxyl group (red or green) on the anomeric carbon relative to the CH_2OH group bound to carbon 5: they either have identical absolute configurations (R,R or S,S) (α), or opposite absolute configurations (R,S or S,R) (β).

Monosaccharides are classified according to three different characteristics: the placement of its carbonyl group, the number of carbon atoms it contains, and its chiral handedness. If the carbonyl group is an aldehyde, the monosaccharide is an aldose; if the carbonyl group is a ketone, the monosaccharide is a ketose. Monosaccharides with three carbon atoms are called trioses, those with four are called tetroses, five are called pentoses, six are hexoses, and so on. These two systems of classification are often combined. For example, glucose is an aldohexose (a six-carbon aldehyde), ribose is an aldopentose (a five-carbon aldehyde), and fructose is a ketohexose (a six-carbon ketone).

Each carbon atom bearing a hydroxyl group (-OH), with the exception of the first and last carbons, are asymmetric, making them stereo centers with two possible configurations each (R or S). Because of this asymmetry, a number of isomers may exist for any given monosaccharide formula. Using Le Bel-van't Hoff rule, the aldohexose D-glucose, for example, has the formula $(C{\cdot}H_2O)_6$, of which four of its six carbons atoms are stereogenic, making D-glucose one of 2^4=16 possible stereoisomers. In the case of glyceraldehydes, an aldotriose, there is one pair of possible stereoisomers, which are enantiomers and epimers. 1, 3-dihydroxyacetone, the ketose corresponding to the aldose glyceraldehydes, is a symmetric molecule with no stereo centers. The assignment of D or L is made according to the orientation of the asymmetric carbon furthest from the carbonyl group: in a standard Fischer projection if the hydroxyl group is on the right the molecule is a D sugar, otherwise it is an L sugar. The "D-" and "L-" prefixes should not be confused with "d-" or "l-", which indicate the direction that the sugar rotates plane polarized light. This usage of "d-" and "l-" is no longer followed in carbohydrate chemistry.

Ring-straight Chain Isomerism

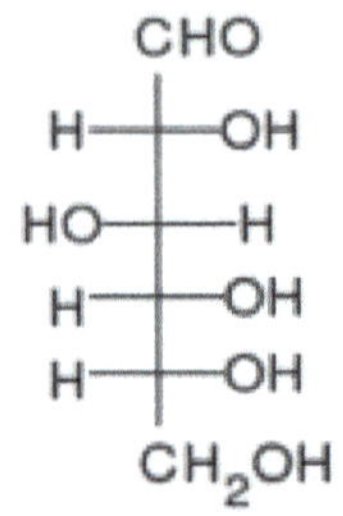

Glucose can exist in both a straight-chain and ring form.

The aldehyde or ketone group of a straight-chain monosaccharide will react reversibly with a hydroxyl group on a different carbon atom to form a hemiacetal or hemiketal, forming a heterocyclic ring with an oxygen bridge between two carbon atoms. Rings with five and six atoms are called furanose and pyranose forms, respectively, and exist in equilibrium with the straight-chain form.

During the conversion from straight-chain form to the cyclic form, the carbon atom containing the carbonyl oxygen, called the anomeric carbon, becomes a stereogenic center with two possible configurations: The oxygen atom may take a position either above or below the plane of the ring. The resulting possible pair of stereoisomers is called anomers. In the *α anomer*, the -OH substituent on the anomeric carbon rests on the opposite side (trans) of the ring from the CH_2OH side branch. The alternative form, in which the CH_2OH substituent and the anomeric hydroxyl are on the same side (cis) of the plane of the ring, is called the *β anomer*.

Use in Living Organisms

Monosaccharides are the major source of fuel for metabolism, being used both as an energy source (glucose being the most important in nature) and in biosynthesis. When monosaccharides are not immediately needed by many cells they are often converted to more space-efficient forms, often polysaccharides. In many animals, including humans, this storage form is glycogen, especially in liver and muscle cells. In plants, starch is used for the same purpose. The most abundant carbohydrate, cellulose, is a structural component of the cell wall of plants and many forms of algae. Ribose is a component of RNA. Deoxyribose is a component of DNA. Lyxose is a component of lyxoflavin found in the human heart. Ribulose and xylulose occur in the pentose phosphate pathway. Galactose, a component of milk sugar lactose, is found in galactolipids in plant cell membranes and in glycoproteins in many tissues. Mannose occurs in human metabolism, especially in the glycosylation of certain proteins. Fructose, or fruit sugar, is found in many plants and in humans, it is metabolized in the liver, absorbed directly into the intestines during digestion, and found in semen. Trehalose, a major sugar of insects, is rapidly hydrolyzed into two glucose molecules to support continuous flight.

Disaccharides

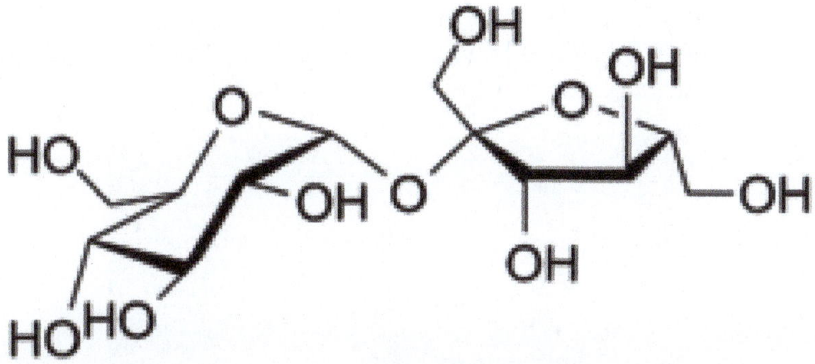

Sucrose, also known as table sugar, is a common disaccharide. It is composed of two monosaccharides: D-glucose (left) and D-fructose (right).

Two joined monosaccharides are called a disaccharide and these are the simplest polysaccharides. Examples include sucrose and lactose. They are composed of two monosaccharide units bound together by a covalent bond known as a glycosidic linkage formed via a dehydration reaction, resulting in the loss of a hydrogen atom from one monosaccharide and a hydroxyl group from the other. The formula of unmodified disaccharides is $C_{12}H_{22}O_{11}$. Although there are numerous kinds of disaccharides, a handful of disaccharides are particularly notable.

Sucrose, pictured to the right, is the most abundant disaccharide, and the main form in which carbohydrates are transported in plants. It is composed of one D-glucose molecule and one D-fructose molecule. The systematic name for sucrose, *O*-α-D-glucopyranosyl-(1→2)-D-fructofuranoside, indicates four things:

- Its monosaccharides: glucose and fructose
- Their ring types: glucose is a pyranose and fructose is a furanose
- How they are linked together: the oxygen on carbon number 1 (C1) of α-D-glucose is linked to the C2 of D-fructose.
- The *-oside* suffix indicates that the anomeric carbon of both monosaccharides participates in the glycosidic bond.

Lactose, a disaccharide composed of one D-galactose molecule and one D-glucose molecule, occurs naturally in mammalian milk. The systematic name for lactose is *O*-β-D-galactopyranosyl-(1→4)-D-glucopyranose. Other notable disaccharides include maltose (two D-glucoses linked α-1,4) and cellulobiose (two D-glucoses linked β-1,4). Disaccharides can be classified into two types: reducing and non-reducing disaccharides. If the functional group is present in bonding with another sugar unit, it is called a reducing disaccharide or biose.

Nutrition

Carbohydrate consumed in food yields 3.87 calories of energy per gram for simple sugars, and 3.57 to 4.12 calories per gram for complex carbohydrate in most other foods. Relatively high levels of carbohydrate are associated with processed foods or refined foods made from plants, including sweets, cookies and candy, table sugar, honey, soft drinks, breads and crackers, jams and fruit products, pastas and breakfast cereals. Lower amounts of carbohydrate are usually associated with unrefined foods, including beans, tubers, rice, and unrefined fruit. Animal-based foods generally have the lowest carbohydrate levels, although milk does contain a high proportion of lactose.

Grain products: rich sources of carbohydrates

Carbohydrates are a common source of energy in living organisms; however, no carbohydrate is an essential nutrient in humans. Humans are able to obtain all of their energy requirement from protein and fats, though the potential for some negative health effects of extreme carbohydrate restriction remains, as the issue has not been studied extensively yet. However, in the case of dietary fiber – indigestible carbohydrates which are not a source of energy – inadequate intake can lead to significant increases in mortality.

Following a diet consisting of very low amounts of daily carbohydrate for several days will usually result in higher levels of blood ketone bodies than an isocaloric diet with similar protein content. This relatively high level of ketone bodies is commonly known as ketosis and is very often confused with the potentially fatal condition often seen in type 1 diabetics known as diabetic ketoacidosis. Somebody suffering ketoacidosis will have much higher levels of blood ketone bodies along with high blood sugar, dehydration and electrolyte imbalance.

Long-chain fatty acids cannot cross the blood–brain barrier, but the liver can break these down to produce ketones. However, the medium-chain fatty acids octanoic and heptanoic acids can cross the barrier and be used by the brain, which normally relies upon glucose for its energy. Gluconeogenesis allows humans to synthesize some glucose from specific amino acids from the glycerol backbone in triglycerides and in some cases from fatty acids.

Organisms typically cannot metabolize all types of carbohydrate to yield energy. Glucose is a nearly universal and accessible source of energy. Many organisms also have the ability to metabolize other monosaccharides and disaccharides but glucose is often metabolized first. In *Escherichia coli*, for example, the lac operon will express enzymes for the digestion of lactose when it is present, but if both lactose and glucose are present the *lac* operon is repressed, resulting in the glucose being used first. Polysaccharides are also common sources of energy. Many organisms can easily break down starches into glucose; most organisms, however, cannot metabolize cellulose or other polysaccharides like chitin and arabinoxylans. These carbohydrate types can be metabolized by some bacteria and protists. Ruminants and termites, for example, use microorganisms to process cellulose. Even though these complex carbohydrates are not very digestible, they represent an important dietary element for humans, called dietary fiber. Fiber enhances digestion, among other benefits.

Based on the effects on risk of heart disease and obesity, the Institute of Medicine recommends that American and Canadian adults get between 45–65% of dietary energy from carbohydrates. The Food and Agriculture Organization and World Health Organization jointly recommend that national dietary guidelines set a goal of 55–75% of total energy from carbohydrates, but only 10% directly from sugars (their term for simple carbohydrates).

Classification

Nutritionists often refer to carbohydrates as either simple or complex. However, the exact distinction between these groups can be ambiguous. The term *complex carbohydrate* was first used in the U.S. Senate Select Committee on Nutrition and Human Needs publication *Dietary Goals for the United States* where it was intended to distinguish sugars from other carbohydrates (which were perceived to be nutritionally superior). However, the report put "fruit, vegetables and whole-grains" in the complex carbohydrate column, despite the fact that these may contain sugars as well as polysaccharides. This confusion persists as today some nutritionists use the term complex carbohydrate to refer to any sort of digestible saccharide present in a whole food, where fiber, vitamins and minerals are also found (as opposed to processed carbohydrates, which provide energy but few other nutrients). The standard usage, however, is to classify carbohydrates chemically: simple if they are sugars (monosaccharides and disaccharides) and complex if they are polysaccharides (or oligosaccharides).

In any case, the simple vs. complex chemical distinction has little value for determining the nutritional quality of carbohydrates. Some simple carbohydrates (e.g. fructose) raise blood glucose slowly, while some complex carbohydrates (starches), especially if processed, raise blood sugar rapidly. The speed of digestion is determined by a variety of factors including which other nutrients are consumed with the carbohydrate, how the food is prepared, individual differences in metabolism, and the chemistry of the carbohydrate.

The USDA's *Dietary Guidelines for Americans 2010* call for moderate- to high-carbohydrate consumption from a balanced diet that includes six one-ounce servings of grain foods each day, at least half from whole grain sources and the rest from enriched.

The glycemic index (GI) and glycemic load concepts have been developed to characterize food behavior during human digestion. They rank carbohydrate-rich foods based on the rapidity and magnitude of their effect on blood glucose levels. Glycemic index is a measure of how quickly food glucose is absorbed, while glycemic load is a measure of the total absorbable glucose in foods. The insulin index is a similar, more recent classification method that ranks foods based on their effects on blood insulin levels, which are caused by glucose (or starch) and some amino acids in food.

Metabolism

Carbohydrate metabolism denotes the various biochemical processes responsible for the formation, breakdown and interconversion of carbohydrates in living organisms.

The most important carbohydrate is glucose, a simple sugar (monosaccharide) that is metabolized by nearly all known organisms. Glucose and other carbohydrates are part of a wide variety of metabolic pathways across species: plants synthesize carbohydrates from carbon dioxide and water by photosynthesis storing the absorbed energy internally, often in the form of starch or lipids. Plant components are consumed by animals and fungi, and used as fuel for cellular respiration. Oxidation of one gram of carbohydrate yields approximately 4 kcal of energy, while the oxidation of one gram of lipids yields about 9 kcal. Energy obtained from metabolism (e.g., oxidation of glucose) is usually stored temporarily within cells in the form of ATP. Organisms capable of aerobic respiration metabolize glucose and oxygen to release energy with carbon dioxide and water as byproducts.

Catabolism

Catabolism is the metabolic reaction which cells undergo to extract energy. There are two major metabolic pathways of monosaccharide catabolism: glycolysis and the citric acid cycle.

In glycolysis, oligo/polysaccharides are cleaved first to smaller monosaccharides by enzymes called glycoside hydrolases. The monosaccharide units can then enter into monosaccharide catabolism. In some cases, as with humans, not all carbohydrate types are usable as the digestive and metabolic enzymes necessary are not present.

Carbohydrate Chemistry

Carbohydrate chemistry is a large and economically important branch of organic chemistry. Some of the main organic reactions that involve carbohydrates are:

- Carbohydrate acetalisation

- Cyanohydrin reaction
- Lobry-de Bruyn-van Ekenstein transformation
- Amadori rearrangement
- Nef reaction
- Wohl degradation
- Koenigs–Knorr reaction
- Carbohydrate digestion

Carbohydrate chemistry is a subdiscipline of chemistry primarily concerned with the synthesis, structure, and function of carbohydrates. Due to the general structure of carbohydrates, their synthesis is often preoccupied with the selective formation of glycosidic linkages and the selective reaction of hydroxyl groups; as a result, it relies heavily on the use of protecting groups.

Glycosidic Bond Formation

- Chemical glycosylation
- Fischer glycosidation
- Glycosyl halide
- Koenigs-Knorr reaction

Protecting Groups

- Carbohydrate acetalisation

Reactions of Carbohydrates

Carbohydrates are reactants in many organic reactions. For example:

- Cyanohydrin reaction
- Lobry-de Bruyn-van Ekenstein transformation
- Amadori rearrangement
- Nef reaction
- Wohl degradation
- Tipson-Cohen reaction
- Ferrier rearrangement
- Ferrier II reaction

Functions of Carbohydrates

Carbohydrates have five major functions within the body:

1. Energy supply, particularly for the brain in the form of glucose
2. Avoiding the breakdown of amino acids for energy
3. Avoiding ketosis from the breakdown of fatty acids
4. Cellular and protein recognition

Cellular and Protein Recognition

Glycoprotein hormones may be removed by the liver from the bloodstream when the passage of time causes the breaking-off of carbohydrates from the glycoproteins.

Amino Sugar

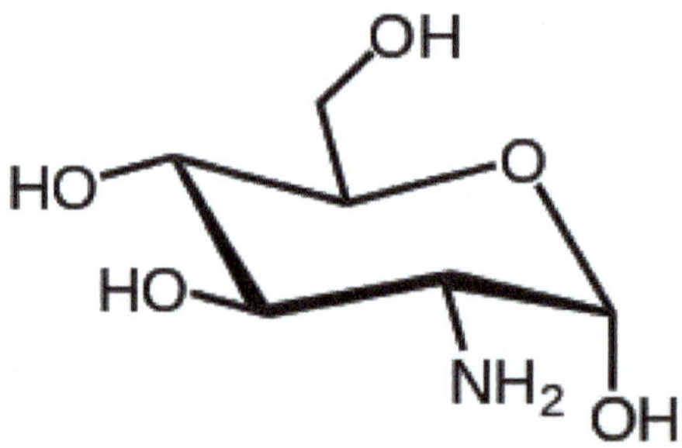

Glucosamine

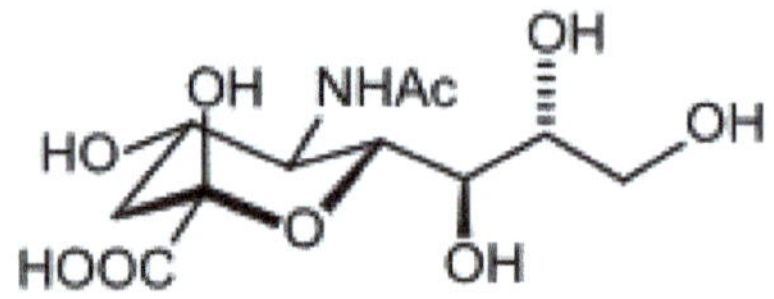

Sialic acid

In chemistry, an amino sugar (or more technically a 2-amino-2-deoxysugar) is a sugar molecule in which a hydroxyl group has been replaced with an amine group. More than 60 amino sugars are known, with one of the most abundant being *N*-Acetyl-D-glucosamine, which is the main component of chitin.

Derivatives of amine containing sugars, such as *N*-acetylglucosamine and sialic acid, whose nitrogens are part of more complex functional groups rather than formally being amines, are also considered amino sugars. Aminoglycosides are a class of antimicrobial compounds that inhibit bacterial protein synthesis. These compounds are conjugates of amino sugars and aminocyclitols.

Synthesis

From Glycals

Glycals are cyclic enol ether derivatives of monosaccharides, having a double bond between carbon atoms 1 and 2 of the ring. *N*-functionalized of glycals at the C2 position, combined with glycosidic bond formation at C1 is a common strategy for the synthesis of amino sugars. This can be achieved using azides with subsequent reduction yielding the amino sugar. One advantage of introducing azide moiety at C-2 lies in its non-participatory ability, which could serve as the basis of stereoseletive synthesis of 1.2-cis-glycosidic linkage.

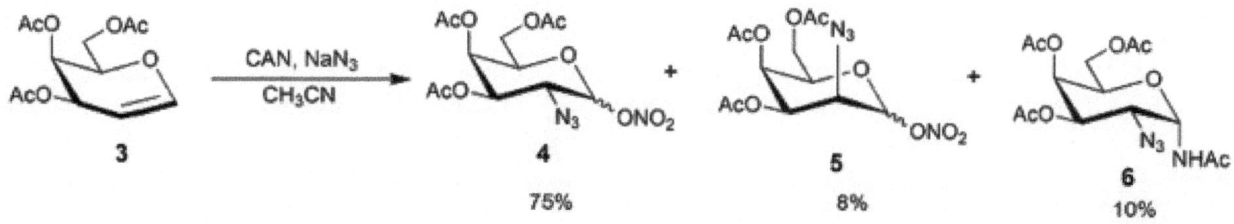

Azides give high regioselectivity, however stereoselectivity both at C-1 and C-2 is generally poor. Usually anomeric mixtures will be obtained and the stereochemistry formed at C-2 is heavily dependent upon the starting substrates. For galactal, addition of azide to the double bond will preferentially occur from equatorial direction because of steric hindrance at the top face caused by axial group at C-4. For glucal, azide could attack from both axial and equatorial directions with almost similar probability, so its selectivity will decrease.

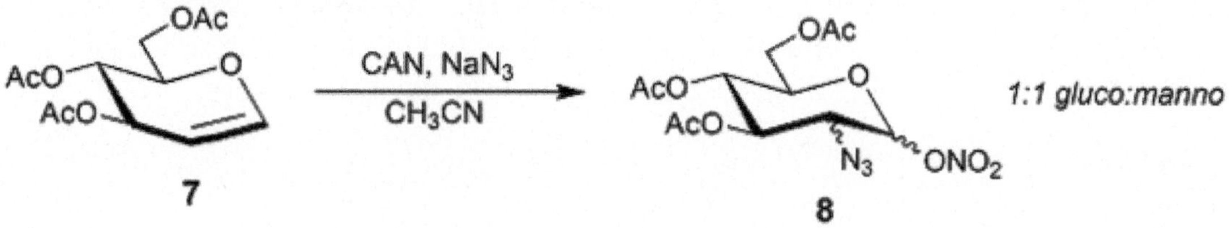

Glycals may also be converted into amino sugars by nitration followed by treatment with thiophenol (Michael addition) to furnish a thioglycoside donor. This is a versatile donor and can react with simple or carbohydrate alcohols to establish the glycosidic linkage, with reduction and *N*-acetylation of nitro group will give the targeted product.

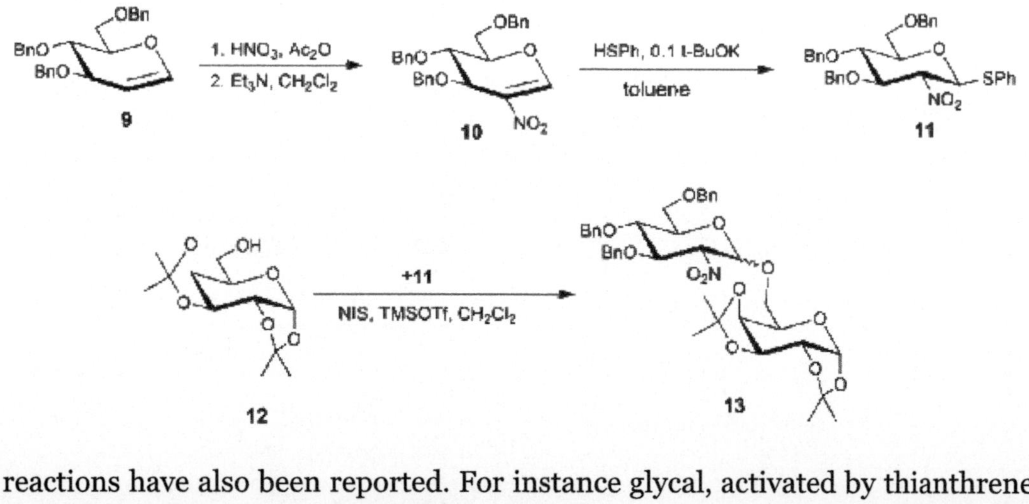

One-pot reactions have also been reported. For instance glycal, activated by thianthrene-5-oxide and Tf_2O is treated with an amide nucleophile and a glycosyl acceptor to produce various 1,2-trans C-2-amidoglycosides. Both the C-2 nitrogen introduction and the glycosidic bond formation precede stereoselectively. This methodology makes the introduction of both natural and non-natural amide functionalities at C-2 possible and more importantly with glycosidic bond formation at the same time in a one-pot procedure.

Via Nucleophilic Displacement

Nucleophilic displacement can be an effective strategy for the synthesis of amino sugars, however success strongly depends upon the nature of nucleophile, the type of leaving group and site of displacements on sugar rings. One aspect of this problem is that displacements at the C2 position tend to be slow as it is adjacent to the anomeric centre; this is particularly true for glycosides with axially-oriented aglycones.

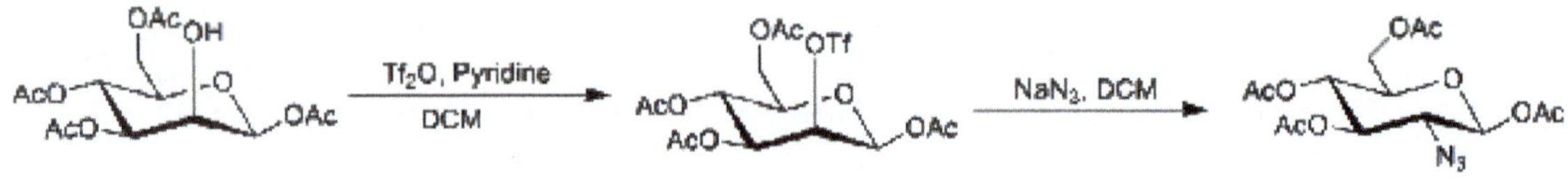

19 20 21

Epoxides are suitable starting materials for realizing nucleophilic displacement reaction to introduce azide into C-2. Anhydrosugar 21 could be transformed into thioglycoside 22, which serves as a donor to react with alcohols to obtain 2-azide-2-deoxy-*O*-glycosides. The subsequent reduction and *N*-acetylation will furnish the desired 2-*N*-acetamido-2-deoxyglycosides.

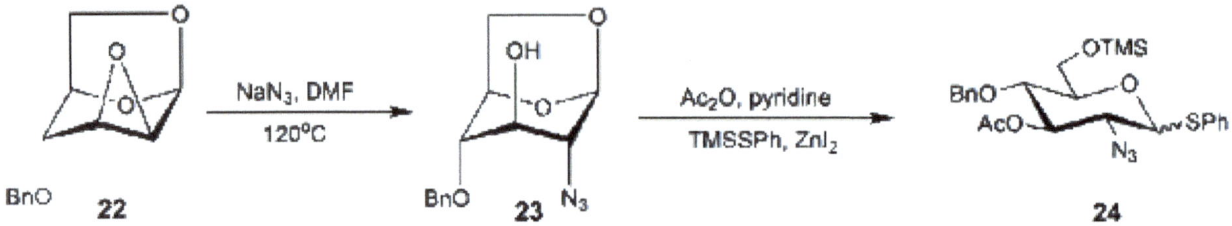

22 23 24

Classification of Carbohydrates

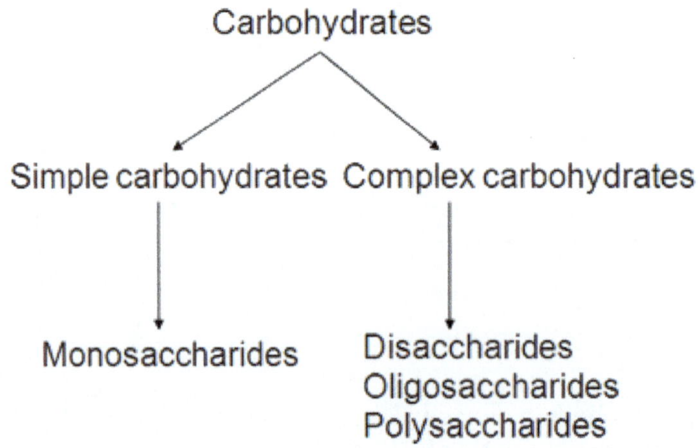

Simple Carbohydrates

- These are made up of a single basic sugar. Simple carbohydrates are responsible for the sweet taste in our food. Fruit sugar, table sugar or corn sugar are all simple sugars. On consumption, these sugars are directly absorbed in the blood and generally used for energy requirements of the body.
- Glucose provides instant energy and reaches different parts of the body via blood, by being quickly metabolized.
- Simple sugars are occur in plenty in natural foods like fruits, vegetables, milk and milk products. Additionally, honey, molasses, corn and maple syrup are also rich sources of simple sugars.

Monosaccharides

- 'Mono' refers to single.
- These are the basic compounds consisting of carbon, hydrogen and oxygen in the ratio 1:2:1 having the emperical formula of $(CH_2O)_n$.
- Monosaccharides are sweet to taste, colourless crystalline solids, freely soluble in water but insoluble in nonpolar solvents.
- Glucose, fructose and galactose are types of monosaccharides.

Basic Structure of Monosaccharides

- An unbranched single bonded carbon chain provides backbone for monosaccharides.
- One of the carbon atom is doubly bonded to an oxygen atom resulting in formation of a carbonyl group.
- The monosaccharides may be an aldehyde (carbonyl group placed at the end of the carbon chain) and is referred to as an ALDOSE or a ketone (when the carbonyl group is placed at any other position in the chain) and is referred to as a KETOSE.
- For example, Glyceraldehyde is an aldose while Dihydroxy acteone is a ketose.
- Monosaccharides containing 3,4,5,6 or 7 carbon atoms, in their backbone are called triose, tetroses, pentoses, hexoses and heptoses respectively.
- For aldehydic or ketonic monosaccharides, they are thus designated as aldotriose (aldehyde + triose) and ketotriose (Ketone+ triose) respectively.
- Likewise, aldopentoses, ketopentoses; aldohexoses, ketohexoses; aldoheptoses and ketoheptoses also exist.
- For example, Glyceraldehyde is an aldotrise while Dihydroxyacteone is a ketotrise.

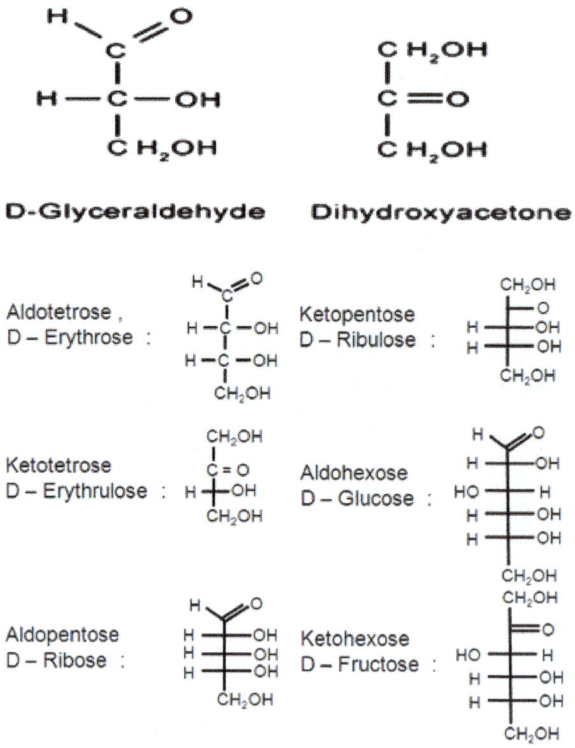

Chain and ring forms: structure and functions of major monosaccharides

Chain and Ring forms

- Many simple sugars can exist in a chain form or a ring form.
- The open-chain form of a monosaccharide often coexists with a closed ring form where the aldehyde/ketonic carbonyl group carbon (C = O) and hydroxylgroup (- OH) react forming a hemiacetal with a new C - O - C bridge. Five and six-membered rings are favoured over other ring sizes because of their low angle and eclipsing strain.
- The cyclic structures are termed furanose(five-membered) or pyranose(six-membered), the nomenclature being derived on the basis of their relationship to common heterocyclic compounds furan and pyran.

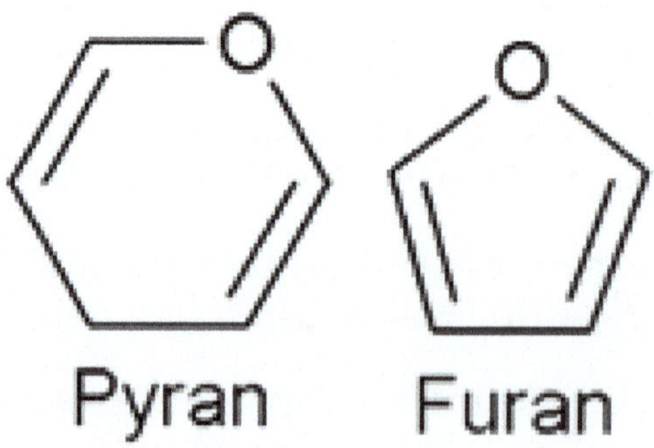

- The cyclic pyranose forms of various monosaccharides drawn in a flat projection are known as a Haworth formula after the British chemist, Norman Harworth. These Haworth formulas make it convenient for displaying stereochemical relationships, but do not represent the true shape of the molecules.

- The glucose ring form is created when the oxygen on carbon number 5 links with the carbon comprising the carbonyl group (carbon number 1) and transfers its hydrogen to the carbonyl oxygen to create a hydroxyl group. The rearrangement produces alpha glucose when the hydroxyl group is on the opposite side of the -CH_2OH group, or beta glucose when the hydroxyl group is on the same side as the -CH_2OH group.

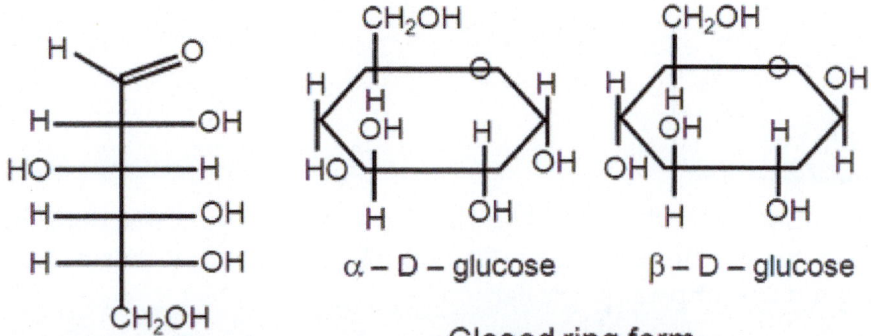

Open chain form

Closed ring form
Haworth projections

- Isomers, such as these, which differ only in their configuration about their carbonyl carbon atom are called anomers.
- Ketoses similarly form hemi-ketal ring leading to furanose like structure. Hence, ketoses are often designated in furanose whereas the aldoses in pyranose form.

Properties of Monosaccharides

- Simple monosaccharides are reducing agents because of their ability to reduce potential oxidising agents like Cu^{2+} and hydrogen peroxide.They are thus called "reducing sugars".
- This reaction forms the basis of Bendict's test for qualitative analysis of simple sugars.
- Glucose, the "blood sugar" and an immediate source of energy for cellular respiration.

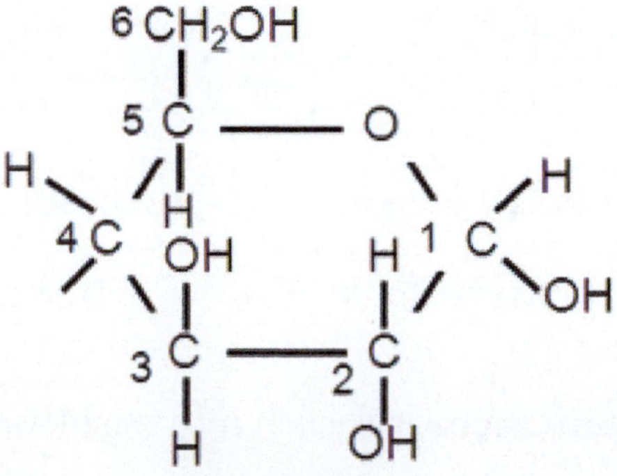

Glucose

Galactose :

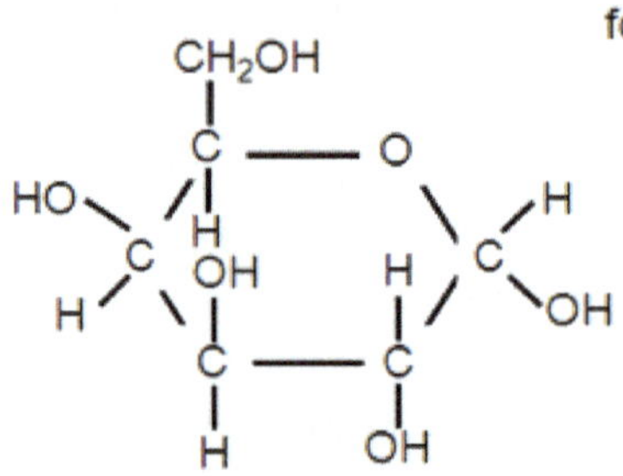

Galactose

Fructose, a ketohexose sugar found in honey fruits.

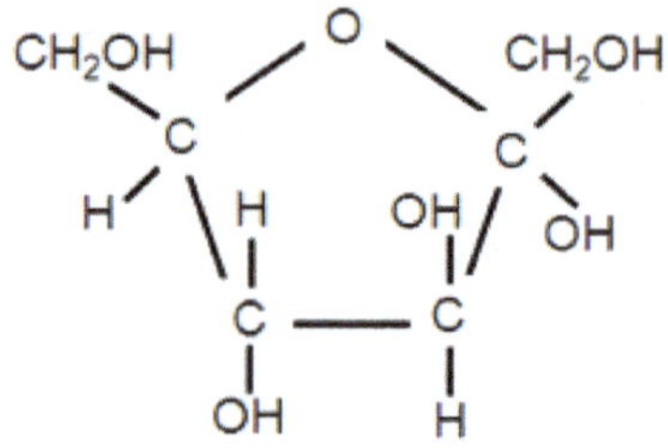

Fructose

Disaccharides

- When two monosaccharides bond together by a condensation reaction, thereby releasing a molecule of water, a disaccharide is formed. The two monosaccharide units are linked by glycosidic bond in α or β anomeric carbon.
- Commonly available disaccahrides are sucrose, maltose and lactose.
- Disaccharides cannot be absorbed through the wall of the small intestine into the blood-stream. They are therefore hydrolyzed to respective monosaccharides by carbohydares present in small intestine, specifically sucrase or invertase, maltase and lactase (β - galactosidase).

Major Disaccharides

Sucrose

- Major carbohydrate present in canesugar, commonly called table sugar.
- Glucose +fructose are linked by α(1 --> 1) glycosidic bonds.

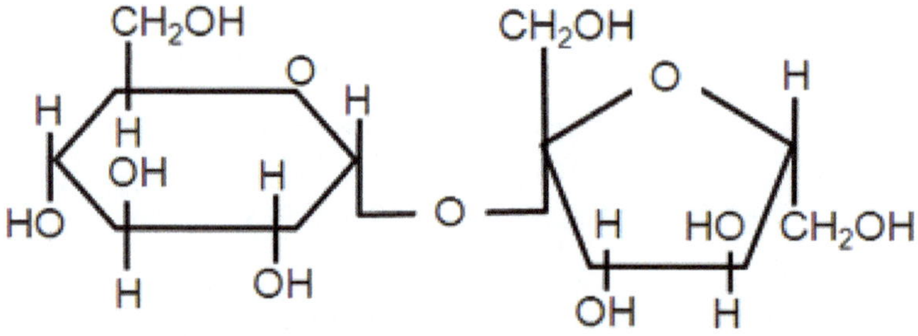

Haworth projection of sucrose

- Sucrose is not a reducing sugar due to the absence of free anomeric carbon from both glucose or fructose units.
- Animals are unable to absorb sucrose as such into the blood stream.
- Enzyme Sucrase, also known as invertase catalyses hydrolysis of sucrose into D - glucose and D - fructose in the intestine, where from they are readily absorbed into blood stream.

Lactose

- A major sugar in milk and milk products.
- Glucose + galactose units linked by α(1 --> 4) glycosidic bonds.

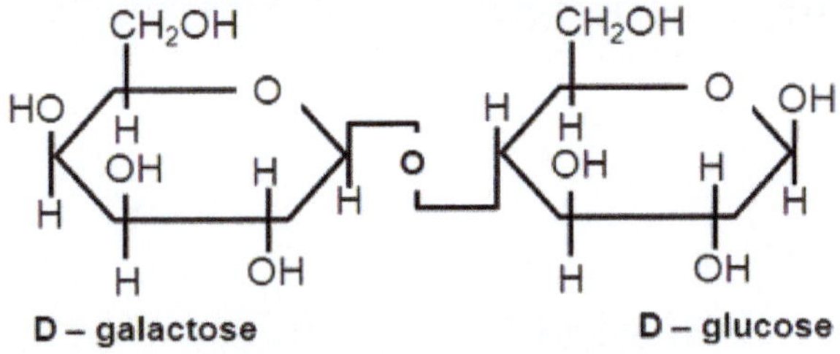

Haworth projection of lactose

- Availability of free carbonyl group on glucose residue makes it a reducing disaccharide.
- Enzyme Lactase catalyzes the hydrolysis of this disaccharide during digestion process in animals.
- Lack of this enzymes leads a clinical condition referred as Lactose intolerance. The subject in such cases are unable to metabolize lactose, because of a lactase deficiency.

Maltose

- Simplest sugar; present in barley malt and also a product of starch digestion.
- Glucose + glucose linked by α (1 --> 4) glycosidic bonds.

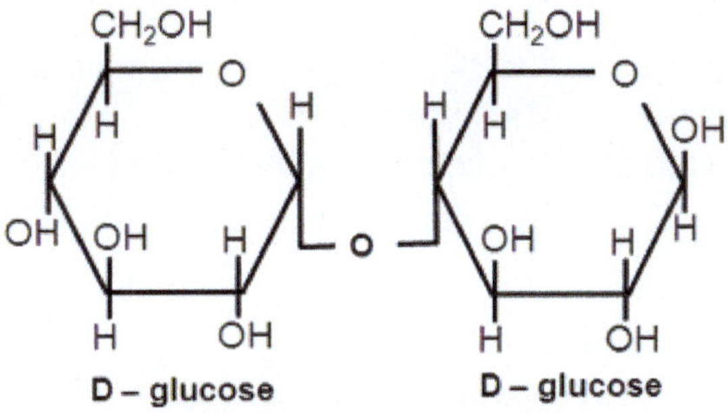

Haworth projection of maltose

- Maltose is a reducing sugar because of the presence of a free carbonyl group which may be oxidized to the free acid.
- Maltose is hydrolyzed to two molecules of D-glucose by the intestinal enzyme maltase, which specifically cleaves the α(1 --> 4). bond.

Cellobiose

The molecule is derived from the condensation of two glucose molecules linked in a β(1 --> 4) fashion. It can be obtained by enzymatic or acidic hydrolysis of cellulose and cellulose rich materials such as cotton, jute or paper.

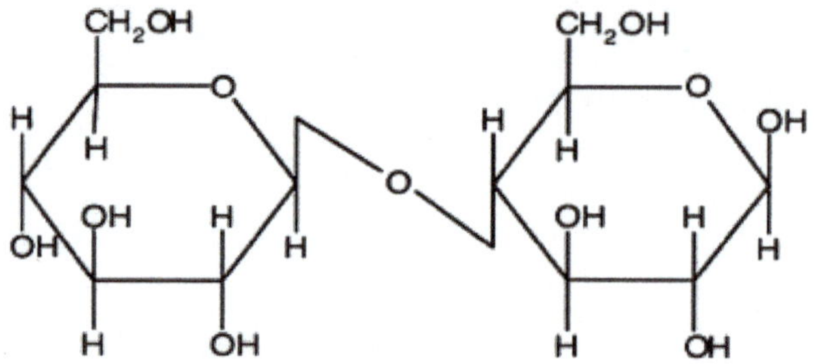

Trehalose

This is another disaccharide which has two α - D - glucose molecules connected through carbon number one in α(1 --> 1) linkage.

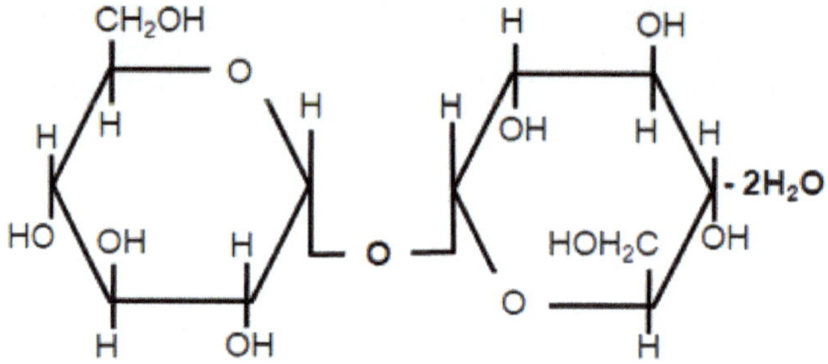

Oligoaccharides and Polysaccharides

Oligosaccharides

- Carbohydrates having more than two or up to ten monosaccharide units are termed as oligosaccharides. Raffinose and stachyose are two major examples of oligosaccharides which consist of repetitive chains of fructose, galactose and glucose.

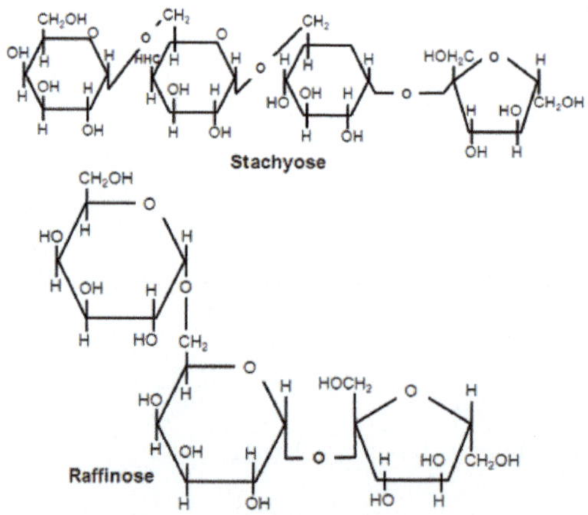

- Raffinose is a trisaccharide widely found in legumes and vegetables like beans, peas, soy, cabbage, brussels sprouts, and broccoli. Here, galactose is bonded to sucrose via a α(1 --> 6) glycosidic linkage. Humans are unable to digest such saccharides Thus undigested saccharides are thus fermented by colonic bacteria resulting into flatulence formation.

Polysaccharides

- Polysaccharides are polymeric carbohydrate structures, formed of repeating units (either mono- or di-saccharides) joined together by glycosidic bonds.
- Polysaccharides have a general formula of $C_x(H_2O)_y$ where x is usually a large number between 200 and 2500.
- These structures are often linear, but may contain various degrees of branching.
- They may be amorphous or even insoluble in water.

Based on the monosccharide units, polysaccharides are broadly put into two categories :

Homopolysaccharides

If the polysaccharides consists of only one type of monosaccharide units, these are called homopolysaccharides.Common example of homopolysaccharide is starch which comprises of only D - glucose units.

- Heteropolysaccharides : When more than one type of monosaccharides are present in a polysaccharide, they are called heteropolysaccharides e.g. pectin (polymer of galactouranic acid and its methylated ester).

Major Storage Polysaccharides

Starch

- Starch constitutes the most predominant storage polysaccharide in nature and is a characteristic storage sugar of all plant cells.
- This is abundant in tubers, like potatoes and corn seeds.
- Starch is a polymer of glucose and contains two types of glucose polymers namely :

(a) Amylose : Amylose consists of linear, unbranched chains of several hundred D-glucose residues. The glucose residues are linked by a α(1 --> 4) glycosidic bond between their C1 and C4 carbon atoms. Molecular weight of such chains may extend upto 50000.

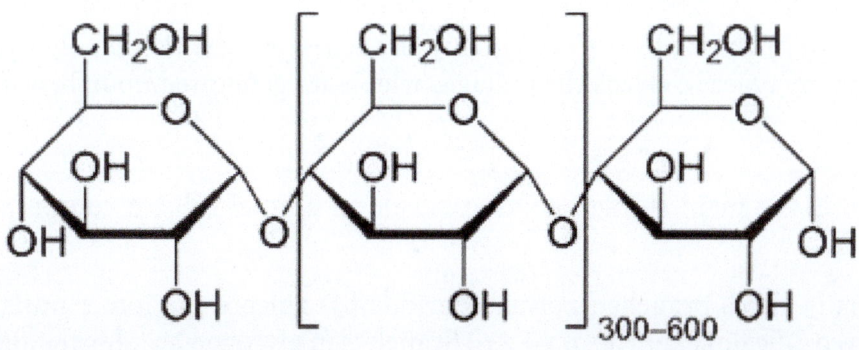

(b) Amylopectin : Amylopectin differs from amylose in being highly branched. The total number of glucose residues in a molecule of amylopectin may run into several thousands. Branching takes place with α(1 --> 6) bonds occurring every 24 to 30 glucose units.

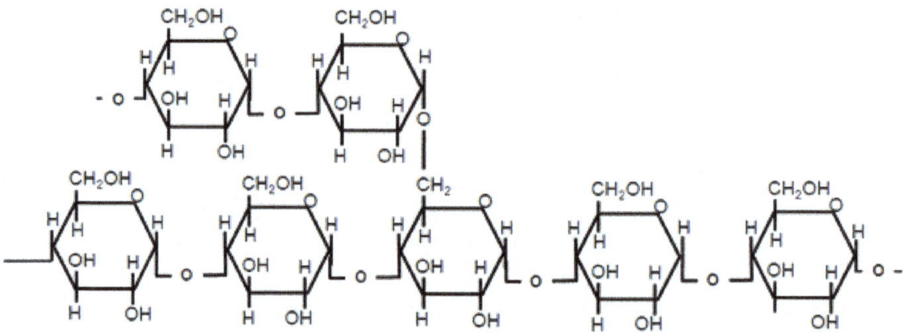

Starch can be transformed into many commercial products by hydrolysis using acids or enzymes as catalysts.

- Produced by the hydrolysis of starch, dextrins are mixtures of polymers of D-glucose units linked by α(1 --> 4) or α(1 --> 6) glycosidic bonds. These are low in molecular weight. When produced by heat, these are termed as pyrodextrins.
- Partial hydrolysis of starch results in a less sweet or almost flavourless Maltodextrin. These are easily digestable and hence act as a source of readily available energy. These can be easily derived from any starch.
- Syrups, such as corn syrup made are made from hydrolysis of corn starch. Corn syrup solids, are mildly sweet semi - crystalline or powdery amorphous products. The latter are not readily digested but partially fermented by intestinal bacteria.
- High Fructose Corn Syrup(HFCS) is another derivative of corn syrup which is as sweet as sugar and is often used in soft drinks.
- When starch is modified by any mechanical or chemical treatments, the resulting product is Modified starch.
- Hydrogenated glucose syrup(HGS) is produced by hydrolyzing starch, followed by its hydrogenation. The resulting syrup is used to produce sugar alcohols as well as hydrogenated oligo- and polysaccharides.
- Polydextrose (poly-D-glucose) is a highly-branched and synthetic polymer. It is known to possess many types of glycosidic linkages which are generated upon heating dextrose.

Glycogen

- Glycogen is the main storage polysaccharide in animal cells, a counterpart of starch in plant cells.
- Glycogen, is also a branched polysaccharide of D glucose monomer units bonded via α(1 --> 4) glycosidic linkage. The α(1 --> 6) branches in glycogen are shorter and more frequent and extensive than those in amylopectin. The overall structure is more compact here.

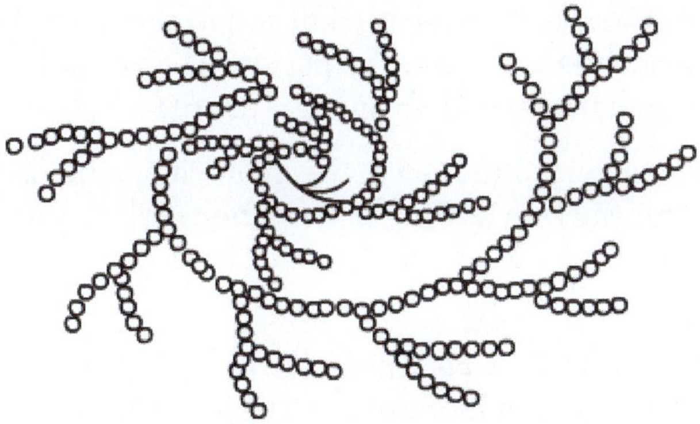

- The glucose chains are organized globularly originating from a pair of molecules of glycogenin, a protein with a molecular weight of 38,000 at the core of the structure.
- Glycogen is easily converted back to glucose to provide energy.
- Both glycogen and starch are hydrolysed in the digestive tract by α - amylases present in the saliva and pancreatic juice which are secreted into the digestive tract.
- α - amylases hydrolyse the α(1 --> 4) glycosidic linkage of the outer branches of glycogen and amylopectin resulting in D-glucose, maltose and a resistant core named as "limit dextrin".
- α - amylases cannot further hydrolyse limit dextrins because of their inability to cleave α(1 --> 6) linkages. This, however, is achieved by a "debranching" enzyme, α(1 --> 6) glucosidase acting on the branch linkages.
- Together α - amylase and α(1 --> 6) glucosidase completely degrade glycogen and starch (amylopectin).
- β - amylase hydrolyses alternate α(1 --> 4) glycosidic linkages yielding mostly maltose with very little glucose.

Cellulose

- Cellulose is the major structural polysaccharide, predominant in the cell wall of the plants. Cellulose is found in cell walls of stalks, stems, trunks, woody portions of plant tissues.
- Like starch, cellulose also comprises of glucose as monomer units which are linked by β(1 --> 4) glycosidic bonds in a linear fashion. The absence of side chains allows cellulose molecules to lie close together and form a rigid structures.

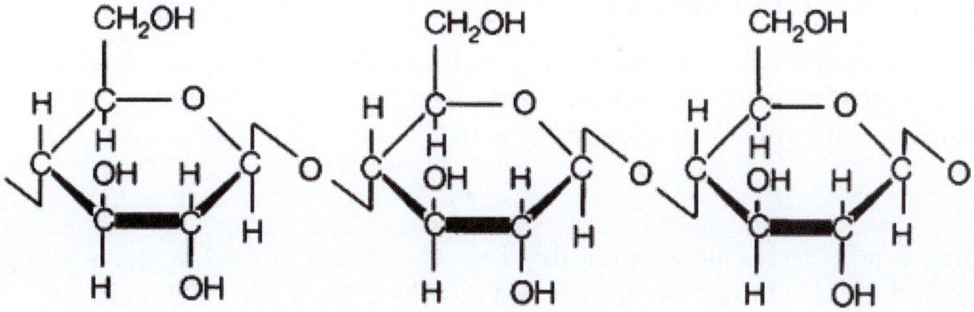

- The basic structural difference between cellulose and starch or glycogen is that in cellulose, the glucose monomer units are linked by β(1 --> 4) glycosidic linkage while in amylose, amylopectin or glycogen the glycosidic linkages are α(1 --> 4).
- On account of this structural difference, the main chains in glycogen or starch assume a coiled and helical conformation leading to formation of dense granules. The main chains of cellulose, on the otherhand take up an extended conformation and allow lateral aggregation to form insoluble fibrils.
- Cellulose are hydrolyzed to its constituent glucose units by microorganisms that inhabit the digestive tract of termites and ruminants. The intestinal tract of termites harbors Triconympha, a parasitic microorganism, that secretes enzyme cellulase which is a cellulose hydrolysing enzyme.
- Cellulose is mostly unavailable as food for vertebrates since they cannot be digested/hydrolysed by any enzyme in them(except cattle and ruminant animals). The digestive system of cattle and ruminant animals like sheep, goats, camels etc is thus abundant with microorganisms that secrete cellulase to degrade cellulose to glucose units.

Function of Starch and Cellulose

- The precise functions of this class of biomolecules in the cells are innumerable.
- Carbohydrates are the source of energy (~4 Kcal/g) in most diets. Potato, rice, wheat, and corn are major sources of starch in the human diet which provide bulk of the calories.
- Cellulose is the major constituent of cell wall. Wood, cotton and paper are forms of cellulose.
- Carbohydrate also forms the part of some glycoproteins.

Biologically Important Polysaccharides

Hemicellulose

- The term "hemicellulose" applies to the polysaccharide components of plant cell walls other than cellulose. This is a polysaccharide in the plant cell walls which is extractable by dilute alkaline solutions.
- Hemicelluloses comprise almost one-third of the carbohydrates in woody plant tissue. This is also found in fruit, plant stems, and grain hulls.
- The chemical structure of hemicelluloses consist of long chains of a variety of pentoses, hexoses, and their corresponding uronic acids.
- Although undigestible, hemicelluloses can be fermented by yeasts and bacteria. The polysaccharides yielding pentoses on hydrolysis are called pentosans. Xylan is major example of a pentosan consisting of D-xylose units with β(1 --> 4) linkages.

Dextran

- Dextran is a polysaccharide wherein the main chains are formed by α(1 --> 6) glycosidic linkages and the side branches are attached by α(1 --> 3) or α(1 --> 4) linkages.

- Dextran is an oral bacterial product that adheres to the teeth, creating a film called plaque.
- It is also used commercially in confections, in lacquers, as food additives, and as plasma volume expanders.

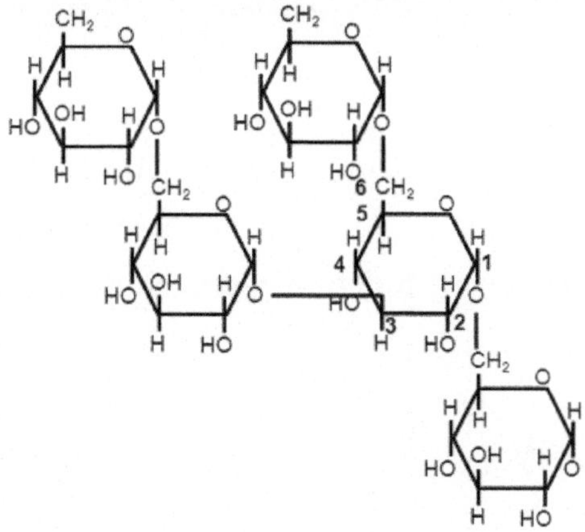

- Some plants store carbohydrates in the form of inulin. Inulins, also called fructans, are polymers consisting of fructose units that have a terminal glucose.
- These are present in many vegetables and fruits, including onions, leeks, garlic, bananas, asparagus, chicory, and Jerusalem artichokes.
- Oligofructose has the same structure as inulin, but the chains consist of 10 or fewer fructose units.
- Oligofructose has approximately 30 to 50% of the sweetness of table sugar. Inulin and oligofructose are nondigestible by human intestinal enzymes, but are fermented by colonic microorganisms.
- Oligofructose are often added to probiotic dairy products like yoghurt. They are consumed by beneficial intestinal bacteria which thereby multiply faster. Such bacteria are important for digestion and the immune system.
- Inulin and oligofructose are used to replace fat or sugar in foods like ice cream, dairy products, confections and baked goods.

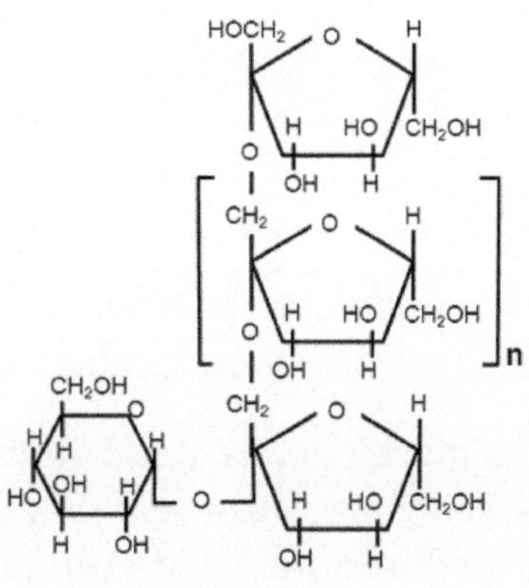

Chitin

- Chitin is third most abundant polysaccharide after starch and cellulose. It is an unbranched polymer of N-Acetyl-D-glucosamine. This is the principal component of arthropod and lower animal exoskeletons, e.g., insect, crab, and shrimp shells and also found in fungi.
- It may be regarded as a derivative of cellulose, in which the hydroxyl groups of the second carbon of each glucose unit have been replaced with acetamido (-NH(C=O)CH_3) groups.

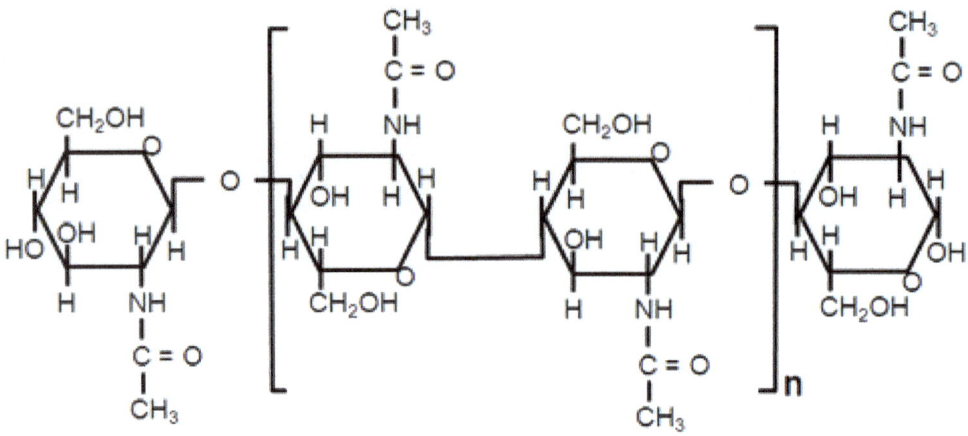

Pectin

- Pectin is a polysaccharide that acts as a binding material for the cell walls of plant tissues. Lemons and oranges contains approximately 30% pectin.
- Pectin is the methylated ester of polygalacturonic acid, which consists of chains of 300 to 1000 galacturonic acid units joined with α(1 --> 4) linkages.
- Pectin is an important ingredient of fruit preserves, jellies, and jams.

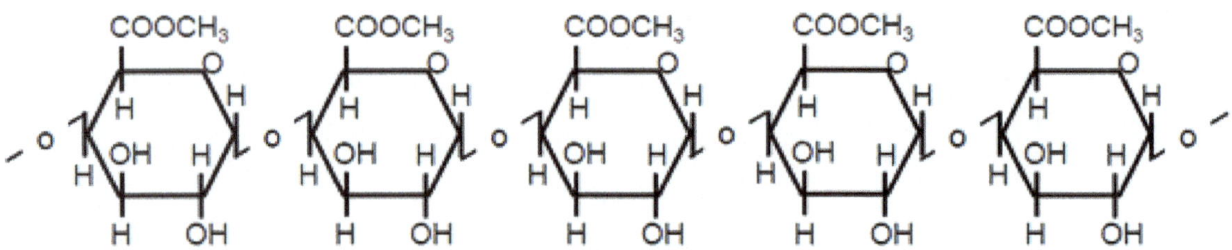

Peptidoglycans

- This is an important polymer found in the bacterial plasma membrane. It consists of polysaccharide and peptide chains in a strong molecular network.
- Peptidoglycan, also known as murein, consists of alternating residues of β(1 --> 4) linked N-acetylglucosamine and N-acetylmuramic acid. A peptide chain of three to five amino acids is attached to the N-acetylmuramic acid. The peptide chain can be cross-linked to

the peptide chain of another strand forming a three dimensional network. Peptidoglycan serves the functional role of providing structural strength and rigidity to the bacterial cell wall.

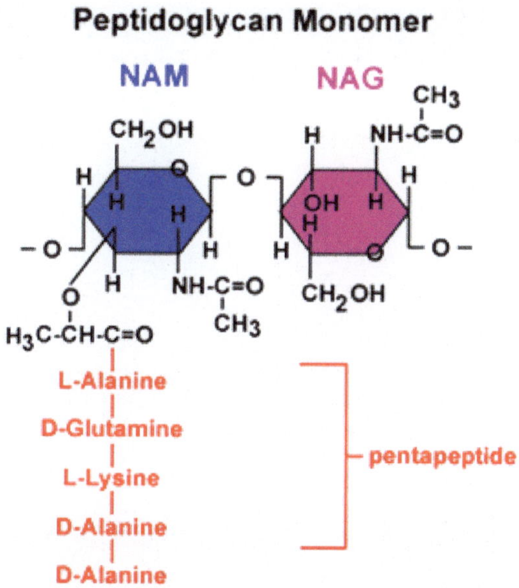

References

- Westman, EC (2002). "Is dietary carbohydrate essential for human nutrition?". The American Journal of Clinical Nutrition. 75 (5): 951–3; author reply 953–4. PMID 11976176
- Maton, Anthea; Jean Hopkins; Charles William McLaughlin; Susan Johnson; Maryanna Quon Warner; David LaHart; Jill D. Wright (1993). Human Biology and Health. Englewood Cliffs, New Jersey, USA: Prentice Hall. pp. 52–59. ISBN 0-13-981176-1
- Jenkins, David; Alexandra L. Jenkins; Thomas M.S. Woleve; Lilian H. Thompson; A. Venkat Rao (February 1986). "Simple and Complex Carbohydrates". Nutrition Reviews. 44 (2): 44–49. doi:10.1111/j.1753-4887.1986.tb07585.x
- "Heptanoate as a neural fuel: energetic and neurotransmitter precursors in normal and glucose transporter I-deficient (G1D) brain". Journal of Cerebral Blood Flow. 33: 175–182. doi:10.1038/jcbfm.2012.151
- Matthews, C. E.; K. E. Van Holde; K. G. Ahern (1999) Biochemistry. 3rd edition. Benjamin Cummings. ISBN 0-8053-3066-6
- Barroca, Nadine; Schmidt, Richard R. (May 2004). "2-Nitro Thioglycoside Donors: Versatile Precursors of β-d-Glycosides of Aminosugars". Organic Letters. 6 (10): 1551–1554. doi:10.1021/ol049729t
- Pichon, L; Huneau, JF; Fromentin, G; Tomé, D (2006). "A high-protein, high-fat, carbohydrate-free diet reduces energy intake, hepatic lipogenesis, and adiposity in rats". The Journal of Nutrition. 136 (5): 1256–60. PMID 16614413
- Campbell, Neil A.; Brad Williamson; Robin J. Heyden (2006). Biology: Exploring Life. Boston, Massachusetts: Pearson Prentice Hall. ISBN 0-13-250882-6
- Pavliak, Viliam; Kováč, Pavol (March 1991). "A short synthesis of 1,3,4,6-tetra-O-acetyl-2-azido-2-deoxy-β-d-glucopyranose and the corresponding α-glucosyl chloride from d-mannose". Carbohydrate Research. 210: 333–337. doi:10.1016/0008-6215(91)80134-9
- Lemieux, R. U.; Ratcliffe, R. M. (15 May 1979). "The azidonitration of tri-O-acetyl-D-galactal". Canadian Journal of Chemistry. 57 (10): 1244–1251. doi:10.1139/v79-203

- Food and Nutrition Board (2002/2005). Dietary Reference Intakes for Energy, Carbohydrate, Fiber, Fat, Fatty Acids, Cholesterol, Protein and Amino Acids. Washington, D.C.: The National Academies Press. Page 769. ISBN 0-309-08537-3
- Liu, Jing; Gin, David Y. (August 2002). "C2-Amidoglycosylation. Scope and Mechanism of Nitrogen Transfer". Journal of the American Chemical Society. 124 (33): 9789–9797. doi:10.1021/ja026281n
- Joint WHO/FAO expert consultation (2003). [1] (PDF). Geneva: World Health Organization. pp. 55–56. ISBN 92-4-120916-X
- Wang, Lai-Xi; Sakairi, Nobuo; Kuzuhara, Hiroyoshi (1990). "1,6-Anhydro-β-D-glucopyranose derivatives as glycosyl donors for thioglycosidation reactions". Journal of the Chemical Society, Perkin Transactions 1 (6): 1677. doi:10.1039/P19900001677

4

An Overview of Amino Acids

Amino acids are the building blocks of proteins comprising amine, carboxyl and a side chain. Out of the many functions of proteins, a few are that it acts as a biocatalyst, helps in transportation of cells, provides mechanical support, and are antibodies to the immune system. The topics discussed in the section are of great importance to broaden the existing knowledge on biomolecular engineering.

Amino Acid

Amino acids are organic compounds containing amine ($-NH_2$) and carboxyl (-COOH) functional groups, along with a side chain (R group) specific to each amino acid. The key elements of an amino acid are carbon, hydrogen, oxygen, and nitrogen, though other elements are found in the side chains of certain amino acids. About 500 amino acids are known (though only 20 appear in the genetic code) and can be classified in many ways. They can be classified according to the core structural functional groups' locations as alpha- (α-), beta- (β-), gamma- (γ-) or delta- (δ-) amino acids; other categories relate to polarity, pH level, and side chain group type (aliphatic, acyclic, aromatic, containing hydroxyl or sulfur, etc.). In the form of proteins, amino acids comprise the second-largest component (water is the largest) of human muscles, cells and other tissues. Outside proteins, amino acids perform critical roles in processes such as neurotransmitter transport and biosynthesis.

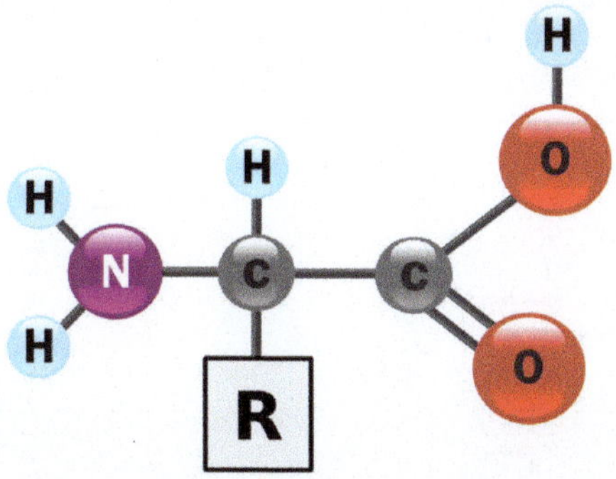

The structure of an alpha amino acid in its un-ionized form

In biochemistry, amino acids having both the amine and the carboxylic acid groups attached to the first (alpha-) carbon atom have particular importance. They are known as 2-, alpha-, or α-amino acids (generic formula $H_2NCHRCOOH$ in most cases, where R is an organic substituent known as a "side chain"); often the term "amino acid" is used to refer specifically to these. They include the 22 proteinogenic ("protein-building") amino acids, which combine into peptide chains ("polypeptides") to form the building-blocks of a vast array of proteins. These are all L-stereoisomers ("left-handed" isomers), although a few D-amino acids ("right-handed") occur in bacterial envelopes, as a neuromodulator (D-serine), and in some antibiotics. Twenty of the proteinogenic amino acids are encoded directly by triplet codons in the genetic code and are known as "standard" amino acids. The other two ("non-standard" or "non-canonical") are selenocysteine (present in many noneukaryotes as well as most eukaryotes, but not coded directly by DNA), and pyrrolysine (found only in some archea and one bacterium). Pyrrolysine and selenocysteine are encoded via variant codons; for example, selenocysteine is encoded by stop codon and SECIS element. *N*-formylmethionine (which is often the initial amino acid of proteins in bacteria, mitochondria, and chloroplasts) is generally considered as a form of methionine rather than as a separate proteinogenic amino acid. Codon–tRNA combinations not found in nature can also be used to "expand" the genetic code and create novel proteins known as alloproteins incorporating non-proteinogenic amino acids.

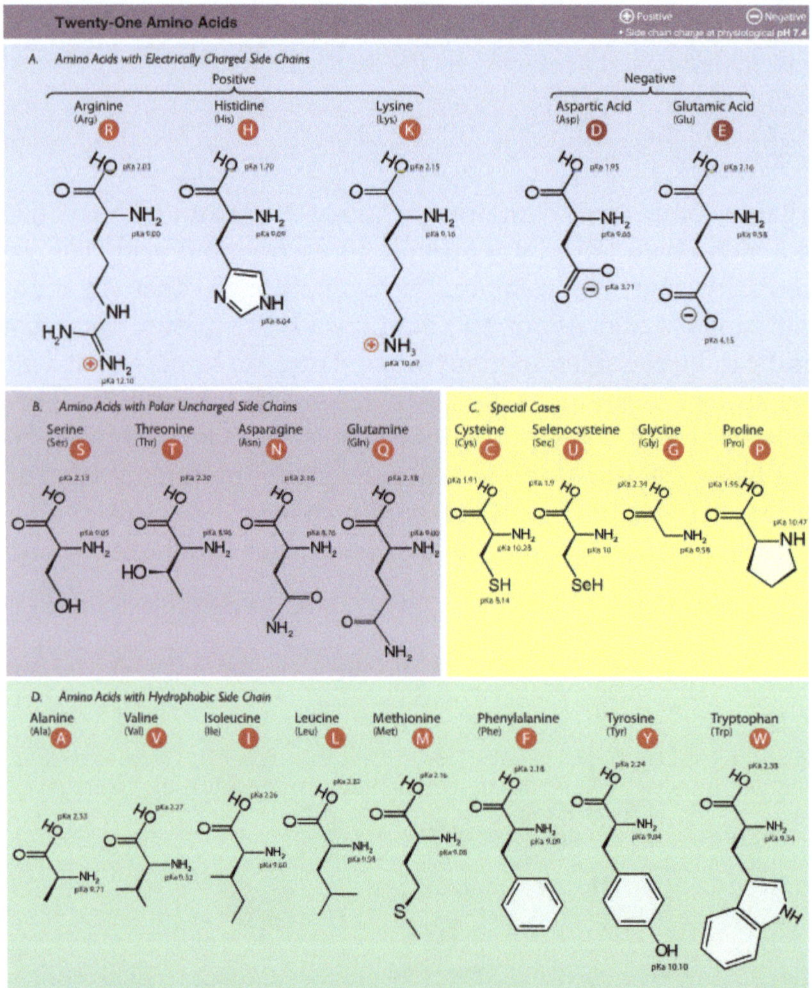

The 21 proteinogenic α-amino acids found in eukaryotes, grouped according to their side chains' pK_a values and charges carried at physiological pH 7.4

Many important proteinogenic and non-proteinogenic amino acids also play critical non-protein roles within the body. For example, in the human brain, glutamate (standard glutamic acid) and gamma-amino-butyric acid ("GABA", non-standard gamma-amino acid) are, respectively, the main excitatory and inhibitory neurotransmitters; hydroxyproline (a major component of the connective tissue collagen) is synthesised from proline; the standard amino acid glycine is used to synthesise porphyrins used in red blood cells; and the non-standard carnitine is used in lipid transport.

Nine proteinogenic amino acids are called "essential" for humans because they cannot be created from other compounds by the human body and so must be taken in as food. Others may be conditionally essential for certain ages or medical conditions. Essential amino acids may also differ between species.

Because of their biological significance, amino acids are important in nutrition and are commonly used in nutritional supplements, fertilizers, and food technology. Industrial uses include the production of drugs, biodegradable plastics, and chiral catalysts.

History

The first few amino acids were discovered in the early 19th century. In 1806, French chemists Louis-Nicolas Vauquelin and Pierre Jean Robiquet isolated a compound in asparagus that was subsequently named asparagine, the first amino acid to be discovered. Cystine was discovered in 1810, although its monomer, cysteine, remained undiscovered until 1884. Glycine and leucine were discovered in 1820. The last of the 20 common amino acids to be discovered was threonine in 1935 by William Cumming Rose, who also determined the essential amino acids and established the minimum daily requirements of all amino acids for optimal growth.

Usage of the term *amino acid* in the English language is from 1898. Proteins were found to yield amino acids after enzymatic digestion or acid hydrolysis. In 1902, Emil Fischer and Franz Hofmeister proposed that proteins are the result of the formation of bonds between the amino group of one amino acid with the carboxyl group of another, in a linear structure that Fischer termed "peptide".

General Structure

In the structure shown at the top of the page, R represents a side chain specific to each amino acid. The carbon atom next to the carboxyl group (which is therefore numbered 2 in the carbon chain starting from that functional group) is called the α–carbon. Amino acids containing an amino group bonded directly to the alpha carbon are referred to as *alpha amino acids*. These include amino acids such as proline which contain secondary amines, which used to be often referred to as "imino acids".

Isomerism

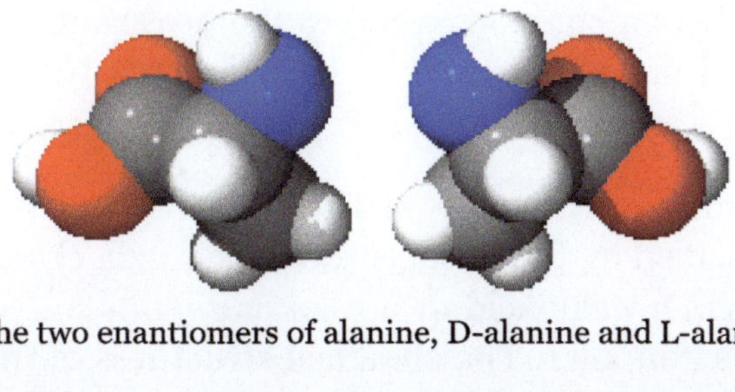

The two enantiomers of alanine, D-alanine and L-alanine

The alpha amino acids are the most common form found in nature, but only when occurring in the L-isomer. The alpha carbon is a chiral carbon atom, with the exception of glycine which has two indistinguishable hydrogen atoms on the alpha carbon. Therefore, all alpha amino acids but glycine can exist in either of two enantiomers, called L or D amino acids, which are mirror images of each other. While L-amino acids represent all of the amino acids found in proteins during translation in the ribosome, D-amino acids are found in some proteins produced by enzyme post-translational modifications after translation and translocation to the endoplasmic reticulum, as in exotic sea-dwelling organisms such as cone snails. They are also abundant components of the peptidoglycan cell walls of bacteria, and D-serine may act as a neurotransmitter in the brain. D-amino acids are used in racemic crystallography to create centrosymmetric crystals, which (depending on the protein) may allow for easier and more robust protein structure determination. The L and D convention for amino acid configuration refers not to the optical activity of the amino acid itself but rather to the optical activity of the isomer of glyceraldehyde from which that amino acid can, in theory, be synthesized (D-glyceraldehyde is dextrorotatory; L-glyceraldehyde is levorotatory). In alternative fashion, the *(S)* and *(R)* designators are used to indicate the absolute stereochemistry. Almost all of the amino acids in proteins are *(S)* at the α carbon, with cysteine being *(R)* and glycine non-chiral. Cysteine has its side chain in the same geometric position as the other amino acids, but the *R/S* terminology is reversed because of the higher atomic number of sulfur compared to the carboxyl oxygen gives the side chain a higher priority, whereas the atoms in most other side chains give them lower priority.

Side Chains

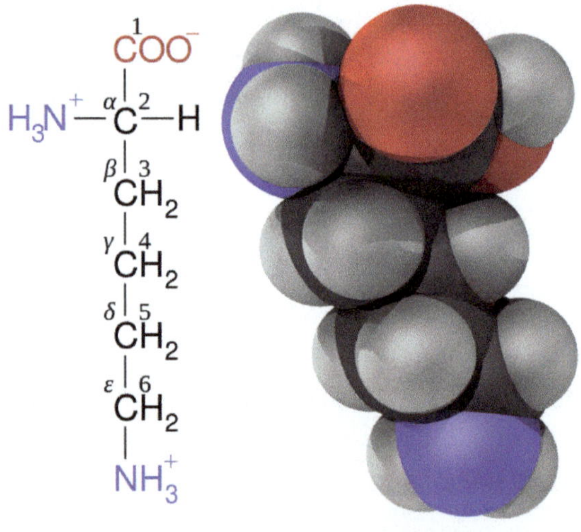

Lysine with carbon atoms labeled

In amino acids that have a carbon chain attached to the α–carbon (such as lysine, shown to the right) the carbons are labeled in order as α, β, γ, δ, and so on. In some amino acids, the amine group is attached to the β or γ-carbon, and these are therefore referred to as *beta* or *gamma amino acids*.

Amino acids are usually classified by the properties of their side chain into four groups. The side chain can make an amino acid a weak acid or a weak base, and a hydrophile if the side chain is polar or a hydrophobe if it is nonpolar. The chemical structures of the 22 standard amino acids, along with their chemical properties, are described more fully in the article on these proteinogenic amino acids.

The phrase "branched-chain amino acids" or BCAA refers to the amino acids having aliphatic side chains that are non-linear; these are leucine, isoleucine, and valine. Proline is the only proteinogenic amino acid whose side-group links to the α-amino group and, thus, is also the only proteinogenic amino acid containing a secondary amine at this position. In chemical terms, proline is, therefore, an imino acid, since it lacks a primary amino group, although it is still classed as an amino acid in the current biochemical nomenclature, and may also be called an "N-alkylated alpha-amino acid".

Zwitterions

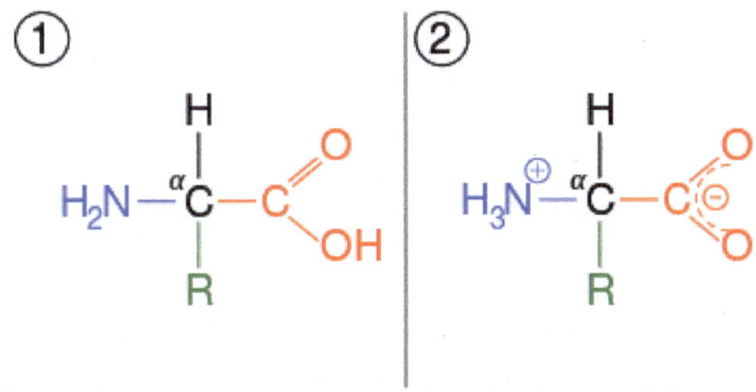

An amino acid in its (1) un-ionized and (2) zwitterionic forms

The α-carboxylic acid group of amino acids is a weak acid, meaning that it releases a hydron (such as a proton) at moderate pH values. In other words, carboxylic acid groups ($-CO_2H$) can be deprotonated to become negative carboxylates ($-CO_2^-$). The negatively charged carboxylate ion predominates at pH values greater than the pKa of the carboxylic acid group. In a complementary fashion, the α-amine of amino acids is a weak base, meaning that it accepts a proton at moderate pH values. In other words, α-amino groups (NH_2-) can be protonated to become positive α-ammonium groups ($^+NH_3-$). The positively charged α-ammonium group predominates at pH values less than the pKa of the α-ammonium group (mean for the 20 common α-amino acids is about 9.4).

Because all amino acids contain amine and carboxylic acid functional groups, they share amphiprotic properties. Below pH 2.2, the predominant form will have a neutral carboxylic acid group and a positive α-ammonium ion (net charge +1), and above pH 9.4, a negative carboxylate and neutral α-amino group (net charge −1). But at pH between 2.2 and 9.4, an amino acid usually contains both a negative carboxylate and a positive α-ammonium group, as shown in structure (2) on the right, so has net zero charge. This molecular state is known as a zwitterion, from the German Zwitter meaning *hermaphrodite* or *hybrid*. The fully neutral form (structure (1) on the right) is a very minor species in aqueous solution throughout the pH range (less than 1 part in 10^7). Amino acids exist as zwitterions also in the solid phase, and crystallize with salt-like properties unlike typical organic acids or amines.

Isoelectric Point

The variation in titration curves when the amino acids are grouped by category can be seen here. With the exception of tyrosine, using titration to differentiate between hydrophobic amino acids is problematic.

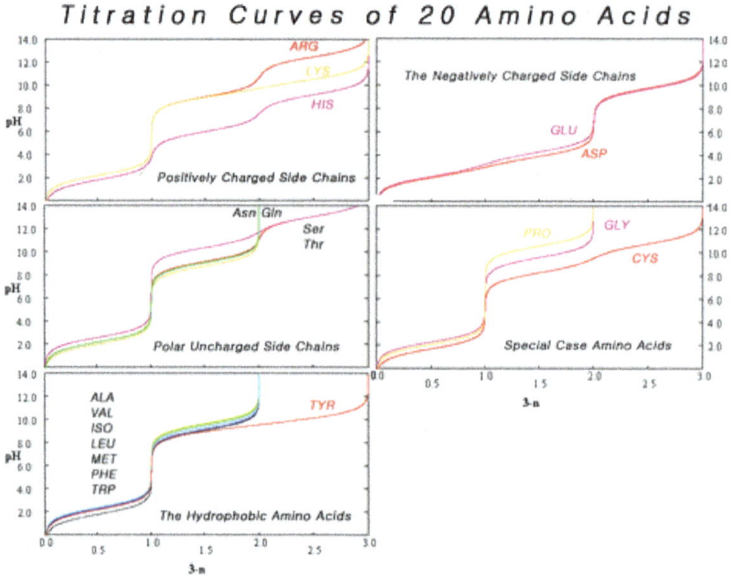

Composite of titration curves of twenty proteinogenic amino acids grouped by side chain category.

At pH values between the two pKa values, the zwitterion predominates, but coexists in dynamic equilibrium with small amounts of net negative and net positive ions. At the exact midpoint between the two pKa values, the trace amount of net negative and trace of net positive ions exactly balance, so that average net charge of all forms present is zero. This pH is known as the isoelectric point pI, so $\text{pI} = \frac{1}{2}(\text{pKa}_1 + \text{pKa}_2)$. The individual amino acids all have slightly different pKa values, so have different isoelectric points. For amino acids with charged side chains, the pKa of the side chain is involved. Thus for Asp, Glu with negative side chains, $\text{pI} = \frac{1}{2}(\text{pKa}_1 + \text{pKa}_R)$, where pKa_R is the side chain pKa. Cysteine also has potentially negative side chain with $\text{pKa}_R = 8.14$, so pI should be calculated as for Asp and Glu, even though the side chain is not significantly charged at neutral pH. For His, Lys, and Arg with positive side chains, $\text{pI} = \frac{1}{2}(\text{pKa}_R + \text{pKa}_2)$. Amino acids have zero mobility in electrophoresis at their isoelectric point, although this behaviour is more usually exploited for peptides and proteins than single amino acids. Zwitterions have minimum solubility at their isoelectric point and some amino acids (in particular, with non-polar side chains) can be isolated by precipitation from water by adjusting the pH to the required isoelectric point.

Occurrence and Functions in Biochemistry

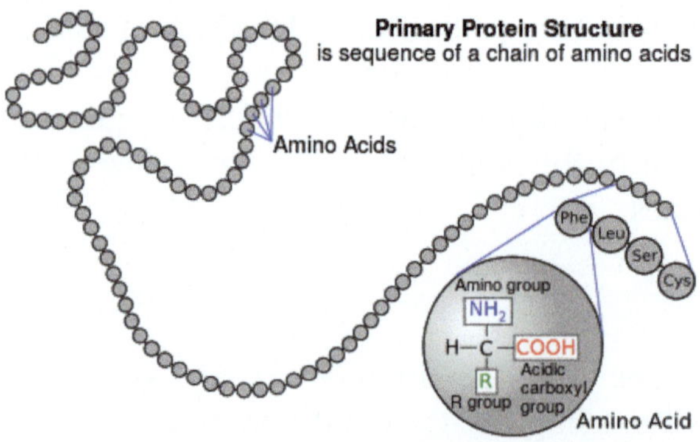

A polypeptide is an unbranched chain of amino acids.

Proteinogenic Amino Acids

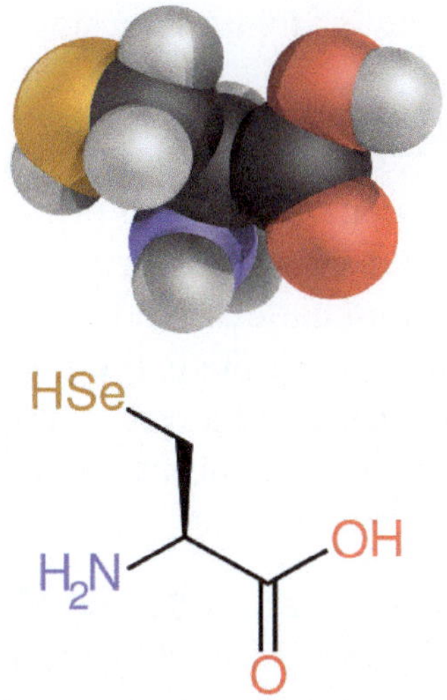

The amino acid selenocysteine

Amino acids are the structural units (monomers) that make up proteins. They join together to form short polymer chains called peptides or longer chains called either polypeptides or proteins. These polymers are linear and unbranched, with each amino acid within the chain attached to two neighboring amino acids. The process of making proteins is called *translation* and involves the step-by-step addition of amino acids to a growing protein chain by a ribozyme that is called a ribosome. The order in which the amino acids are added is read through the genetic code from an mRNA template, which is a RNA copy of one of the organism's genes.

Twenty-two amino acids are naturally incorporated into polypeptides and are called proteinogenic or natural amino acids. Of these, 20 are encoded by the universal genetic code. The remaining 2, selenocysteine and pyrrolysine, are incorporated into proteins by unique synthetic mechanisms. Selenocysteine is incorporated when the mRNA being translated includes a SECIS element, which causes the UGA codon to encode selenocysteine instead of a stop codon. Pyrrolysine is used by some methanogenic archaea in enzymes that they use to produce methane. It is coded for with the codon UAG, which is normally a stop codon in other organisms. This UAG codon is followed by a PYLIS downstream sequence.

Non-proteinogenic Amino Acids

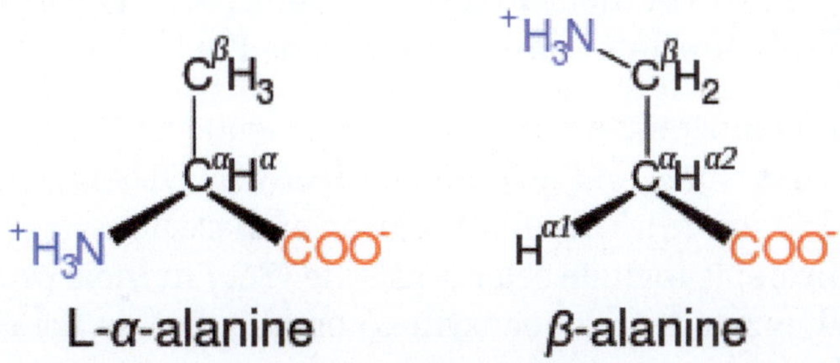

β-alanine and its α-alanine isomer

Aside from the 22 proteinogenic amino acids, there are many other amino acids that are called *non-proteinogenic*. Those either are not found in proteins (for example carnitine, GABA, Levothyroxine) or are not produced directly and in isolation by standard cellular machinery (for example, hydroxyproline and selenomethionine).

Non-proteinogenic amino acids that are found in proteins are formed by post-translational modification, which is modification after translation during protein synthesis. These modifications are often essential for the function or regulation of a protein; for example, the carboxylation of glutamate allows for better binding of calcium cations, and the hydroxylation of proline is critical for maintaining connective tissues. Another example is the formation of hypusine in the translation initiation factor EIF5A, through modification of a lysine residue. Such modifications can also determine the localization of the protein, e.g., the addition of long hydrophobic groups can cause a protein to bind to a phospholipid membrane.

Some non-proteinogenic amino acids are not found in proteins. Examples include 2-aminoisobutyric acid and the neurotransmitter gamma-aminobutyric acid. Non-proteinogenic amino acids often occur as intermediates in the metabolic pathways for standard amino acids – for example, ornithine and citrulline occur in the urea cycle, part of amino acid catabolism. A rare exception to the dominance of α-amino acids in biology is the β-amino acid beta alanine (3-aminopropanoic acid), which is used in plants and microorganisms in the synthesis of pantothenic acid (vitamin B_5), a component of coenzyme A.

D-amino Acid Natural Abundance

D-isomers are uncommon in live organisms. For instance, gramicidin is a polypeptide made up from mixture of D- and L-amino acids. Other compounds containing D-amino acid are tyrocidine and valinomycin. These compounds disrupt bacterial cell walls, particularly in Gram-positive bacteria. Only 837 D-amino acids were found in Swiss-Prot database (187 million amino acids analysed).

Non-standard Amino Acids

The 20 amino acids that are encoded directly by the codons of the universal genetic code are called *standard* or *canonical* amino acids. A modified form of methionine (*N*-formylmethionine) is often incorporated in place of methionine as the initial amino acid of proteins in bacteria, mitochondria and chloroplasts. Other amino acids are called *non-standard* or *non-canonical.* Most of the non-standard amino acids are also non-proteinogenic (i.e. they cannot be incorporated into proteins during translation), but two of them are proteinogenic, as they can be incorporated translationally into proteins by exploiting information not encoded in the universal genetic code.

The two non-standard proteinogenic amino acids are selenocysteine (present in many non-eukaryotes as well as most eukaryotes, but not coded directly by DNA) and pyrrolysine (found only in some archaea and one bacterium). The incorporation of these non-standard amino acids is rare. For example, 25 human proteins include selenocysteine (Sec) in their primary structure, and the structurally characterized enzymes (selenoenzymes) employ Sec as the catalytic moiety in their active sites. Pyrrolysine and selenocysteine are encoded via variant codons. For example, selenocysteine is encoded by stop codon and SECIS element.

In Human Nutrition

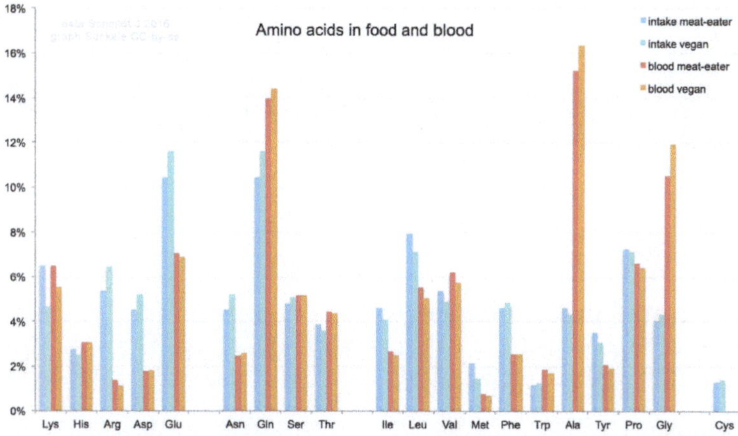

Share of amino acid in different human diets and the resulting mix of amino acids in human blood serum. Glutamate and glutamine are the most frequent in food at over 10%, while alanine, glutamine, and glycine are the most common in blood.

When taken up into the human body from the diet, the 20 standard amino acids either are used to synthesize proteins and other biomolecules or are oxidized to urea and carbon dioxide as a source of energy. The oxidation pathway starts with the removal of the amino group by a transaminase; the amino group is then fed into the urea cycle. The other product of transamidation is a keto acid that enters the citric acid cycle. Glucogenic amino acids can also be converted into glucose, through gluconeogenesis. Of the 20 standard amino acids, nine (His, Ile, Leu, Lys, Met, Phe, Thr, Trp and Val), are called essential amino acids because the human body cannot synthesize them from other compounds at the level needed for normal growth, so they must be obtained from food. In addition, cysteine, taurine, tyrosine, and arginine are considered semiessential amino-acids in children (though taurine is not technically an amino acid), because the metabolic pathways that synthesize these amino acids are not fully developed. The amounts required also depend on the age and health of the individual, so it is hard to make general statements about the dietary requirement for some amino acids. Dietary exposure to the non-standard amino acid BMAA has been linked to human neurodegenerative diseases, including ALS.

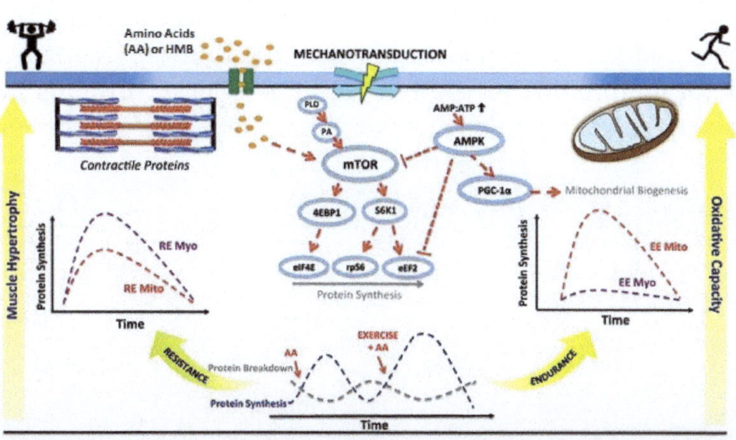

Diagram of the molecular signaling cascades that are involved in myofibrillar muscle protein synthesis and mitochondrial biogenesis in response to physical exercise and specific amino acids or their derivatives (primarily L-leucine and HMB). Many amino acids derived from food protein promote the activation of mTORC1 and increase protein synthesis by signaling through Rag GTPases.

Abbreviations and representations:

- PLD: phospholipase D
- PA: phosphatidic acid
- mTOR: mechanistic target of rapamycin
- AMP: adenosine monophosphate
- ATP: adenosine triphosphate
- AMPK: AMP-activated protein kinase
- PGC-1α: peroxisome proliferator-activated receptor gamma coactivator-1α
- S6K1: p70S6 kinase
- 4EBP1: eukaryotic translation initiation factor 4E-binding protein 1
- eIF4E: eukaryotic translation initiation factor 4E
- RPS6: ribosomal protein S6
- eEF2: eukaryotic elongation factor 2
- RE: resistance exercise; EE: endurance exercise
- Myo: myofibrillar; Mito: mitochondrial
- AA: amino acids
- HMB: β-hydroxy β-methylbutyric acid
- ↑ represents activation
- T represents inhibition

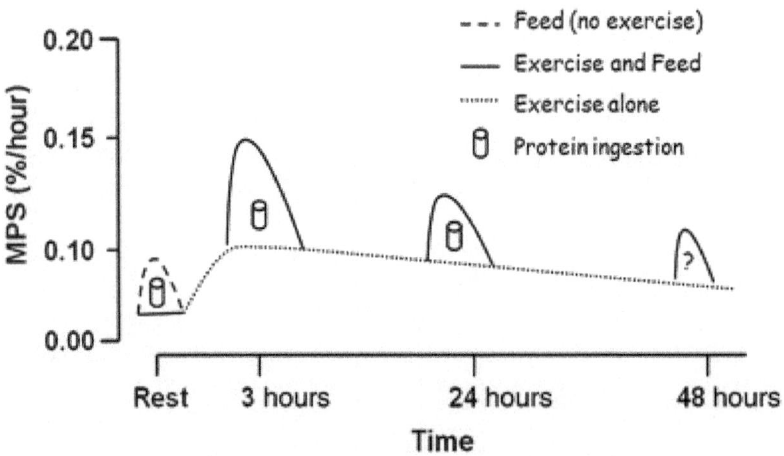

Resistance training stimulates muscle protein synthesis (MPS) for a period of up to 48 hours following exercise (shown by dotted line). Ingestion of a protein-rich meal at any point during this period will augment the exercise-induced increase in muscle protein synthesis (shown by solid lines).

Non-protein Functions

In humans, non-protein amino acids also have important roles as metabolic intermediates, such as in the biosynthesis of the neurotransmitter gamma-amino-butyric acid (GABA). Many amino acids are used to synthesize other molecules, for example:

- Tryptophan is a precursor of the neurotransmitter serotonin.
- Tyrosine (and its precursor phenylalanine) are precursors of the catecholamine neurotransmitters dopamine, epinephrine and norepinephrine and various trace amines.
- Phenylalanine is a precursor of phenethylamine and tyrosine in humans. In plants, it is a precursor of various phenylpropanoids, which are important in plant metabolism.
- Glycine is a precursor of porphyrins such as heme.
- Arginine is a precursor of nitric oxide.
- Ornithine and S-adenosylmethionine are precursors of polyamines.
- Aspartate, glycine, and glutamine are precursors of nucleotides. However, not all of the functions of other abundant non-standard amino acids are known.

Some non-standard amino acids are used as defenses against herbivores in plants. For example, canavanine is an analogue of arginine that is found in many legumes, and in particularly large amounts in *Canavalia gladiata* (sword bean). This amino acid protects the plants from predators such as insects and can cause illness in people if some types of legumes are eaten without processing. The non-protein amino acid mimosine is found in other species of legume, in particular *Leucaena leucocephala*. This compound is an analogue of tyrosine and can poison animals that graze on these plants.

Uses in Industry

Amino acids are used for a variety of applications in industry, but their main use is as additives to animal feed. This is necessary, since many of the bulk components of these feeds, such as soybeans, either have low levels or lack some of the essential amino acids: lysine, methionine, threonine, and tryptophan are most important in the production of these feeds. In this industry, amino acids are also used to chelate metal cations in order to improve the absorption of minerals from supplements, which may be required to improve the health or production of these animals.

The food industry is also a major consumer of amino acids, in particular, glutamic acid, which is used as a flavor enhancer, and aspartame (aspartyl-phenylalanine-1-methyl ester) as a low-calorie artificial sweetener. Similar technology to that used for animal nutrition is employed in the human nutrition industry to alleviate symptoms of mineral deficiencies, such as anemia, by improving mineral absorption and reducing negative side effects from inorganic mineral supplementation.

The chelating ability of amino acids has been used in fertilizers for agriculture to facilitate the delivery of minerals to plants in order to correct mineral deficiencies, such as iron chlorosis. These fertilizers are also used to prevent deficiencies from occurring and improving the overall health of the plants. The remaining production of amino acids is used in the synthesis of drugs and cosmetics.

Similarly, some amino acids derivatives are used in pharmaceutical industry. They include 5-HTP (5-hydroxytryptophan) used for experimental treatment of depression, L-DOPA (L-dihydroxyphenylalanine) for Parkinson's treatment, and eflornithine drug that inhibits ornithine decarboxylase and used in the treatment of sleeping sickness.

Expanded Genetic Code

Since 2001, 40 non-natural amino acids have been added into protein by creating a unique codon (recoding) and a corresponding transfer-RNA:aminoacyl – tRNA-synthetase pair to encode it with diverse physicochemical and biological properties in order to be used as a tool to exploring protein structure and function or to create novel or enhanced proteins.

Nullomers

Nullomers are codons that in theory code for an amino acid, however in nature there is a selective bias against using this codon in favor of another, for example bacteria prefer to use CGA instead of AGA to code for arginine. This creates some sequences that do not appear in the genome. This characteristic can be taken advantage of and used to create new selective cancer-fighting drugs and to prevent cross-contamination of DNA samples from crime-scene investigations.

Chemical Building Blocks

Amino acids are important as low-cost feedstocks. These compounds are used in chiral pool synthesis as enantiomerically pure building-blocks.

Amino acids have been investigated as precursors chiral catalysts, e.g., for asymmetric hydrogenation reactions, although no commercial applications exist.

Biodegradable Plastics

Amino acids are under development as components of a range of biodegradable polymers. These materials have applications as environmentally friendly packaging and in medicine in drug delivery and the construction of prosthetic implants. These polymers include polypeptides, polyamides, polyesters, polysulfides, and polyurethanes with amino acids either forming part of their main chains or bonded as side chains. These modifications alter the physical properties and reactivities of the polymers. An interesting example of such materials is polyaspartate, a water-soluble biodegradable polymer that may have applications in disposable diapers and agriculture. Due to its solubility and ability to chelate metal ions, polyaspartate is also being used as a biodegradeable anti-scaling agent and a corrosion inhibitor. In addition, the aromatic amino acid tyrosine is being developed as a possible replacement for toxic phenols such as bisphenol A in the manufacture of polycarbonates.

Reactions

As amino acids have both a primary amine group and a primary carboxyl group, these chemicals can undergo most of the reactions associated with these functional groups. These include nucleophilic addition, amide bond formation, and imine formation for the amine group, and esterification, amide bond formation, and decarboxylation for the carboxylic acid group. The combination of these functional groups allow amino acids to be effective polydentate ligands for metal-amino acid chelates. The multiple side chains of amino acids can also undergo chemical reactions. The types of these reactions are determined by the groups on these side chains and are, therefore, different between the various types of amino acid.

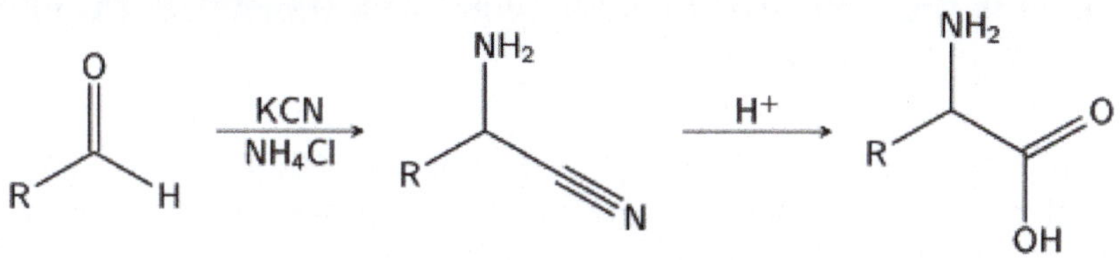

The Strecker amino acid synthesis

Chemical Synthesis

Several methods exist to synthesize amino acids. One of the oldest methods begins with the bromination at the α-carbon of a carboxylic acid. Nucleophilic substitution with ammonia then converts the alkyl bromide to the amino acid. In alternative fashion, the Strecker amino acid synthesis involves the treatment of an aldehyde with potassium cyanide and ammonia, this produces an α-amino nitrile as an intermediate. Hydrolysis of the nitrile in acid then yields a α-amino acid. Using ammonia or ammonium salts in this reaction gives unsubstituted amino acids, whereas substituting primary and secondary amines will yield substituted amino acids. Likewise, using ketones, instead of aldehydes, gives α,α-disubstituted amino acids. The classical synthesis gives racemic mixtures of α-amino acids as products, but several alternative procedures using asymmetric auxiliaries or asymmetric catalysts have been developed.

At the current time, the most-adopted method is an automated synthesis on a solid support (e.g., polystyrene beads), using protecting groups (e.g., Fmoc and t-Boc) and activating groups (e.g., DCC and DIC).

Peptide Bond Formation

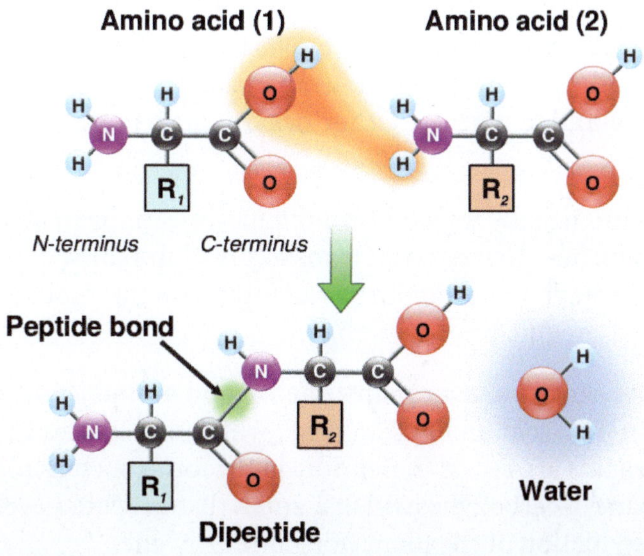

The condensation of two amino acids to form a *dipeptide* through a *peptide bond*

As both the amine and carboxylic acid groups of amino acids can react to form amide bonds, one amino acid molecule can react with another and become joined through an amide linkage. This polymerization of amino acids is what creates proteins. This condensation reaction yields the newly formed peptide bond and a molecule of water. In cells, this reaction does not occur directly; instead, the amino acid is first activated by attachment to a transfer RNA molecule through an ester bond. This aminoacyl-tRNA is produced in an ATP-dependent reaction carried out by an aminoacyl tRNA synthetase. This aminoacyl-tRNA is then a substrate for the ribosome, which catalyzes the attack of the amino group of the elongating protein chain on the ester bond. As a result of this mechanism, all proteins made by ribosomes are synthesized starting at their N-terminus and moving toward their C-terminus.

However, not all peptide bonds are formed in this way. In a few cases, peptides are synthesized by specific enzymes. For example, the tripeptide glutathione is an essential part of the defenses of cells against oxidative stress. This peptide is synthesized in two steps from free amino acids. In the first step, gamma-glutamylcysteine synthetase condenses cysteine and glutamic acid through a peptide bond formed between the side chain carboxyl of the glutamate (the gamma carbon of this side chain) and the amino group of the cysteine. This dipeptide is then condensed with glycine by glutathione synthetase to form glutathione.

In chemistry, peptides are synthesized by a variety of reactions. One of the most-used in solid-phase peptide synthesis uses the aromatic oxime derivatives of amino acids as activated units. These are added in sequence onto the growing peptide chain, which is attached to a solid resin support. The ability to easily synthesize vast numbers of different peptides by varying the types and order of amino acids (using combinatorial chemistry) has made peptide synthesis particularly important in creating libraries of peptides for use in drug discovery through high-throughput screening.

Biosynthesis

In plants, nitrogen is first assimilated into organic compounds in the form of glutamate, formed from alpha-ketoglutarate and ammonia in the mitochondrion. In order to form other amino acids, the plant uses transaminases to move the amino group to another alpha-keto carboxylic acid. For example, aspartate aminotransferase converts glutamate and oxaloacetate to alpha-ketoglutarate and aspartate. Other organisms use transaminases for amino acid synthesis, too.

Nonstandard amino acids are usually formed through modifications to standard amino acids. For example, homocysteine is formed through the transsulfuration pathway or by the demethylation of methionine via the intermediate metabolite S-adenosyl methionine, while hydroxyproline is made by a posttranslational modification of proline.

Microorganisms and plants can synthesize many uncommon amino acids. For example, some microbes make 2-aminoisobutyric acid and lanthionine, which is a sulfide-bridged derivative of alanine. Both of these amino acids are found in peptidic lantibiotics such as alamethicin. However, in plants, 1-aminocyclopropane-1-carboxylic acid is a small disubstituted cyclic amino acid that is a key intermediate in the production of the plant hormone ethylene.

Catabolism

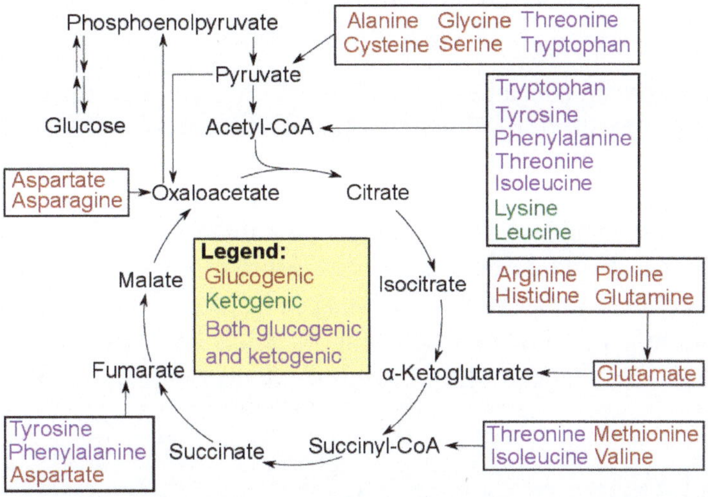

Catabolism of proteinogenic amino acids. Amino acids can be classified according to the properties of their main products as either of the following:
* Glucogenic, with the products having the ability to form glucose by gluconeogenesis * Ketogenic, with the products not having the ability to form glucose. These products may still be used for ketogenesis or lipid synthesis.
* Amino acids catabolized into both glucogenic and ketogenic products.

Amino acids must first pass out of organelles and cells into blood circulation via amino acid transporters, since the amine and carboxylic acid groups are typically ionized. Degradation of an amino acid, occurring in the liver and kidneys, often involves deamination by moving its amino group to alpha-ketoglutarate, forming glutamate. This process involves transaminases, often the same as those used in amination during synthesis. In many vertebrates, the amino group is then removed through the urea cycle and is excreted in the form of urea. However, amino acid degradation can produce uric acid or ammonia instead. For example, serine dehydratase converts serine to pyruvate and ammonia. After removal of one or more amino groups, the remainder of the molecule can sometimes be used to synthesize new amino acids, or it can be used for energy by entering glycolysis or the citric acid cycle, as detailed in image at right.

Physicochemical Properties of Amino Acids

The 20 amino acids encoded directly by the genetic code can be divided into several groups based on their properties. Important factors are charge, hydrophilicity or hydrophobicity, size, and functional groups. These properties are important for protein structure and protein–protein interactions. The water-soluble proteins tend to have their hydrophobic residues (Leu, Ile, Val, Phe, and Trp) buried in the middle of the protein, whereas hydrophilic side chains are exposed to the aqueous solvent. (Note that in biochemistry, a residue refers to a specific monomer within the polymeric chain of a polysaccharide, protein or nucleic acid.) The integral membrane proteins tend to have outer rings of exposed hydrophobic amino acids that anchor them into the lipid bilayer. In the case part-way between these two extremes, some peripheral membrane proteins have a patch of hydrophobic amino acids on their surface that locks onto the membrane. In similar fashion, proteins that have to bind to positively charged molecules have surfaces rich with negatively charged amino acids like glutamate and aspartate, while proteins binding to negatively charged molecules have surfaces rich with positively charged chains like lysine and arginine. There are different hydrophobicity scales of amino acid residues.

Some amino acids have special properties such as cysteine, that can form covalent disulfide bonds to other cysteine residues, proline that forms a cycle to the polypeptide backbone, and glycine that is more flexible than other amino acids.

Many proteins undergo a range of posttranslational modifications, when additional chemical groups are attached to the amino acids in proteins. Some modifications can produce hydrophobic lipoproteins, or hydrophilic glycoproteins. These type of modification allow the reversible targeting of a protein to a membrane. For example, the addition and removal of the fatty acid palmitic acid to cysteine residues in some signaling proteins causes the proteins to attach and then detach from cell membranes.

Table of Standard Amino Acid Abbreviations and Properties

Amino Acid	3-Letter	1-Letter	Side chain class	Side chain polarity	Side chain charge (pH 7.4)	Hydropathy index	Absorbance λ_{max}(nm)	ε at λ_{max} (mM^{-1} cm^{-1})	MW (Weight)	Occurrence in proteins (%)
Alanine	Ala	A	aliphatic	nonpolar	neutral	1.8			89.094	8.76
Arginine	Arg	R	basic	basic polar	positive	−4.5			174.203	5.78
Asparagine	Asn	N	amide	polar	neutral	−3.5			132.119	3.93
Aspartic acid	Asp	D	acid	acidic polar	negative	−3.5			133.104	5.49
Cysteine	Cys	C	sulfur-containing	nonpolar	neutral	2.5	250	0.3	121.154	1.38
Glutamic acid	Glu	E	acid	acidic polar	negative	−3.5			147.131	6.32
Glutamine	Gln	Q	amide	polar	neutral	−3.5			146.146	3.9
Glycine	Gly	G	aliphatic	nonpolar	neutral	−0.4			75.067	7.03
Histidine	His	H	basic aromatic	basic polar	positive(10% neutral(90%)	−3.2	211	5.9	155.156	2.26
Isoleucine	Ile	I	aliphatic	nonpolar	neutral	4.5			131.175	5.49
Leucine	Leu	L	aliphatic	nonpolar	neutral	3.8			131.175	9.68
Lysine	Lys	K	basic	basic polar	positive	−3.9			146.189	5.19
Methionine	Met	M	sulfur-containing	nonpolar	neutral	1.9			149.208	2.32
Phenylalanine	Phe	F	aromatic	nonpolar	neutral	2.8	257, 206, 188	0.2, 9.3, 60.0	165.192	3.87
Proline	Pro	P	cyclic	nonpolar	neutral	−1.6			115.132	5.02

Amino Acid	3-Letter	1-Letter	Side chain class	Side chain polarity	Side chain charge (pH 7.4)	Hydropathy index	Absorbance λ_{max}(nm)	ε at λ_{max} ($mM^{-1} cm^{-1}$)	MW (Weight)	Occurrence in proteins (%)
Serine	Ser	S	hydroxyl-containing	polar	neutral	−0.8			105.093	7.14
Threonine	Thr	T	hydroxyl-containing	polar	neutral	−0.7			119.119	5.53
Tryptophan	Trp	W	aromatic	nonpolar	neutral	−0.9	280, 219	5.6, 47.0	204.228	1.25
Tyrosine	Tyr	Y	aromatic	polar	neutral	−1.3	274, 222, 193	1.4, 8.0, 48.0	181.191	2.91
Valine	Val	V	aliphatic	nonpolar	neutral	4.2			117.148	6.73

Two additional amino acids are in some species coded for by codons that are usually interpreted as stop codons:

21st and 22nd amino acids	3-Letter	1-Letter	MW(Weight)
Selenocysteine	Sec	U	168.064
Pyrrolysine	Pyl	O	255.313

In addition to the specific amino acid codes, placeholders are used in cases where chemical or crystallographic analysis of a peptide or protein cannot conclusively determine the identity of a residue. They are also used to summarise conserved protein sequence motifs. The use of single letters to indicate sets of similar residues is similar to the use of abbreviation codes for degenerate bases.

Ambiguous Amino Acids	3-Letter	1-Letter	
Any / unknown	Xaa	X	All
Asparagine or aspartic acid	Asx	B	D, N
Glutamine or glutamic acid	Glx	Z	E, Q
Leucine or Isoleucine	Xle	J	I, L
Hydrophobic		Φ	V, I, L, F, W, Y, M
Aromatic		Ω	F, W, Y, H
Aliphatic (Non-Aromatic)		Ψ	V, I, L, M
Small		π	P, G, A, S
Hydrophilic		ζ	S, T, H, N, Q, E, D, K, R
Positively charged		+	K, R, H
Negatively charged		−	D, E

Unk is sometimes used instead of Xaa, but is less standard.

In addition, many non-standard amino acids have a specific code. For example, several peptide drugs, such as Bortezomib and MG132, are artificially synthesized and retain their protecting groups, which have specific codes. Bortezomib is Pyz-Phe-boroLeu, and MG132 is Z-Leu-Leu-Leu-al. To aid in the analysis of protein structure, photo-reactive amino acid analogs are available. These include photoleucine (pLeu) and photomethionine (pMet).

Amino Acids-building Blocks of Proteins

- Proteins are one of the most important class of biomolecules which play pivotal roles in a wide array of cellular activities.
- Proteins are constituted by amino acids, as monomeric unit or building blocks. A typical amino acid has the amino, carboxyl moieties and "R" group (also called as side chain).

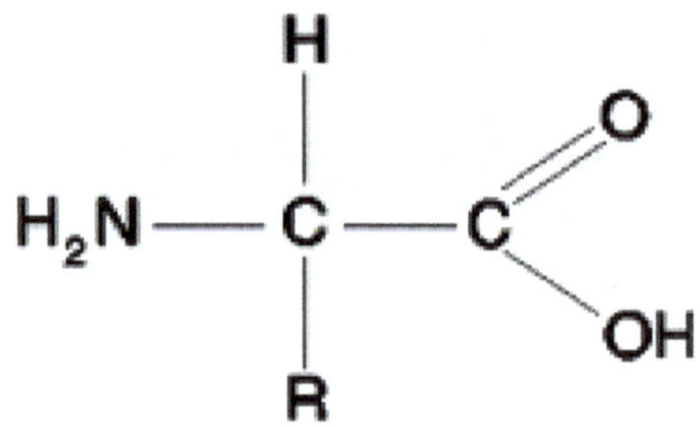

Amino Acids

- When dissolved in water, amino acids exists in solution as dipolar ion called "zwitter ion". It can act as an acid (proton donor) or a base (proton acceptor). Because of this dual nature, they are often called as ampholytes.
- Amino acids in the protein are linked by peptide bonds. Thus, two amino acids joined by peptide bond are called di-peptide, three- tripeptide, four- tetrapeptide. The longer chain is called polypeptide. Proteins are made of single or more polypeptide chains.

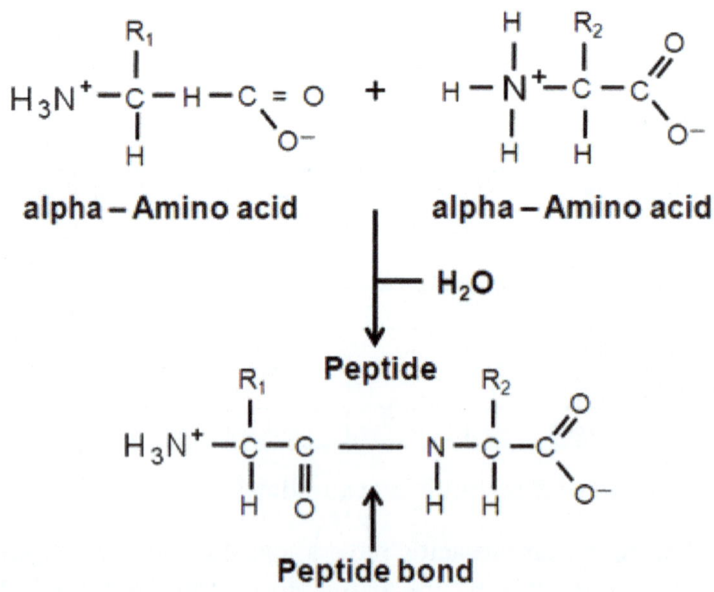

- There are twenty different kind of amino acids found in biological systems beside some unusual amino acids. The nature of R-group varies from amino acid to amino acid.
- Each functional group of amino acid has a fixed pKa value. pKa values of amino acid side chains play an important role in defining the pH-dependent characteristics of a protein. Thus, the ionization state of amino acids will be pH dependent.
- Each amino acid has a standard three letter and one letter abbreviations which are used instead of full name.

Structure of Amino Acids

Functional Groups and pKa Values of Different Amino Acids

Amino acid	Symbol	Structure	*pK1(COOH)	pk2(NH2)	pK R Group
Amino Acids with aliphatic R – Groups					
Glycine	Gly – G	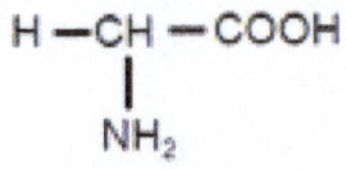	2.4	9.8	
Alanine	Ala – A	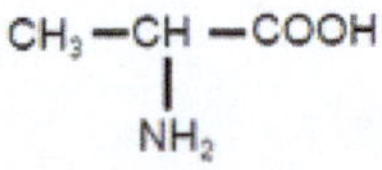	2.4	9.9	
Valine	Val – V	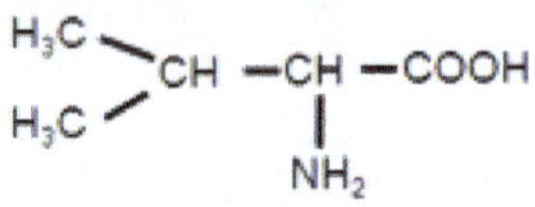	2.2	9.7	
Leucine	Leu – L	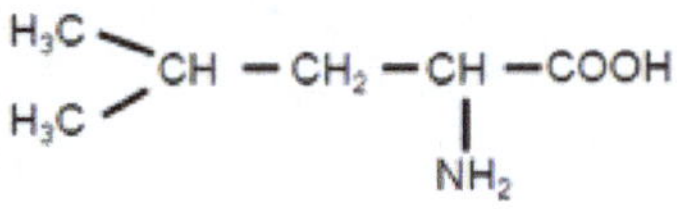	2.3	9.7	
Isoleucine	Ile - I	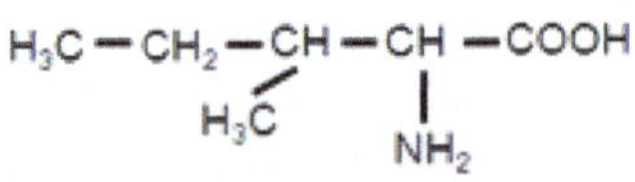	2.3	9.8	

Non-Aromatic Amino Acids with Hydroxyl R-Groups

Amino acid	Symbol	Structure	*pK1(COOH)	pk2(NH2)	pK R Group
Serine	Ser – S	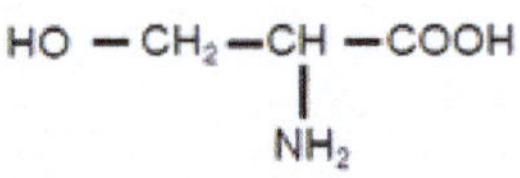	2.2	9.2	~13
Threonine	Thr – T	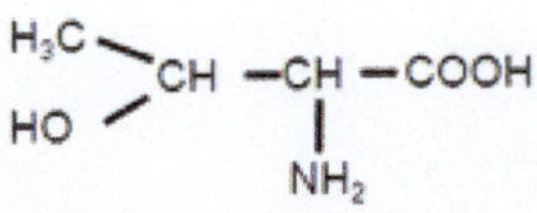	2.1	9.1	~13

Amino Acids with Sulfur-Containing R-Groups

Amino acid	Symbol	Structure	*pK1(COOH)	pk2(NH2)	pK R Group
Cysteine	Ser – S	$HS-CH_2-CH(NH_2)-COOH$	1.9	10.8	8.3
Methionine	Met – M	$H_3C-S-(CH_2)_2-CH(NH_2)-COOH$	2.1	9.3	

Acidic Amino Acids and their Amides

Amino acid	Symbol	Structure	*pK1(COOH)	pk2(NH2)	pK R Group
Aspartic Acid	Asp – D	$HOOC-CH_2-CH(NH_2)-COOH$	2.0	9.9	3.9
Asparagine	Asn – N	$H_2N-C(=O)-CH_2-CH(NH_2)-COOH$	2.1	8.8	
Glutamic Acid	Glu – E	$HOOC-CH_2-CH_2-CH(NH_2)-COOH$	2.1	9.5	4.1
Glutamine	Gin – Q	$H_2N-C(O)-CH_2-CH_2-CH(NH_2)-COOH$	2.2	9.1	

Basic Amino Acids

Amino acid	Symbol	Structure	*pK1(COOH)	pk2(NH2)	pK R Group
Arginine	Arg – R	$H_2N-C(=NH)-HN-CH_2-CH_2-CH_2-CH(NH_2)-COOH$	1.8	9.0	12.5
Lysine	Lys – K	$H_2N-(CH_2)_4-CH(NH_2)-COOH$	2.2	9.2	10.8
Histidine	His – H	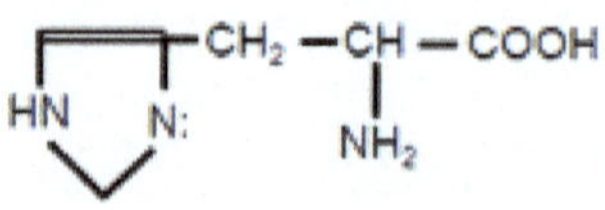	1.8	9.2	6.0

Amino Acids with Aromatic Rings

Amino acid	Symbol	Structure	*pK1(COOH)	pk2(NH2)	pK R Group
Phenylalanine	Phe – F	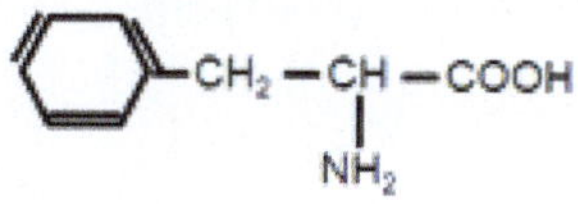	2.2	9.2	
Tyrosine	Tyr – Y	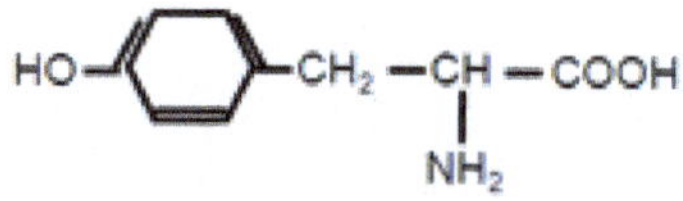	2.2	9.1	10.1
Tryptophan T	Trp – W	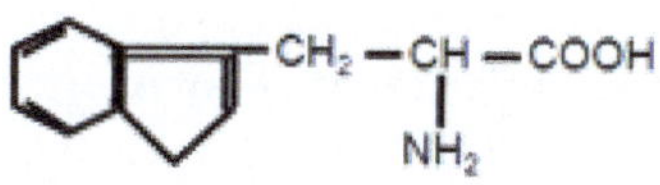	2.4	9.4	

Imino Acids

Amino acid	Symbol	Structure	*pK1(COOH)	pk2(NH2)	pK R Group
Proline	Pro – P	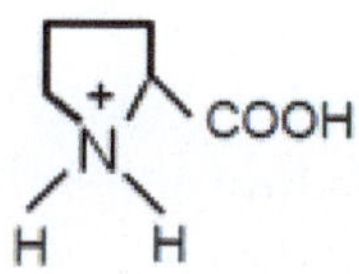	2.0		10.6

- All amino acids except glycine (R = H) are chiral, existing in D and L form. However, all the amino acids in biological system exists in the L-configuration, where "L" implies that the amino acid confirmation similar to L-glyceraldehyde.

Properties

Properties of Amino Acids

- The properties of each amino acid are mainly dictated by the side chain, which can vary in size, shape, charge, reactivity and ability to hydrogen bond. The amino acids are grouped according to the properties of their side chains :

1. Amino acids with non-polar or hydrophobic R group :-

- Aliphatic : The first six amino acids, glycine (GLY, G), alanine (ALA, A), Methionine (Met, M), valine (VAL, V) leucine (LEU, L), and isoleucine (ILE, I), proline (PRO) and are aliphatic in nature. Glycine is smallest. Glycine and alanine are too small to have a hydrophobic effect. Methionine is sulphur containing amino acid. Valine, leucine and isoleucine are considerably hydrophobic.
- Aromatic: Phenylalanine (PHE, F), tryptophan (TRP, Y) and tyrosine (TYR, W) are aromatic in nature. These contain aromatic side chain. They are specifically absorbs at 280 nm thus form the basis of quantitative estimation of protein by ultra Violet (UV) method.

2. Amino acids with polar but uncharged R group :

- These are sulfur containing, namely cysteine (CYS, C), two hydroxyl-containing serine (SER S) threonine (THR, T) and amide containing Aspargine (ASN, N) and glutamine (GLN, Q).
- One typical imino acid, Proline (PRO, P) is also found in this category. Because of its cyclic structure, it leads to bending of protein chain. Proline is an imine and usual in that its nitrogen atom present as secondary.
- Cysteine is involved in inter molecular di sulfide bond (called cystine) with other cysteine of the poly peptide chain. These disulphide bonds are the only covalent bond beside peptide bond in the protein and impart stability to the protein.
- Ser and Thr have side chains which can hydrogen bond to water or to other groups on neighbouring macromolecules. Asn and Gln are amide of acidic amino acids- aspartic and glutamic acid.

3. Polar positively charged amino acids :

- Amino acids lysine (LYS, K), arginine (ARG, R) and histidine (HIS, H) are considered basic hydrophilic, since they contain basic side chain groups that will have a positive charge at pH 7.4.

4. Polar negatively charged amino acids :

- Aspartic acid (ASP, D) and glutamic acid (GLU, E) are considered acidic hydrophilic, since they contain acidic side chain groups that will have a negative charge at pH 7.4.

Peptide

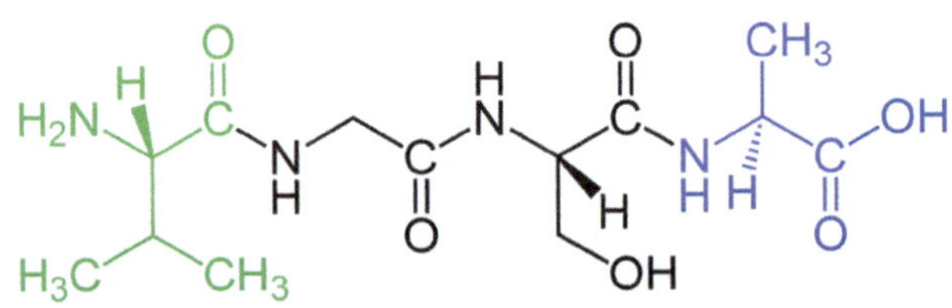

A tetrapeptide (example Val-Gly-Ser-Ala) with **green** marked amino end (**L-Valine**) and **blue** marked carboxyl end (**L-Alanine**).

Peptides are biologically occurring short chains of amino acid monomers linked by peptide (amide) bonds.

The covalent chemical bonds are formed when the carboxyl group of one amino acid reacts with the amine group of another. The shortest peptides are dipeptides, consisting of 2 amino acids joined by a single peptide bond, followed by tripeptides, tetrapeptides, etc. A polypeptide is a long, continuous, and unbranched peptide chain. Hence, peptides fall under the broad chemical classes of biological oligomers and polymers, alongside nucleic acids, oligosaccharides and polysaccharides, etc.

Peptides are distinguished from proteins on the basis of size, and as an arbitrary benchmark can be understood to contain approximately 50 or fewer amino acids. Proteins consist of one or more polypeptides arranged in a biologically functional way, often bound to ligands such as coenzymes and cofactors, or to another protein or other macromolecule (DNA, RNA, etc.), or to complex macromolecular assemblies. Finally, while aspects of the lab techniques applied to peptides versus polypeptides and proteins differ (e.g., the specifics of electrophoresis, chromatography, etc.), the size boundaries that distinguish peptides from polypeptides and proteins are not absolute: long peptides such as amyloid beta have been referred to as proteins, and smaller proteins like insulin have been considered peptides.

Amino acids that have been incorporated into peptides are termed "residues" due to the release of either a hydrogen ion from the amine end or a hydroxyl ion from the carboxyl end, or both, as a water molecule is released during formation of each amide bond. All peptides except cyclic peptides have an N-terminal and C-terminal residue at the end of the peptide (as shown for the tetrapeptide in the image).

Peptide Classes

Peptides are divided into several classes, depending on how they are produced:

Milk Peptides

Two naturally occurring milk peptides are formed from the milk protein casein when digestive enzymes break this down; they can also arise from the proteinases formed by lactobacilli during the fermentation of milk.

Ribosomal Peptides

Ribosomal peptides are synthesized by translation of mRNA. They are often subjected to proteolysis to generate the mature form. These function, typically in higher organisms, as hormones and signaling molecules. Some organisms produce peptides as antibiotics, such as microcins. Since they are translated, the amino acid residues involved are restricted to those utilized by the ribosome.

However, these peptides frequently have posttranslational modifications such as phosphorylation, hydroxylation, sulfonation, palmitoylation, glycosylation and disulfide formation. In general, they are linear, although lariat structures have been observed. More exotic manipulations do occur, such as racemization of L-amino acids to D-amino acids in platypus venom.

Nonribosomal Peptides

Nonribosomal peptides are assembled by enzymes that are specific to each peptide, rather than by the ribosome. The most common non-ribosomal peptide is glutathione, which is a component of the antioxidant defenses of most aerobic organisms. Other nonribosomal peptides are most common in unicellular organisms, plants, and fungi and are synthesized by modular enzyme complexes called *nonribosomal peptide synthetases*.

These complexes are often laid out in a similar fashion, and they can contain many different modules to perform a diverse set of chemical manipulations on the developing product. These peptides are often cyclic and can have highly complex cyclic structures, although linear nonribosomal peptides are also common. Since the system is closely related to the machinery for building fatty acids and polyketides, hybrid compounds are often found. The presence of oxazoles or thiazoles often indicates that the compound was synthesized in this fashion.

Peptones

Peptones are derived from animal milk or meat digested by proteolysis. In addition to containing small peptides, the resulting material includes fats, metals, salts, vitamins and many other biological compounds. Peptones are used in nutrient media for growing bacteria and fungi.

Peptide Fragments

Peptide fragments refer to fragments of proteins that are used to identify or quantify the source protein. Often these are the products of enzymatic degradation performed in the laboratory on a controlled sample, but can also be forensic or paleontological samples that have been degraded by natural effects.

Peptide Synthesis

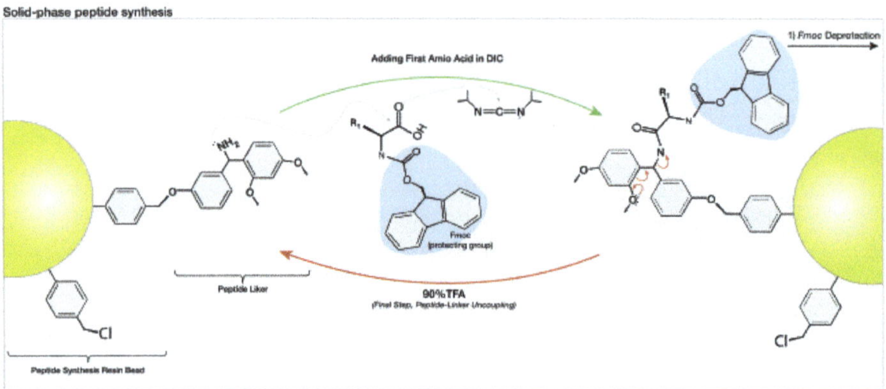

Solid-phase peptide synthesis on a rink amide resin using Fmoc-α-amine-protected amino acid

Peptides in Molecular Biology

Peptides received prominence in molecular biology for several reasons. The first is that peptides allow the creation of *peptide antibodies* in animals without the need of purifying the protein of interest. This involves synthesizing antigenic peptides of sections of the protein of interest. These will then be used to make antibodies in a rabbit or mouse against the protein.

Another reason is that peptides have become instrumental in mass spectrometry, allowing the identification of proteins of interest based on peptide masses and sequence.

Peptides have recently been used in the study of protein structure and function. For example, synthetic peptides can be used as probes to see where protein-peptide interactions occur.

Inhibitory peptides are also used in clinical research to examine the effects of peptides on the inhibition of cancer proteins and other diseases. For example, one of the most promising application is through peptides that target LHRH. These particular peptides act as an agonist, meaning that they bind to a cell in a way that regulates LHRH receptors. The process of inhibiting the cell receptors suggests that peptides could be beneficial in treating prostate cancer. However, additional investigations and experiments are required before the cancer-fighting attributes, exhibited by peptides, can be considered definitive.

Well-known Peptide Families

The peptide families in this section are ribosomal peptides, usually with hormonal activity. All of these peptides are synthesized by cells as longer "propeptides" or "proproteins" and truncated prior to exiting the cell. They are released into the bloodstream where they perform their signaling functions.

Antimicrobial Peptides

- Magainin family
- Cecropin family
- Cathelicidin family
- Defensin family

Tachykinin Peptides

- Substance P
- Kassinin
- Neurokinin A
- Eledoisin
- Neurokinin B

Vasoactive Intestinal Peptides

- VIP (*Vasoactive Intestinal Peptide*; PHM27)
- PACAP *Pituitary Adenylate Cyclase Activating Peptide*
- Peptide PHI 27 (*Peptide Histidine Isoleucine 27*)
- GHRH 1-24 *(Growth Hormone Releasing Hormone 1-24)*
- Glucagon
- Secretin

Pancreatic Polypeptide-related Peptides

- NPY *(NeuroPeptide Y)*

- PYY *(Peptide YY)*
- APP *(Avian Pancreatic Polypeptide)*
- PPY *Pancreatic PolYpeptide*

Opioid Peptides

- Proopiomelanocortin (POMC) peptides
- Enkephalin pentapeptides
- Prodynorphin peptides

Calcitonin Peptides

- Calcitonin
- Amylin
- AGG01

Other Peptides

- B-type Natriuretic Peptide (BNP) - produced in myocardium & useful in medical diagnosis
- Lactotripeptides - Lactotripeptides might reduce blood pressure, although the evidence is mixed.

Notes on Terminology

Length:

- A *polypeptide* is a single linear chain of many amino acids, held together by amide bonds.
- A *protein* consists of one or more polypeptides (more than about 50 amino acids long).
- An *oligopeptide* consists of only a few amino acids (between two and twenty).

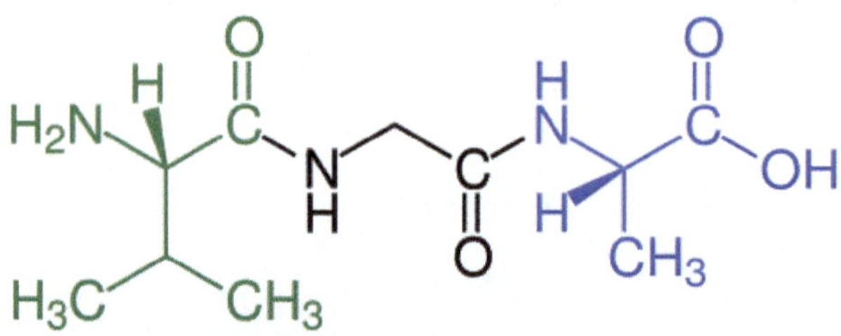

A tripeptide (example Val-Gly-Ala) with **green** marked amino end (**L-Valine**) and **blue** marked carboxyl end (**L-Alanine**)

Number of amino acids:

- A *monopeptide* has one amino acid.

- A *dipeptide* has two amino acids.
- A *tripeptide* has three amino acids.
- A *tetrapeptide* has four amino acids.
- A *pentapeptide* has five amino acids.
- A *hexapeptide* has six amino acids.
- A *heptapeptide* has seven amino acids.
- An *octapeptide* has eight amino acids (e.g., angiotensin II).
- A *nonapeptide* has nine amino acids (e.g., oxytocin).
- A *decapeptide* has ten amino acids (e.g., gonadotropin-releasing hormone & angiotensin I).
- An *undecapeptide* (or *monodecapeptide*) has eleven amino acids, a *dodecapeptide* (or *didecapeptide*) has twelve amino acids, a *tridecapeptide* has thirteen amino acids, and so forth.
- An *icosapeptide* has twenty amino acids, a *tricontapeptide* has thirty amino acids, a *tetracontapeptide* has forty amino acids, and so forth.

Function:

- A *neuropeptide* is a peptide that is active in association with neural tissue.
- A *lipopeptide* is a peptide that has a lipid connected to it, and *pepducins* are lipopeptides that interact with GPCRs.
- A *peptide hormone* is a peptide that acts as a hormone.
- A proteose is a mixture of peptides produced by the hydrolysis of proteins. The term is somewhat archaic.

Doping in Sports

The term *peptide* has been used to mean *secretagogue peptides* and *peptide hormones* in sports doping matters: secretagogue peptides are classified as Schedule 2 (S2) prohibited substances on the World Anti-Doping Agency (WADA) Prohibited List, and are therefore prohibited for use by professional athletes both in and out of competition. Such secretagogue peptides have been on the WADA prohibited substances list since at least 2008. The Australian Crime Commission cited the alleged misuse of secretagogue peptides in Australian sport including growth hormone releasing peptides CJC-1295, GHRP-6, and GHSR (gene) hexarelin. There is ongoing controversy on the legality of using secretagogue peptides in sports.

Glutathione

Glutathione (GSH) is an important antioxidant in plants, animals, fungi, and some bacteria and archaea. Glutathione is capable of preventing damage to important cellular components caused by reactive oxygen species such as free radicals, peroxides, lipid peroxides, and heavy metals. It is a tripeptide with a gamma peptide linkage between the carboxyl group of the glutamate side chain and the amine group of cysteine, and the carboxyl group of cysteine is attached by normal peptide linkage to a glycine.

Thiol groups are reducing agents, existing at a concentration around 5 mM in animal cells. Glutathione reduces disulfide bonds formed within cytoplasmic proteins to cysteines by serving as an electron donor. In the process, glutathione is converted to its oxidized form, glutathione disulfide (GSSG), also called L-(−)-glutathione.

Once oxidized, glutathione can be reduced back by glutathione reductase, using NADPH as an electron donor. The ratio of reduced glutathione to oxidized glutathione within cells is often used as a measure of cellular oxidative stress.

Biosynthesis

The biosynthesis pathway for glutathione is found in some bacteria, such as cyanobacteria and proteobacteria, but is missing in many other bacteria. Most eukaryotes, including humans, synthesize glutathione, but some do not, such as Leguminosae, *Entamoeba*, and *Giardia*. The only archaea that make glutathione are halobacteria.

Glutathione is not an essential nutrient for humans, since it can be synthesized in the body from the amino acids L-cysteine, L-glutamic acid, and glycine; it does not have to be present as a supplement in the diet. The sulfhydryl group (SH) of cysteine serves as a proton donor and is responsible for its biological activity. Cysteine is the rate-limiting factor in cellular glutathione biosynthesis, since this amino acid is relatively rare in foods.

Cells make glutathione in two adenosine triphosphate-dependent steps:

- First, *gamma*-glutamylcysteine is synthesized from L-glutamate and cysteine via the enzyme gamma-glutamylcysteine synthetase (glutamate cysteine ligase, GCL). This reaction is the rate-limiting step in glutathione synthesis.
- Second, glycine is added to the C-terminal of *gamma*-glutamylcysteine via the enzyme glutathione synthetase.

Animal glutamate cysteine ligase (GCL) is a heterodimeric enzyme composed of a catalytic and a modulatory subunit. The catalytic subunit is necessary and sufficient for all GCL enzymatic activity, whereas the modulatory subunit increases the catalytic efficiency of the enzyme. Mice lacking the catalytic subunit (i.e., lacking all *de novo* GSH synthesis) die before birth. Mice lacking the modulatory subunit demonstrate no obvious phenotype, but exhibit marked decrease in GSH and increased sensitivity to toxic insults.

While all animal cells are capable of synthesizing glutathione, glutathione synthesis in the liver has been shown to be essential. GCLC knockout mice die within a month of birth due to the absence of hepatic GSH synthesis. Major transport into the blood stream is driven by an electrochemical gradient, specifically through the transport proteins RcGshT and RsGshT. Similarly, glutathione *S*-conjugates, synthesized hepatically, feature preferential secretion into bile.

The plant glutamate cysteine ligase (GCL) is a redox-sensitive homodimeric enzyme, conserved in the plant kingdom. In an oxidizing environment, intermolecular disulfide bridges are formed and the enzyme switches to the dimeric active state. The midpoint potential of the critical cysteine pair is -318 mV. In addition to the redox-dependent control, the plant GCL enzyme is feedback inhibited by glutathione. GCL is exclusively located in plastids, and glutathione synthetase (GS) is dual-targeted to plastids and cytosol, thus GSH and gamma-glutamylcysteine are exported from the plastids. Both glutathione biosynthesis enzymes are essential in plants; knock-outs of GCL and GS are lethal to embryo and seedling.

Function

Glutathione exists in both reduced (GSH) and oxidized (GSSG) states. In the reduced state, the thiol group of cysteine is able to donate a reducing equivalent (H^{+}+ e^{-}) to other molecules, such as reactive oxygen species to neutralize them, or to protein cysteines to maintain their reduced forms. With donating an electron, glutathione itself becomes reactive and readily reacts with another reactive glutathione to form glutathione disulfide (GSSG). Such a reaction is probable due to the relatively high concentration of glutathione in cells (up to 7 mM in the liver).

Generally, interactions between GSH and other molecules with higher relative electrophilicity deplete GSH levels within the cell. An exception to this case involves the sensitivity of GSH to the electrophilic compound's relative concentration. In high concentrations, the organic molecule Diethyl maleate fully depleted GSH levels in cells. However, in low concentrations, a minor decrease in cellular GSH levels was followed by a two-fold increase.

GSH can be regenerated from GSSG by the enzyme glutathione reductase (GSR): NADPH reduces FAD present in GSR to produce a transient FADH-anion. This anion then quickly breaks a disulfide bond (Cys58 - Cys63) and leads to Cys63's nucleophilically attacking the nearest sulfide unit in the GSSG molecule (promoted by His467), which creates a mixed disulfide bond (GS-Cys58) and a GS-anion. His467 of GSR then protonates the GS-anion to form the first GSH. Next, Cys63 nucleophilically attacks the sulfide of Cys58, releasing a GS-anion, which, in turn, picks up a solvent proton and is released from the enzyme, thereby creating the second GSH. So, for every GSSG and NADPH, two reduced GSH molecules are gained, which can again act as antioxidants scavenging reactive oxygen species in the cell.

In healthy cells and tissue, more than 90% of the total glutathione pool is in the reduced form (GSH) and less than 10% exists in the disulfide form (GSSG). An increased GSSG-to-GSH ratio is considered indicative of oxidative stress.

Glutathione has multiple functions:

- It maintains levels of reduced glutaredoxin and glutathione peroxidase
- It is one of the major endogenous antioxidants produced by the cells, participating directly in the neutralization of free radicals and reactive oxygen compounds, as well as maintaining exogenous antioxidants such as vitamins C and E in their reduced (active) forms.
- Regulation of the nitric oxide cycle is critical for life, but can be problematic if unregulated.

- It is used in metabolic and biochemical reactions such as DNA synthesis and repair, protein synthesis, prostaglandin synthesis, amino acid transport, and enzyme activation. Thus, every system in the body can be affected by the state of the glutathione system, especially the immune system, the nervous system, the gastrointestinal system, and the lungs.
- It has a vital function in iron metabolism. Yeast cells depleted of GSH or containing toxic levels of GSH show an intense iron starvation-like response and impairment of the activity of extramitochondrial ISC enzymes thus inhibiting oxidative endoplasmic reticulum folding, followed by death.
- It has roles in progression of the cell cycle, including cell death. GSH levels regulate redox changes to nuclear proteins necessary for the initiation of cell differentiation. Differences in GSH levels also determine the expressed mode of cell death, being either apoptosis or cell necrosis. Manageably low levels result in the systematic breakage of the cell whereas excessively low levels result in rapid cell death.

Function in Animals

GSH is known as a substrate in conjugation reactions, which is catalyzed by glutathione S-transferase enzymes in cytosol, microsomes, and mitochondria. However, GSH is also capable of participating in nonenzymatic conjugation with some chemicals.

In the case of *N*-acetyl-*p*-benzoquinone imine (NAPQI), the reactive cytochrome P450-reactive metabolite formed by paracetamol (acetaminophen), which becomes toxic when GSH is depleted by an overdose of acetaminophen, glutathione is an essential antidote to overdose. Glutathione conjugates to NAPQI and helps to detoxify it. In this capacity, it protects cellular protein thiol groups, which would otherwise become covalently modified; when all GSH has been spent, NAPQI begins to react with the cellular proteins, killing the cells in the process. The preferred treatment for an overdose of this painkiller is the administration (usually in atomized form) of *N*-acetyl-L-cysteine (often as a preparation called Mucomyst), which is processed by cells to L-cysteine and used in the *de novo* synthesis of GSH.

Glutathione (GSH) participates in leukotriene synthesis and is a cofactor for the enzyme glutathione peroxidase. It is also important as a hydrophilic molecule that is added to lipophilic toxins and waste in the liver during biotransformation before they can become part of the bile. Glutathione is also needed for the detoxification of methylglyoxal, a toxin produced as a byproduct of metabolism.

This detoxification reaction is carried out by the glyoxalase system. Glyoxalase I (EC 4.4.1.5) catalyzes the conversion of methylglyoxal and reduced glutathione to *S*-D-lactoyl-glutathione. Glyoxalase II (EC 3.1.2.6) catalyzes the hydrolysis of *S*-D-lactoyl-glutathione to glutathione and D-lactic acid.

Glutathione, along with oxidized glutathione (GSSG) and S-nitrosoglutathione (GSNO), have been found to bind to the glutamate recognition site of the NMDA and AMPA receptors (via their γ-glutamyl moieties), and may be endogenous neuromodulators. At millimolar concentrations, they may also modulate the redox state of the NMDA receptor complex. In addition, glutathione has been found to bind to and activate ionotropic receptors that are different from any other excitatory amino acid receptor, and which may constitute *glutathione receptors*, potentially making it a neurotransmitter.

Function in Plants

In plants, glutathione is crucial for biotic and abiotic stress management. It is a pivotal component of the glutathione-ascorbate cycle, a system that reduces poisonous hydrogen peroxide. It is the precursor of phytochelatins, glutathione oligomers that chelate heavy metals such as cadmium. Glutathione is required for efficient defence against plant pathogens such as *Pseudomonas syringae* and *Phytophthora brassicae*. Adenylyl-sulfate reductase, an enzyme of the sulfur assimilation pathway, uses glutathione as an electron donor. Other enzymes using glutathione as a substrate are glutaredoxins. These small oxidoreductases are involved in flower development, salicylic acid, and plant defence signalling.

Supplementation

Calcitriol (1,25-dihydroxyvitamin D_3), the active metabolite of vitamin D_3, after being synthesized from calcifediol in the kidney, increases glutathione levels in the brain and appears to be a catalyst for glutathione production. It takes about ten days for the body to process vitamin D_3 into calcitriol.

S-adenosylmethionine (SAMe), a cosubstrate involved in methyl group transfer, has also been shown to increase cellular glutathione content in persons suffering from a disease-related glutathione deficiency.

Low glutathione is commonly observed in wasting and negative nitrogen balance, as seen in cancer, HIV/AIDS, sepsis, trauma, burns, and athletic overtraining. Low levels are also observed in periods of starvation. These effects are hypothesized to be influenced by the higher glycolytic activity associated with cachexia, which result from reduced levels of oxidative phosphorylation.

Bioavailability

Glutathione is only to a small extent bioavailable to humans; the human body is capable of maintaining a consistent level of GSH. Oral introduction of GSH into the body is, in fact, scarcely effective to increase its plasma and/or intracellular concentration. At the base of its poor bioavailability is the nature of glutathione which, being a tripeptide, is the substrate of proteases (peptidases) of the alimentary canal, and the absence of a specific *carrier* of glutathione at the level of cell membrane.

Methods to Determine Glutathione

Ellman's Reagent and Monobromobimane

Reduced glutathione may be visualized using Ellman's reagent or bimane derivatives such as monobromobimane. The monobromobimane method is more sensitive. In this procedure, cells are lysed and thiols extracted using a HCl buffer. The thiols are then reduced with dithiothreitol and labelled by monobromobimane. Monobromobimane becomes fluorescent after binding to GSH. The thiols are then separated by HPLC and the fluorescence quantified with a fluorescence detector.

Monochlorobimane

Monochlorobimane can be used to quantify glutathione in vivo. The quantification is done by confocal laser scanning microscopy after application of the dye to living cells. This quantification process relies on measuring the rates of fluorescence changes and is limited to plant cells.

5-Chloromethylfluorescein Diacetate (CMFDA)

CMFDA was initially used as a cell tracker. Unfortunately, it has also been mistakenly used as a glutathione probe. Unlike monochlorobimane, whose fluorescence increases upon reacting with glutathione, the fluorescence increase of CMFDA is due to the hydrolysis of the acetate groups inside cells. Although CMFDA may react with glutathione in cells, the fluorescence increase does not reflect the reaction. Therefore, studies using CMFDA as a glutathione probe should be revisited and re-interpreted.

ThiolQuant Green

The major limitation of these bimane based probes and many other reported probes is that these probes are based on irreversible chemical reactions with glutathione, which renders these probes incapable of monitoring the real-time glutathione dynamics. Recently, the first reversible reaction based fluorescent probe-ThiolQuant Green (TQG)-for glutathione was reported. ThiolQuant Green can not only perform high resolution measurements of glutathione levels in single cells using a confocal microscope, but also be applied in flow cytometry to perform bulk measurements.

Protein based Glutathione Probes

Another approach, which allows measurement of the glutathione redox potential at a high spatial and temporal resolution in living cells is based on redox imaging using the redox-sensitive green fluorescent protein (roGFP) or redox sensitive yellow fluorescent protein (rxYFP) GSSG because its very low physiological concentration is difficult to measure accurately unless the procedure is carefully executed and monitored and the occurrence of interfering compounds is properly addressed. GSSG concentration ranges from 10 to 50 μM in all solid tissues, and from 2 to 5 μM in blood (13–33 nmol per gram Hb). GSH-to-GSSG ratio ranges from 100 to 700.

Other Biological Implications

Cancer

Once a tumor has been established, elevated levels of glutathione may act to protect cancerous cells by conferring resistance to chemotherapeutic drugs. The antineoplastic mustard drug canfosfamide was modelled on the structure of glutathione.

Cystic Fibrosis

Several studies have been completed on the effectiveness of introducing inhaled glutathione to people with cystic fibrosis with mixed results.

Alzheimer's Disease

Whilst extracellular Aβ plaques, NFT, inflammation in the form or reactive astrocytes and microglia, and neuronal loss are all consistent pathological features of AD, a mechanistic link between these factors is yet to be clarified. Although the majority of past research has focused on fibrillar Aβ, soluble oligomeric Aβ species are now considered to be of major pathological importance in AD. Up-regulation of GSH may be protective against the oxidative and neurotoxic effects of oligomeric Aβ.

Uses

Winemaking

The content of glutathione in must, the first raw form of wine, determines the browning, or caramelizing effect, during the production of white wine by trapping the caffeoyltartaric acid quinones generated by enzymic oxidation as grape reaction product. Its concentration in wine can be determined by mass spectrometry.

Cosmetics

Glutathione plays an important role in preventing oxidative damage to the skin. In addition to its many recognized biological functions, glutathione has also been associated with skin lightening ability. The role of glutathione as a skin whitener was discovered as a side effect of large doses of glutathione. Glutathione utilizes different mechanisms to exert its action as a skin whitening agent at various levels of melanogenesis. It inhibits melanin synthesis by means of stopping the neurotransmitter precursor L-DOPA's ability to interact with tyrosinase in the process of melanin production. Glutathione inhibits the actual production as well as agglutination of melanin by interrupting the function of L-DOPA. Another study found that glutathione inhibits melanin formation by direct inactivation of the enzyme tyrosinase by binding and chelating copper within the enzyme's active site. Glutathione's antioxidant property allows it to inhibit melanin synthesis by quenching of free radicals and peroxides that contribute to tyrosinase activation and melanin formation. Its antioxidant property also protects the skin from UV radiation and other environmental as well as internal stressors that generate free radicals that cause skin damage and hyperpigmentation. In most mammals, melanin formation consists of eumelanin (brown-black pigment) and pheomelanin (yellow-red pigment) as either mixtures or co-polymers. Increase in glutathione level may induce the pigment cell to produce pheomelanin instead of eumelanin pigments. A research by Te-Sheng Chang found lowest levels of reduced glutathione to be associated with eumelanin type pigmentation, whereas the highest ones were associated with the pheomelanin. As a result, it is reasonable to assume that depletion of glutathione would result in eumelanin formation. Prota observed that decreased glutathione concentration led to the conversion of L-Dopaquinone to Dopachrome, increasing the formation of brown-black pigment (eumelanin).

Importance of Gamma-glutamylcysteine as a Precursor for Glutathione synthesis

Gamma-glutamylcysteine (GGC) is the immediate precursor to GSH. GGC supplementation would circumvent feedback inhibitory control of GCL by the end product GSH. Accordingly, a method of elevating GSH levels with the notable advantage of bypassing negative feedback inhibition has been described. Because of this, GGC has been the focus of therapeutic efforts since Puri and Meister 1983. The first documented use of GGC in brains appears to be Pileblad and Magnusson, 1992. Astroglia cells are capable of utilising GGC. Direct delivery of the GSH precursor GCC to brain has been reported to effectively replenish levels of GSH in the brain.

Most of the work done on GGC has been preclinical, based on in vivo animal models, or in vitro brain cultures. In order for the therapeutic value of GGC elevation against AD to be vindicated, three empirical hurdles have to be cleared. The first is to demonstrate that delivery of GCC into the brain can indeed increase GSH. The second is to demonstrate that the increase in GGC can indeed reduce oxidative stress in the brain, a condition frequently linked with cognitive decline.

Gramicidin

Gramicidin is a heterogeneous mixture of three antibiotic compounds, gramicidins A, B and C, making up 80%, 6%, and 14%, respectively, all of which are obtained from the soil bacterial species *Bacillus brevis* and called collectively gramicidin D. Gramicidin D contains linear pentadecapeptides, that is chains made up of 15 amino acids. This is in contrast to gramicidin S, which is a cyclic peptide chain.

Gramicidin was discovered in the United States in 1939.

Medical uses

Gramicidin is active against Gram-positive bacteria, except for the Gram-positive bacilli, and against select Gram-negative organisms, such as *Neisseria* bacteria. Its therapeutic use is limited to topical application, as it induces hemolysis in lower concentrations than bacteria cell death, so it cannot be administered internally. Since the exterior epidermis is composed of dead cells, applying it to the surface of the skin will not cause harm.

It is used primarily as a topical antibiotic and is one of the three constituents of consumer antibiotic polysporin ophthalmic solution.

History

In 1939, the French microbiologist René Dubos isolated the substance tyrothricin and later showed it was composed of two substances, gramicidin (20%) and tyrocidine (80%). These were the first antibiotics to be manufactured commercially.

Composition and Structure

Gramicidin is a polypeptide with alternating L- and D-amino acids, sharing the general formula: formyl-L-X-Gly-L-Ala-D-Leu-L-Ala-D-Val-L-Val-D-Val-L-Trp-D-Leu-L-Y-D-Leu-L-Trp-D-Leu-L-Trp-ethanolamine

X and Y depend upon the gramicidin molecule. There exist valine and isoleucine variants of all three gramicidin species, and 'X' can be either. Y determines which is which; as Y gramicidin A contains tryptophan, B phenylalanine, and C tyrosine. Also note the alternating stereochemical configurations (in the form of D and L) of the amino acids; this is vital to the formation of the β-helix.

The chain assembles inside of the hydrophobic interior of the cellular lipid bilayer to form a β-helix. The helix itself is not long enough to span the membrane, but it dimerizes to form the elongated channel needed to span the whole membrane.

The structure of the gramicidin head-to-head dimer in micelles and lipid bilayers was determined by solution and solid-state NMR. The structure was first proposed by D. W. Urry in 1971. In organic solvents and crystals, this peptide forms different types of non-native double helices.

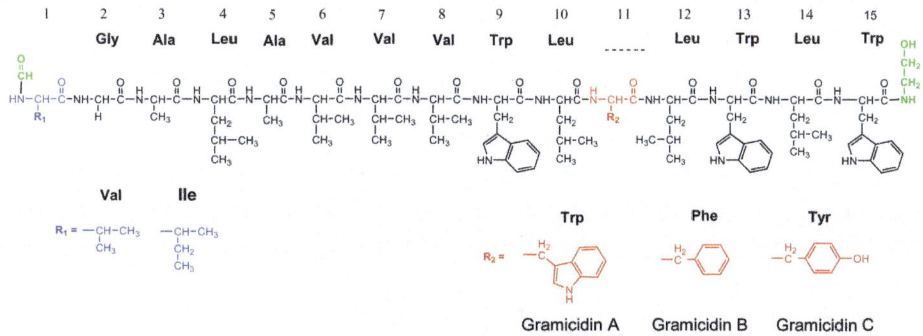

Structures of gramicidins A, B and C

Pharmacological Effect

Gramicidin's bactericidal activity is a result of increasing the permeability of the bacterial cell membrane, allowing inorganic monovalent cations (e.g. Na^+) to travel through unrestricted and thereby destroying the ion gradient between the cytoplasm and the extracellular environment.

That gramicidin D functions as a channel was demonstrated by Hladky and Haydon, who investigated the unitary conductance of the channel. In general, gramicidin channels are completely selective for monovalent cations and the single-channel conductances for the alkali cations are ranked in the same order as the aqueous mobilities of these ions. Divalent cations like Ca^{2+} block the channel by binding near its mouth so it is essentially impermeable to divalent cations and also excludes anions. Cl^- in particular is excluded from the channel because its hydration shell is thermodynamically stronger than that of most monovalent cations. The channel is permeable to most monovalent cations, which move through the channel in single file. The channel is filled with about six water molecules, almost all of which must be displaced when an ion is transported. Thus, ions moving through the gramicidin pore carry along a single file of water molecules. Such a flux of ion and water molecules is known as flux coupling. In the presence of a second type of permeable ion, the two ions couple their flux, as well. Like valinomycin and nonactin, the gramicidin channel is selective for potassium over sodium, but only slightly so. It has a permeability ratio of 2.9. Though it is impermeable to anions, conditions exist under which some anion permeation may be observed. Its ability to bind and transport cations is due to the presence of cation-binding sites, one strong and the other weak, in the channel.

Aspartame

Aspartame is an artificial, non-saccharide sweetener used as a sugar substitute in some foods and beverages. In the European Union, it is codified as E951. Aspartame is a methyl ester of the aspartic acid/phenylalanine dipeptide. It was first sold under the brand name NutraSweet. It was first synthesized in 1965, and the patent expired in 1992.

The safety of aspartame has been the subject of several political and medical controversies, United States congressional hearings and Internet hoaxes since its initial approval for use in food products by the U.S. Food and Drug Administration (FDA) in 1981. The European Food Safety Authority concluded in its 2013 re-evaluation that aspartame and its breakdown products are safe for human consumption at current levels of exposure, corroborating other medical reviews. However, because its breakdown products include phenylalanine, aspartame must be avoided by people with the genetic condition phenylketonuria (PKU).

Chemistry

Aspartame is a methyl ester of the dipeptide of the natural amino acids L-aspartic acid and L-phenylalanine. Under strongly acidic or alkaline conditions, aspartame may generate methanol by hydrolysis. Under more severe conditions, the peptide bonds are also hydrolyzed, resulting in free amino acids.

While known aspects of synthesis are covered by patents, many details are proprietary. Two approaches to synthesis are used commercially. In the chemical synthesis, the two carboxyl groups of aspartic acid are joined into an anhydride, and the amino group is protected by converting it to a functional group that will not interfere in the next reaction. Phenylalanine is converted to its methyl ester and combined with the *N*-protected aspartic anhydride; then the protecting group is removed from aspartic nitrogen by acid hydrolysis. The drawback of this technique is that a byproduct, the bitter-tasting β-form, is produced when the wrong carboxyl group from aspartic acid links to phenylalanine. A process using an enzyme from *Bacillus thermoproteolyticus* to catalyze the condensation of the chemically altered amino acids will produce high yields without the β-form byproduct. A variant of this method, which has not been used commercially, uses unmodified aspartic acid, but produces low yields. Methods for directly producing aspartyl-phenylalanine by enzymatic means, followed by chemical methylation, have also been tried, but not scaled for industrial production.

Mechanism of Action

The perceived sweetness of aspartame (and other sweet substances like acesulfame K) in humans is due to its binding of the heterodimer G-protein coupled receptor formed by the proteins TAS1R2 and TAS1R3.

Properties and use

Aspartame is approximately 200 times sweeter than sucrose (table sugar). Due to this property, even though aspartame produces four kilocalories of energy per gram (17 kJ/g) when metabolized, the quantity of aspartame needed to produce a sweet taste is so small that its caloric contribution is negligible. The taste of aspartame and other artificial sweeteners differs from that of table sugar in the times of onset and how long the sweetness lasts, though aspartame comes closest to sugar's taste profile among approved artificial sweeteners. The sweetness of aspartame lasts longer than that of sucrose, so it is often blended with other artificial sweeteners such as acesulfame potassium to produce an overall taste more like that of sugar.

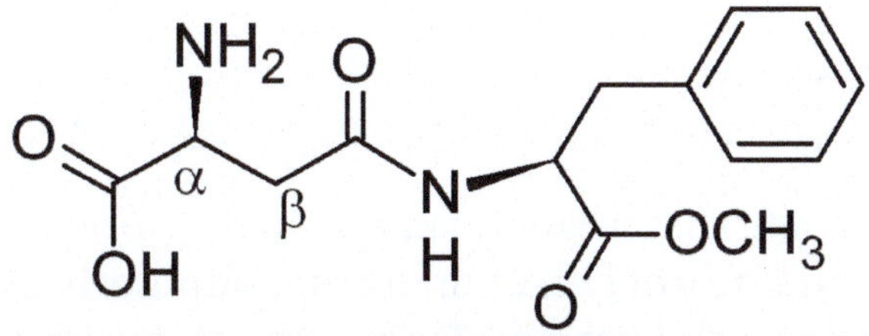

Beta-aspartame differs from aspartame based upon which carboxyl group of aspartate binds to the nitrogen of phenylalanine.

Like many other peptides, aspartame may hydrolyze (break down) into its constituent amino acids under conditions of elevated temperature or high pH. This makes aspartame undesirable as a baking sweetener, and prone to degradation in products hosting a high pH, as required for a long shelf life. The stability of aspartame under heating can be improved to some extent by encasing it in fats or in maltodextrin. The stability when dissolved in water depends markedly on pH. At room temperature, it is most stable at pH 4.3, where its half-life is nearly 300 days. At pH 7, however, its half-life is only a few days. Most soft-drinks have a pH between 3 and 5, where aspartame is reasonably stable. In products that may require a longer shelf life, such as syrups for fountain beverages, aspartame is sometimes blended with a more stable sweetener, such as saccharin.

Aspartame's major decomposition products are its cyclic dipeptide (in a 2,5-diketopiperazine, or DKP, form), the de-esterified dipeptide (aspartyl-phenylalanine), and its constituent components, phenylalanine, aspartic acid, and methanol. At 180 °C, aspartame undergoes decomposition to form a diketopiperazine derivative.

In products such as powdered beverages, the amine in aspartame can undergo a Maillard reaction with the aldehyde groups present in certain aroma compounds. The ensuing loss of both flavor and sweetness can be prevented by protecting the aldehyde as an acetal.

Descriptive analyses of solutions containing aspartame report a sweet aftertaste as well as bitter and off-flavor aftertastes.

Discovery and Approval

Aspartame was discovered in 1965 by James M. Schlatter, a chemist working for G.D. Searle & Company. Schlatter had synthesized aspartame as an intermediate step in generating a tetrapeptide of the hormone gastrin, for use in assessing an anti-ulcer drug candidate. He discovered its sweet taste when he licked his finger, which had become contaminated with aspartame, to lift up a piece of paper. Torunn Atteraas Garin participated in the development of aspartame as an artificial sweetener.

In 1975, prompted by issues regarding Flagyl and Aldactone, a U.S. FDA task force team reviewed 25 studies submitted by the manufacturer, including 11 on aspartame. The team reported "serious deficiencies in Searle's operations and practices". The FDA sought to authenticate 15 of the submitted studies against the supporting data. In 1979, the Center for Food Safety and Applied Nutrition (CFSAN) concluded, since many problems with the aspartame studies were minor and did not affect the conclusions, the studies could be used to assess aspartame's safety.

In 1980, the FDA convened a Public Board of Inquiry (PBOI) consisting of independent advisors charged with examining the purported relationship between aspartame and brain cancer. The PBOI concluded aspartame does not cause brain damage, but it recommended against approving aspartame at that time, citing unanswered questions about cancer in laboratory rats.

Citing data from a Japanese study that had not been available to the members of the PBOI, and after seeking advice from an expert panel that found fault with statistical analyses underlying the PBOI's hesitation, yet argued against approval, FDA commissioner Hayes approved aspartame for use in dry goods. In 1983, the FDA further approved aspartame for use in carbonated beverages, and for use in other beverages, baked goods, and confections in 1993. In 1996, the FDA removed all restrictions from aspartame, allowing it to be used in all foods.

Several European Union countries approved aspartame in the 1980s, with EU-wide approval in 1994. The European Commission Scientific Committee on Food reviewed subsequent safety studies and reaffirmed the approval in 2002. The European Food Safety Authority reported in 2006 that the previously established Acceptable daily intake was appropriate, after reviewing yet another set of studies.

Compendial Status

- British Pharmacopoeia
- United States Pharmacopeia

Safety and Approval Controversies

Aspartame has been found to be safe for human consumption by more than ninety countries worldwide, with FDA officials describing aspartame as "one of the most thoroughly tested and studied food additives the agency has ever approved" and its safety as "clear cut", but has been the subject of several controversies, hoaxes and health scares.

Initially aspartame was approved by the U.S. Food and Drug Administration (FDA) in 1974; however, problems with Searle's safety testing program, including testing of aspartame, were discovered subsequently. The approval was rescinded the following year, but after outside reviews of the problematic tests and additional testing, final approval was granted in 1981. Because allegations of conflicts of interest marred the FDA's approval of aspartame, the U.S. Government Accountability Office reviewed the actions of involved officials in 1986 and the approval process in 1987; neither the allegations of conflict of interest nor problems in the final approval process were substantiated.

In addition, the Centers for Disease Control investigated in 1984 and was unable to find any significant epidemiological associations to serious risk or harm.

Since December 1998, a widely circulated email hoax has cited aspartame as the cause of numerous diseases.

The weight of existing scientific evidence indicates that aspartame is safe at current levels of consumption as a non-nutritive sweetener. Reviews conducted by regulatory agencies decades after aspartame was first approved have supported its continued availability.

Safety and Health Effects

The safety of aspartame has been studied extensively since its discovery with research that includes animal studies, clinical and epidemiological research, and postmarketing surveillance. Aspartame is one of the most rigorously tested food ingredients. Peer-reviewed comprehensive review articles and independent reviews by governmental regulatory bodies have analyzed the published research on the safety of aspartame and have found aspartame is safe for consumption at current levels. Aspartame has been deemed safe for human consumption by over 100 regulatory agencies in their respective countries, including the UK Food Standards Agency, the European Food Safety Authority (EFSA) and Health Canada.

Intake

The acceptable daily intake (ADI) value for aspartame, as well as other food additives studied, is defined as the "amount of a food additive, expressed on a body weight basis, that can be ingested daily over a lifetime without appreciable health risk." The Joint FAO/WHO Expert Committee on Food Additives (JECFA) and the European Commission's Scientific Committee on Food has determined this value is 40 mg/kg of body weight for aspartame, while FDA has set its ADI for aspartame at 50 mg/kg.

The primary source for exposure to aspartame in the United States is diet soft drinks, though it can be consumed in other products, such as pharmaceutical preparations, fruit drinks, and chewing gum among others in smaller quantities. A 12 US fluid ounce (355 ml) can of diet soda contains 180 milligrams (0.0063 oz) of aspartame, and for a 75 kg (165 lb) adult, it takes approximately 21 cans of diet soda daily to consume the 3,750 milligrams (0.132 oz) of aspartame that would surpass the FDA's 50 milligrams per kilogram of body weight ADI of aspartame from diet soda alone.

Reviews have analyzed studies which have looked at the consumption of aspartame in countries worldwide, including the United States, countries in Europe, and Australia, among others. These reviews have found that even the high levels of intake of aspartame, studied across multiple countries and different methods of measuring aspartame consumption, are well below the ADI for safe consumption of aspartame. Reviews have also found that populations that are believed to be especially high consumers of aspartame such as children and diabetics are below the ADI for safe consumption, even considering extreme worst-case scenario calculations of consumption.

In a report released on 10 December 2013, the EFSA said that, after an extensive examination of evidence, it ruled out the "potential risk of aspartame causing damage to genes and inducing cancer," and deemed the amount found in diet sodas an amount safe to consume.

Metabolites

Aspartame is rapidly hydrolyzed in the small intestines. Even with ingestion of very high doses of aspartame (over 200 mg/kg), no aspartame is found in the blood due to the rapid breakdown.

These metabolites have been studied in a wide range of populations including infants, children, adolescents, and healthy adults. In healthy adults and children, even enormous doses of aspartame do not lead to plasma levels of metabolites that are a concern for safety.

Upon ingestion, aspartame breaks down into residual components, including aspartic acid, phenylalanine, methanol, in ratio of 4:5:1 by mass and further breakdown products including formaldehyde and formic acid. Human studies show that formic acid is excreted faster than it is formed after ingestion of aspartame. In some fruit juices, *higher* concentrations of methanol can be found than the amount produced from aspartame in beverages.

Aspartate

Aspartic acid (aspartate) is one of the most common amino acids in the typical diet. As with methanol and phenylalanine, intake of aspartic acid from aspartame is less than would be expected from other dietary sources. At the 90th percentile of intake, aspartame provides only between 1% and 2% of the daily intake of aspartic acid. There has been some speculation that aspartame, in conjunction with other amino acids like glutamate, may lead to excitotoxicity, inflicting damage on brain and nerve cells. However, clinical studies have shown no signs of neurotoxic effects, and studies of metabolism suggest it is not possible to ingest enough aspartic acid and glutamate through food and drink to levels that would be expected to be toxic.

Methanol

The methanol produced by the metabolism of aspartame is absorbed and quickly converted into formaldehyde and then completely oxidized to formic acid. The methanol from aspartame is unlikely to be a safety concern for several reasons. The amount of methanol in aspartame is less than that found in fruit juices and citrus fruits, and there are other dietary sources for methanol such as fermented beverages. Therefore, the amount of methanol produced from aspartame is likely to be less than that from natural sources. With regard to formaldehyde, it is rapidly converted in the body, and the amounts of formaldehyde from the metabolism of aspartame are trivial when compared to the amounts produced routinely by the human body and from other foods and drugs. At the highest expected human doses of consumption of aspartame, there are no increased blood levels of methanol or formic acid, and ingesting aspartame at the 90th percentile of intake would produce 25 times less methanol than what would be considered toxic.

Phenylalanine and Phenylketonuria

High levels of the naturally occurring essential amino acid phenylalanine are a health hazard to those born with phenylketonuria (PKU), a rare inherited disease that prevents phenylalanine from being properly metabolized. Since individuals with PKU must consider aspartame as an additional source of phenylalanine, foods containing aspartame sold in the United States must state "Phenylketonurics: Contains Phenylalanine" on their product labels.

In the UK, foods that contain aspartame are legally required by the country's Food Standards Agency to list the substance among the product's ingredients and carry the warning "Contains a source of phenylalanine" – this is usually at the foot of the list of ingredients. Manufacturers are also required to print '"with sweetener(s)" on the label close to the main product name on foods that contain “sweeteners such as aspartame” or “with sugar and sweetener(s)” on “foods that contain both sugar and sweetener”.

In Canada, foods that contain aspartame are legally required by the country to list the substance among the product’s ingredients and include a measure of the amount of aspartame per serving. As well, labels must state that the product contains phenylalanine – this is usually in the order of ingredients, contained in brackets.

Phenylalanine is one of the essential amino acids and is required for normal growth and maintenance of life. Concerns about the safety of phenylalanine from aspartame center largely on hypothetical changes in neurotransmitter levels as well as ratios of neurotransmitters to each other in the blood and brain that could lead to neurological symptoms. Reviews of the literature have found no consistent findings to support such concerns, and while high doses of aspartame consumption may have some biochemical effects, these effects are not seen in toxicity studies to suggest aspartame can adversely affect neuronal function. Like methanol, the typical diet will lead to ingestion of significantly higher amounts of phenylalanine than would be expected from aspartame consumption. People with the genetic disorder phenylketonuria are advised to avoid aspartame as they have a decreased ability to metabolize phenylalanine. Common foods such as milk, meat, and fruits provide far greater amounts of these metabolites in a diet than does aspartame.

Lactation

In a study done in 1979, the effect of aspartame ingestion on blood and milk amino acid levels in lactating women was tested. In this study, six women from the ages of 20 to 29 with established lactation were studied after oral administration of aspartame or lactose (50 mg/kg body weight) in a random order, with the intent to study the differences in breast milk between the two. The study resulted with the conclusion that aspartame administration at 50 mg/kg body weight has a small effect upon the milk aspartate levels and although a small increase in aspartate time-effect scores was noted over the four-hour postabsorptive period, no significant difference was noted over the entire 24-hour watching period.

Cancer

Reviews have found no association between aspartame and cancer. These reviews have looked at numerous carcinogenicity studies in animals, epidemiologic studies in humans, as well as *in vitro* genotoxicity studies. These studies have found no significant evidence that aspartame causes cancer in animals, damages the genome, or causes cancer in humans at doses currently used. This position is supported by multiple regulatory agencies like the FDA and EFSA as well as scientific bodies such as the National Cancer Institute. Aspartame did not show any DNA-damaging properties either.

Concern about possible carcinogenic properties of aspartame was originally raised and popularized in the mainstream media by John Olney in the 1970s and again in 1996 by suggesting that aspartame may be related to brain tumors. Reviews have found that these concerns were flawed, due to reliance on the ecological fallacy and the purported mechanism of causing tumors being unlikely to actually cause cancer. Independent agencies such as the FDA and National Cancer Institute have reanalyzed multiple studies based on these worries and found no association between aspartame and brain cancer.

As discussed in the article on controversies around aspartame, the Cesare Maltoni Cancer Research Center of the European Ramazzini Foundation of Oncology and Environmental Sciences released several studies which claimed that aspartame can increase several malignancies in rodents, concluding that aspartame is a potential carcinogen at normal dietary doses. The EFSA and the FDA discounted the study results due to lack of transparency and numerous flaws in the study, finding no reason to revise their previously established acceptable daily intake levels for aspartame.

Neurological and Psychiatric Symptoms

Numerous allegations have been made via the Internet and in consumer magazines purporting neurotoxic effects of aspartame leading to neurological or psychiatric symptoms such as seizures, headaches, and mood changes. Review of the biochemistry of aspartame has found no evidence that the doses consumed would plausibly lead to neurotoxic effects. Comprehensive reviews have not found any evidence for aspartame as a cause for these symptoms. One review did provide a theoretical biochemical background of neurotoxicity and suggested further testing. However, a panel of EFSA experts noted that this review's conclusions were partially based on Internet sources and therefore were not scientifically robust. These experts also concurred with a critique that significant scientific errors were made in the critical review that led to unsubstantiated and misleading interpretations. A review of the pediatric literature did not show any significant findings for safety concerns with regard to neuropsychiatric conditions such as panic attacks, mood changes, hallucinations or with ADHD or seizures.

Headaches

Headaches are the most common symptom reported by consumers. While one small review noted aspartame is likely one of many dietary triggers of migraines, in a list that includes "cheese, chocolate, citrus fruits, hot dogs, monosodium glutamate, aspartame, fatty foods, ice cream, caffeine withdrawal, and alcoholic drinks, especially red wine and beer," other reviews have noted conflicting studies about headaches and still more reviews lack any evidence and references to support this claim.

Weight Change and Hunger

Since the caloric contribution of aspartame is negligible, it has been used as a means for weight loss through its role as a sugar substitute, with reviews finding that aspartame may aid in weight loss as part of a multidisciplinary weight loss program. On its own, aspartame is not known by medical literature to cause weight gain or weight loss. Although some researchers have stated that aspartame contributes to weight gain, hunger and increase in appetite, broad reviews and regulators conclude that aspartame has no appreciable effect on appetite.

Commercial uses

Under the trade names Equal, NutraSweet, and Canderel, aspartame is an ingredient in approximately 6,000 consumer foods and beverages sold worldwide, including (but not limited to) diet sodas and other soft drinks, instant breakfasts, breath mints, cereals, sugar-free chewing gum, cocoa mixes, frozen desserts, gelatin desserts, juices, laxatives, chewable vitamin supplements, milk drinks, pharmaceutical drugs and supplements, shake mixes, tabletop sweeteners, teas, instant coffees, topping mixes, wine coolers and yogurt. It is provided as a table condiment in some countries. Aspartame is less suitable for baking than other sweeteners, because it breaks down when heated and loses much of its sweetness.

NutraSweet Company

In 1985, Monsanto Company bought G.D. Searle, and the aspartame business became a separate Monsanto subsidiary, the NutraSweet Company. In March 2000, Monsanto sold it to J.W. Childs Equity Partners II L.P. European use patents on aspartame expired starting in 1987, and the U.S. patent expired in 1992. Since then, the company has competed for market share with other manufacturers, including Ajinomoto, Merisant and the Holland Sweetener Company.

Ajinomoto

Many aspects of industrial synthesis of aspartame were established by Ajinomoto. In 2004, the market for aspartame, in which Ajinomoto, the world's largest aspartame manufacturer, had a 40 percent share, was 14,000 metric tons a year, and consumption of the product was rising by 2 percent a year. Ajinomoto acquired its aspartame business in 2000 from Monsanto for $67M.

In 2008, Ajinomoto sued British supermarket chain Asda, part of Wal-Mart, for a malicious falsehood action concerning its aspartame product when the substance was listed as excluded from the chain's product line, along with other "nasties". In July 2009, a British court found in favour of Asda. In June 2010, an appeals court reversed the decision, allowing Ajinomoto to pursue a case against Asda to protect aspartame's reputation. Asda said that it would continue to use the term "no nasties" on its own-label products, but the suit was settled in 2011 with Asda choosing to remove references to aspartame from its packaging.

In November 2009, Ajinomoto announced a new brand name for its aspartame sweetener – AminoSweet.

Holland Sweetener Company

A joint venture of DSM and Tosoh, the Holland Sweetener Company manufactured aspartame using the enzymatic process developed by Toyo Soda (Tosoh) and sold as the brand Sanecta. Additionally, they developed a combination aspartame-acesulfame salt under the brand name Twinsweet. They left the sweetener industry in late 2006, because "global aspartame markets are facing structural oversupply, which has caused worldwide strong price erosion over the last five years", making the business "persistently unprofitable".

Competing Products

Because sucralose, unlike aspartame, retains its sweetness after being heated, and has at least twice the shelf life of aspartame, it has become more popular as an ingredient. This, along with differences in marketing and changing consumer preferences, caused aspartame to lose market share to sucralose. In 2004, aspartame traded at about $30/kg and sucralose, which is roughly three times sweeter by weight, at around $300/kg.

Ant-killer Hoax

Aspartame has been falsely claimed to have been originally developed as ant poison. The source for this was a satirical article posted on "thespoof" website. Further claims that the substance actually is poisonous to ants were inferred from that online article being quoted as fact by various anti-aspartame websites, and videos of numerous trials of this rumor have been shown on YouTube, or posted on social networks, some even claiming success in eradicating ants with Aspartame or with other sweeteners.

Protein Conformation

- The spatial arrangement of atoms in a protein is called its conformation.
- The possible conformations of a protein include any structural state that can be achieved without breaking covalent bonds. A change in conformation could occur, for example, by rotation about single bonds. Of the numerous conformations that are theoretically possible in a protein containing hundreds of peptide bonds, one or (more commonly) a few generally predominate under biological conditions.
- The need for multiple stable conformations reflects the changes that must occur in most proteins as they bind to other molecules or catalyze reactions.
- The conformations existing under a given set of conditions are usually the ones that are thermodynamically the most stable, having the lowest Gibbs free energy (G). Proteins in any of their functional, folded conformations are called native proteins.

Structure

- Proteins have a total of four levels of structures.
- Primary structure.
- Secondary structure.
- Tertiary structure.
- Quaternary structure.

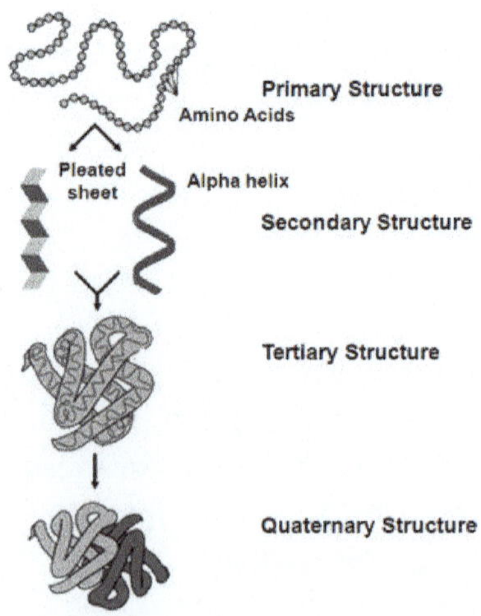

Primary Structure

- The simple amino acid sequence of a protein is called as its primary structure.
- Since the possible way of arrangement of the chain will depend on the sequence of amino acid residues leading to proper protein folding, the primary structure dictate three dimensional structure of the proteins.

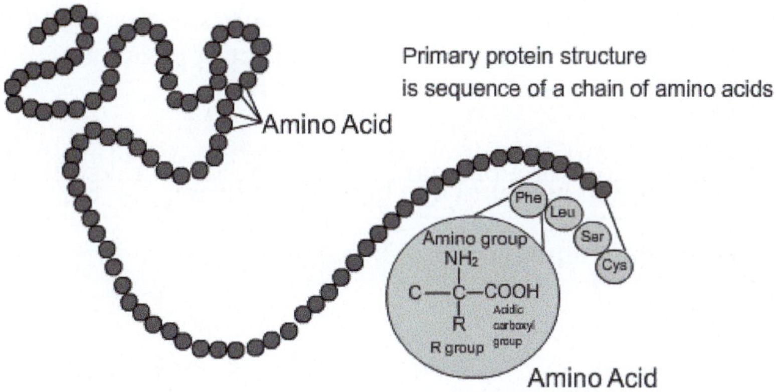

Secondary Structure

- This defines the interaction of closely located amino acids in a chain. Two main types of secondary structures observed in the proteins are helices (α helices) and pleated sheets (β pleated sheets).
- Alpha helix is a helical structure around an axis.
- This is coiled in clockwise (right handed) manner. It has an average of 3.6 amino acids per turn. Please write other dimensions too.
- There are 3.6 residue per turn with in a fix pitch of 5.4°A
- Thus the rise per residue come out to be 1.5°A. In a typical α - helix φ value ranges from 113 to 132° and ψ from 123 to 136°.

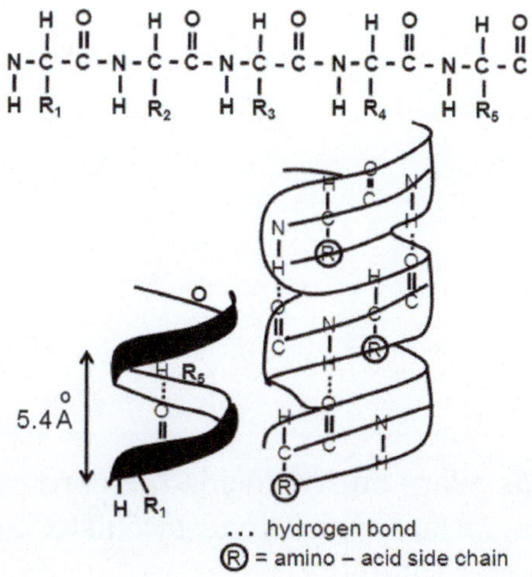

The reason as to why alpha helices form more readily in proteins than any other possible conformations is that these arrangements make optimal use of internal hydrogen bonds to attain stability. The helix is stabilized by hydrogen bonding between the carbonyl of each first amino acid of the chain to the NH of the amino acid four residues away. All main chain amino and carboxyl groups are thus hydrogen bonded, and the R groups stick out from the structure in a spiral arrangement.

Beta pleated sheet is composed of two or more straight chains that are hydrogen bonded side by side. If the amino termini are on the same end of each chain, the sheet is termed parallel, and if the chains run in the opposite direction (amino terminal on opposite ends), the sheet is termed antiparallel.

Pleated sheets may be formed from a single chain if it contains a beta turn, which forms a hairpin loop structure. Often a proline can be found in a beta turn, since it places a "kink" in the chain.

Glycine and Ala are predominant amino acids in beta sheet.

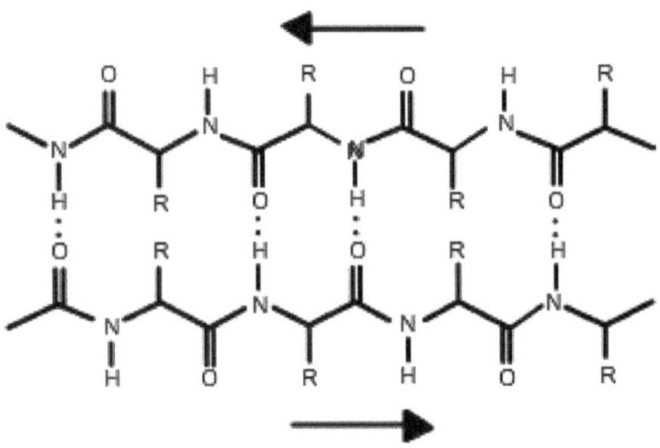

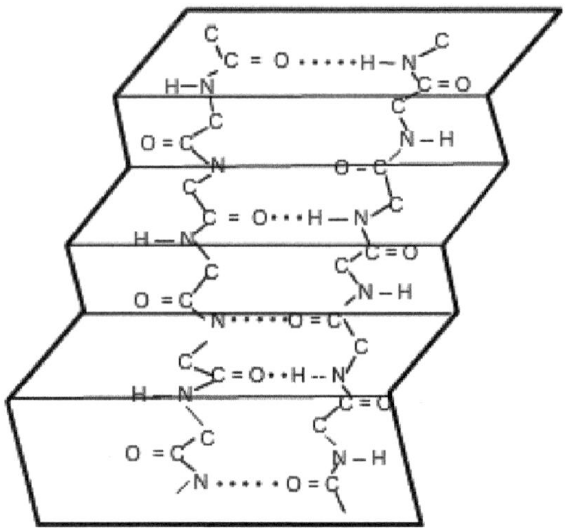

- Tertiary structure :- refers to the arrangement of amino acids in the space i.e. in three dimensional form.
- Distinct amino acid are brought closer in chain are further linked by :

  - polar-polar interaction,

  - hydrophobic interaction,

  - ionic interaction,

  - disulfide bonds,

  - Van der Waals forces.

  - hydrogen bonds.
- Hydrophobic amino acids :- are buried inside the core of protein and charged and polar group are located on the surface which tend to cluster and exclude water. This allows a protein to have greater water solubility.

- Quaternary structure :- If protein consists of more than one polypeptide chains, their association with each other – implies the Quaternary structure.

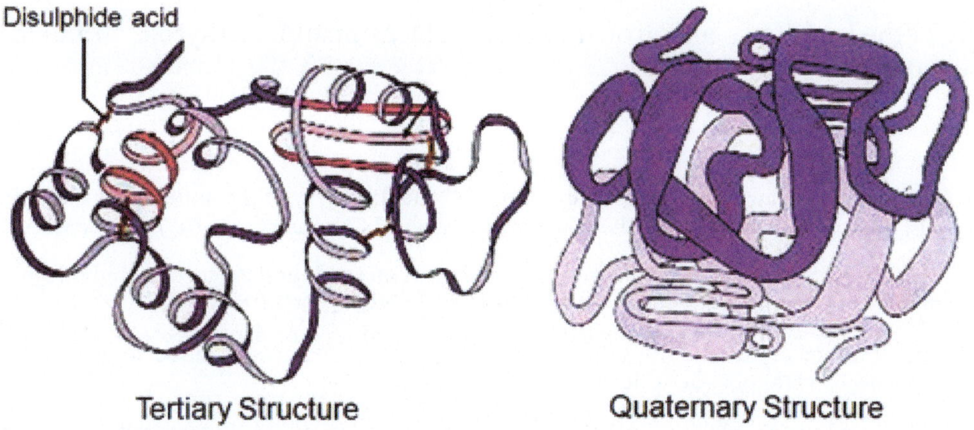

- Accordingly protein are termed as dimeric (wherein one chain is referred as monomeric unit), trimeric or oligomeric. If the chains are similar i.e. have same amino acid sequence these are called homomeric or heteromeric if chains are different.
- Proteins have also been put into two major groups :

(a) Fibrous proteins, having polypeptide chains arranged in long strands or sheets.

(b) Globular proteins, having polypeptide chains folded into a spherical or globular shape.

- The two groups are structurally distinct : fibrous proteins usually consist largely of a single type of secondary structure; globular proteins often contain several types of secondary structure e.g a- keratins are predominantly alpha helix whereas silk proteins are beta –sheets.
- The two groups differ functionally in that the structures that provide structure, support, shape, and external protection to vertebrates are made of fibrous proteins, whereas most enzymes and regulatory proteins are globular proteins.

Functions of Proteins

Proteins carry out most diverse and possibly the largest volumes of cellular functions. Some of the key functions are summarized as below :

- Biocatalysis- Almost all the biological reactions are catalyzed by the enzymes. These are substrate specific and carry out reactions at very high rates under mild physiological conditions. Several thousand enzymes have been identified to date.
- Membrane are constitute of lipoprotein and some proteins are integral part of membrane. Receptors found on the membrane are also protein in nature.
- Transport and storage proteins - small molecules are often carried by proteins in the physiological setting e.g. haemoglobin is responsible for the transport of oxygen to tissues.
- Muscle are made up of proteins and their contraction is done by actin and myosin protein.

- Mechanical support - skin and bone are strengthened by the protein collagen.
- Antibodies of immune system are protein structures.
- Many of the hormones and growth factors such as insulin or thyroid stimulating hormone are proteins.

References

- Simoni RD, Hill RL, Vaughan M (September 2002). "The discovery of the amino acid threonine: the work of William C. Rose [classical article]". The Journal of Biological Chemistry. 277 (37): E25. PMID 12218068
- Pollegioni L, Servi S, eds. (2012). Unnatural Amino Acids: Methods and Protocols. Methods in Molecular Biology – Volume 794. Humana Press. p. v. ISBN 978-1-61779-331-8. OCLC 756512314
- "Chapter 1: Proteins are the Body's Worker Molecules". The Structures of Life. National Institute of General Medical Sciences. 27 October 2011. Retrieved 20 May 2008
- Michal G, Schomburg D, eds. (2012). Biochemical Pathways: An Atlas of Biochemistry and Molecular Biology (2nd ed.). Oxford: Wiley-Blackwell. p. 5. ISBN 978-0-470-14684-2
- Elzanowski A, Ostell J (7 April 2008). "The Genetic Codes". National Center for Biotechnology Information (NCBI). Retrieved 10 March 2010
- Magee T, Seabra MC (April 2005). "Fatty acylation and prenylation of proteins: what's hot in fat". Current Opinion in Cell Biology. 17 (2): 190–6. PMID 15780596. doi:10.1016/j.ceb.2005.02.003
- Wolosker H, Dumin E, Balan L, Foltyn VN (July 2008). "D-amino acids in the brain: D-serine in neurotransmission and neurodegeneration". The FEBS Journal. 275 (14): 3514–26. PMID 18564180. doi:10.1111/j.1742-4658.2008.06515
- Simon M (2005). Emergent computation: emphasizing bioinformatics. New York: AIP Press/Springer Science+Business Media. pp. 105–106. ISBN 0-387-22046-1
- Strecker A (1850). "Ueber die künstliche Bildung der Milchsäure und einen neuen, dem Glycocoll homologen Körper". Justus Liebigs Annalen der Chemie. 75 (1): 27–45. doi:10.1002/jlac.18500750103
- Jodidi, S. L. (1 March 1926). "The Formol Titration of Certain Amino Acids". Journal of the American Chemical Society. 48 (3): 751–753. doi:10.1021/ja01414a033
- Hatem, Salama Mohamed Ali (2006). "Gas chromatographic determination of Amino Acid Enantiomers in tobacco and bottled wines". University of Giessen. Retrieved 17 November 2008
- Gutteridge A, Thornton JM (November 2005). "Understanding nature's catalytic toolkit". Trends in Biochemical Sciences. 30 (11): 622–9. PMID 16214343. doi:10.1016/j.tibs.2005.09.006
- Vermeer C (March 1990). "Gamma-carboxyglutamate-containing proteins and the vitamin K-dependent carboxylase". The Biochemical Journal. 266 (3): 625–36. PMC 1131186. PMID 2183788. doi:10.1042/bj2660625
- Creighton TH (1993). "Chapter 1". Proteins: structures and molecular properties. San Francisco: W. H. Freeman. ISBN 978-0-7167-7030-5
- "Nomenclature and Symbolism for Amino Acids and Peptides". IUPAC-IUB Joint Commission on Biochemical Nomenclature. 1983. Archived from the original on 9 October 2008. Retrieved 17 November 2008
- Kostrzewa RM, Nowak P, Kostrzewa JP, Kostrzewa RA, Brus R (March 2005). "Peculiarities of L: -DOPA treatment of Parkinson's disease". Amino Acids. 28 (2): 157–64. PMID 15750845. doi:10.1007/s00726-005-0162-4
- Brosnan JT (April 2000). "Glutamate, at the interface between amino acid and carbohydrate metabolism". The Journal of Nutrition. 130 (4S Suppl): 988S–90S. PMID 10736367
- Liebecq C, ed. (1992). Biochemical Nomenclature and Related Documents (2nd ed.). Portland Press. pp. 39–69. ISBN 978-1-85578-005-7

- "U.S. GAO – HRD-87-46 Food and Drug Administration: Food Additive Approval Process Followed for Aspartame, 18 June 1987". Retrieved 5 September 2008.
- Rosenthal GA (2001). "L-Canavanine: a higher plant insecticidal allelochemical". Amino Acids. 21 (3): 319–30. PMID 11764412. doi:10.1007/s007260170017
- Turner BL, Harborne JB (1967). "Distribution of canavanine in the plant kingdom". Phytochemistry. 6 (6): 863–66. doi:10.1016/S0031-9422(00)86033-1
- Fürst P, Stehle P (June 2004). "What are the essential elements needed for the determination of amino acid requirements in humans?". The Journal of Nutrition. 134 (6 Suppl): 1558S–1565S. PMID 15173430
- Elmore DT, Barrett GC (1998). Amino acids and peptides. Cambridge, UK: Cambridge University Press. pp. 48–60. ISBN 0-521-46827-2
- Jones RC, Buchanan BB, Gruissem W (2000). Biochemistry & molecular biology of plants. Rockville, Md: American Society of Plant Physiologists. pp. 371–2. ISBN 0-943088-39-9
- Shemin D, Rittenberg D (December 1946). "The biological utilization of glycine for the synthesis of the protoporphyrin of hemoglobin". The Journal of Biological Chemistry. 166 (2): 621–5. PMID 20276176
- Hausman, Robert E., Cooper, Geoffrey M. (2004). The cell: a molecular approach. Washington, D.C: ASM Press. p. 51. ISBN 0-87893-214-3

5

An Overview of Lipids

Among the four major biomolecules, one of them is lipids. It is insoluble in a water solution but soluble in organic solvents. It has many important biological functions. It acts as the energy source in animals, humans, birds, etc. It activates important mitochondrial enzymes. It is a major component of electron systems in mitochondria. The aspects elucidated in this chapter are of vital importance, and provide a better understanding of biomolecular engineering.

Lipids

- Lipids are one among the four major biomolecules of living systems.
- By definition, these are the class of biomolecules which are insoluble or sparingly soluble in aqueous solutions and soluble in organic solvents.
- Fatty acids are major constituents of lipids. Fatty acids are mono carboxylic acid containing short/ long-chain hydrocarbon molecules. Some important fatty acids are enlisted below.
- The numbering of carbons in fatty acids begins with the carbon of the carboxylate group. Fatty acid represented by the total number of carbons e.g, palmitic acid a 16-carbon fatty acid $CH3(CH_2)_{14}COOH$ is designated as C16.
- It is customary to write it as C 16:0 where zero represent that there is no double bond in the fatty acid). If there is one double bond, then it will be written as C:16:1

Representation	Common Name	Structure
4:0	Butyric acid	$CH3(CH_2)_2COOH$
6:0	Caproic acid	$CH3(CH_2)_4COOH$
10:0	Decanoic acid	$CH3(CH_2)_8COOH$
12:0	Lauric acid	$CH3(CH_2)_{10}COOH$
14:0	Myristic acid	$CH3(CH_2)_{12}COOH$
16:0	Palmitic acid	$CH3(CH_2)_{14}COOH$
18:0	Stearic acid	$CH3(CH_2)_{16}COOH$
20:0	Arachidic acid	$CH3(CH_2)_{18}COOH$

Types of Fatty Acids

Saturated Fatty Acids

All sets of examples in the previous table were fatty acids that contained no carbon-carbon double bonds. These are called saturated fatty acids. Saturated fatty acids having short carbon chain are liquid at room temperature, whereas long carbon chain fatty acids are solid.

Unsaturated Fatty Acids

These have carbon-carbon double bonds in between, thus leading to unsatuaration. The representations for these fatty acids consists of the number of carbon atoms, followed by the number double bond and the place of unsaturation. The place of unsaturation in a fatty acid is indicated by the symbol (Δ) and the number of the first carbon of the double bond in superscript form. Thus oleic acid a 16-carbon fatty acid with one site of unsaturation between carbons 9 and 10, and will be represented by $C16{:}1\Delta^9$.

Some commonly occurring unsaturated fatty acids are:

$18{:}1\Delta^{9}$	Oleic acid	$CH_3(CH_2)_7C{=}C(CH_2)_7COOH$
$18{:}2\ \Delta^{9,12}$	Linoleic acid	$CH_3(CH_2)_4C{=}CCH_2C{=}C(CH_2)_7COOH$
$18{:}3\ \Delta^{9,12,15}$	Linolenic acid	$CH_3CH_2C{=}CCH_2C{=}CCH_2C{=}C(CH_2)_7COOH$
$20{:}4\ \Delta^{5,8,11,14}$	Arachidonic acid	$CH_3(CH_2)_3(CH_2C{=}C)_4(CH_2)_3COOH$

The presence of double bonds in unsaturated fatty acids significantly lowers the melting point making them liquid. More the no. of bonds lower the MP.

Biological Functions of Lipids

Lipids perform and are involved in variety of important cellular functions. However, following are some of the major physiological functions attributed to lipids :

- Energy source in animals, insects, birds and high lipid seeds e.g. triacyl glycerols.
- Activators of enzymes namely glucose-6-phosphatase, stearoyl CoA desaturase, monooxygenases which are important mitochondrial enzymes.
- Some of the lipids derivatives serve as vitamins and hormones e.g. Prostaglandins.
- Arachidonic acid is the specific precursor for all prostaglandins and leukotrienes. It is bound as an acyl moiety to the 2 position of a number of phospholipids and this is inactive as a substrate. Upon release by the action of phospholpiase A2, the free arachidonic acid is converted by a cyclooxygenase to form two very prostaglandin derivatives, prostacyclin I2, an active physiological vasodilator and Thromboxane A2, an active vasoconstrictor.
- Essential components of biological membranes e.g. shingolipids and glycoloipids.
- As lipoproteins in protein modification and recognitions.
- Components of the electron transport system in the inner membrane of mitochondria.

Structure and Function of Various Classes of Lipids

Classification of Lipids

Lipids are generally classified into following seven groups :

1. Acyl glycerols.
2. Phosholipids.
3. Sphingolipids.
4. Glycolipids.
5. Alkyl glyceryl ethers.
6. Terpenoids.
7. Wax

1. Acylglycerols

- These are also called as Triacylglycerides or neutral lipids, being composed of a glycerol backbone, in which each alcoholic group is esterified by fatty acids.
- Following is the typical triglyceride structure in which fatty acids are indicated by R. These are most commonly occurring form of lipids in cell, stored in adipose or fat depot. Their function is to serve as major energy source.

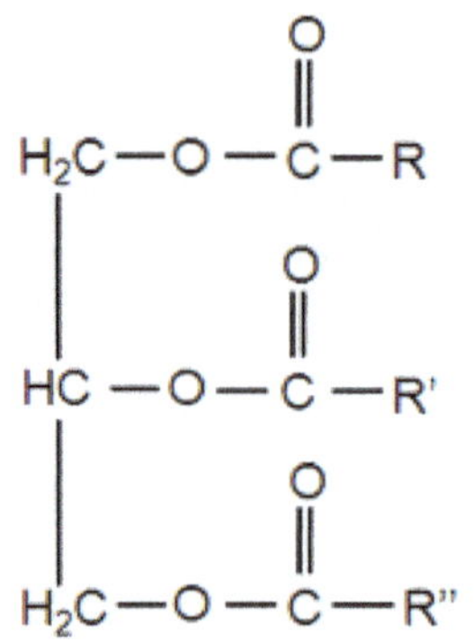

R1 R2 and R3 can be same or different fatty acids.
By nomenclature the top carbon position is called Sn1, middle Sn2 And bottom Sn3

2. Phospholipids

- They are the major component of membrane.
- The basic structure of phospholipids is very similar to that of the triacylglycerides except that Carbon -3 (Sn3), of the glycerol backbone is esterified by phosphoric acid.
- This basic phospholipids is called phosphatidic acid. Phosphotidic acid have further substitution as –X shown in adjoining figure.
- Phospholipids are amphipathic in nature due to presence of both hydrophilic / charged substitution (PO4- and X at Sn-3) and hydrophobic fatty acid chains at Sn1 and Sn2.

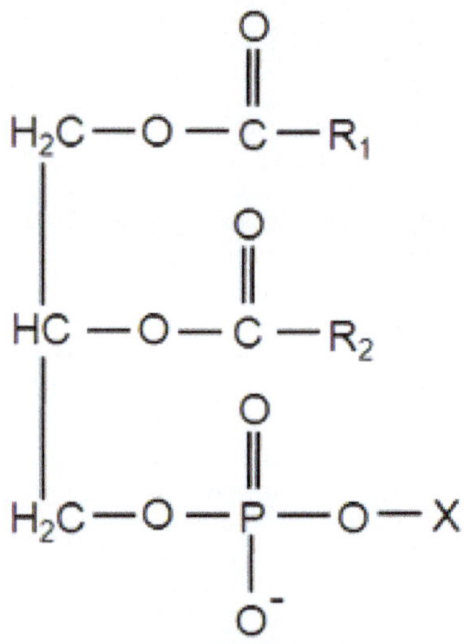

Different Types of Phospholipids

- Several different types of phosopholipids are formed by further attachment of different groups (X) at Phosphatidic acid Sn-3 phosphoric acid.
- Some representative examples are :

X = Ethanolamine, the phospholipid is called phosphatidylethanolamine.

X = Choline ---> phosphatidylcholine, also called lecithins.

X = Serine ---> phosphatidylserine.

X = Glycerol ---> phosphatidylglycerol.

X = myo-inositol ---> phosphatidylinositol.

X = diphosphatidylglycerol ---> cardiolipins.

Their presence and functions have been briefly enumerated in the following slide.

Name of phospholipid	Source	Function
phosphatidylethanolamine	main lipid component of microbial membranes	Forms the main building block of membrane bilayers.
Phosphatidylcholine, also called lecithins	main lipid component of plant and animal membranes; present heavily on the outer leaflet of the plasma membrane	Integral component of the lipoproteins circulating in the plasma
phosphatidylserine	minor membrane lipid; forms a small percentage of the total phospholipid in the plasma membrane and endoplasmic reticulum of the cell	Forms an intermediate in phosphatidylethanolamine biosynthesis

Name of phospholipid	Source	Function
phosphatidylglycerol	main component of some bacterial membranes, also found in plants and animal membranes	Involved in proper functioning of the enzymes associated with oxidative phosphorylation
phosphatidylinositol	important membrane constituent	Component of building blocks of membranes; crucial roles in interfacial binding of proteins; regulation of proteins at the cell interface
cardiolipins	component of the inner mitochondrial membrane	Triggers apoptosis Serves as proton trap for oxidative phosphorylation. Cholesterol translocation from outer to the inner mitochondrial membrane

3. Sphingolipids

- Sphingolipids are composed of a backbone of sphingosine, which is derived from glycerol.
- The structure of sphingosine is shown below :

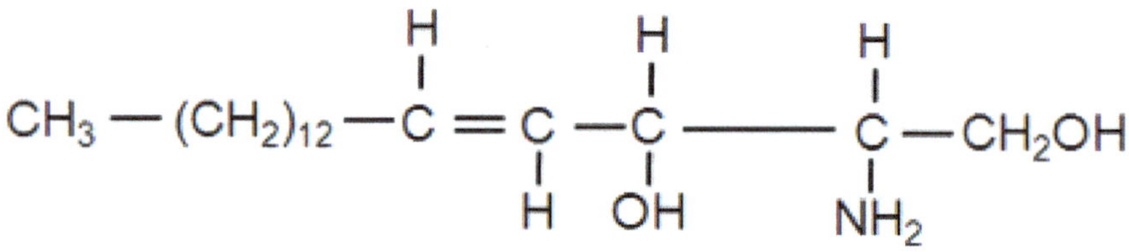

- Sphingolipids are predominantly present in the myelin sheath of nerve fibers.

Some of the important sphingolipids are :

- Ceramides :– In this case, the sphingosine is N-acetylated at CH2OH by a variety of fatty acids generating different types of ceramides.
- Sphingomyelin :- It is an abundant sphingolipid in which CH2OH is esterified by phosphoric acid and choline instead of fatty acid.
- Glycosphingolipids other major class of sphingolipids are generated by substitution of carbohydrates at CH2OH. Cerbrosides and Gangliosides are major classes of glycosphingolipids.
- Cerebrosides :- It is also called galactocerebrosides because galactose is the carbohydrate present in it.
- Gangliosides :- It also contains sialic acid.

Structure and Function of Various Classes of Lipids

Glycolipids

- They are carbohydrate containing derivative of triglycerides. Galactose is the predominant carbohydrate present in glycolipids. Sn-3 monogalactosyl galactosyl diacyl glycerol and Sn-3 di galactosyl diacyl glycerol are commonly present in membrane structures especially in the chloroplast membrane.

- A representative structure of glycolipid is given below:

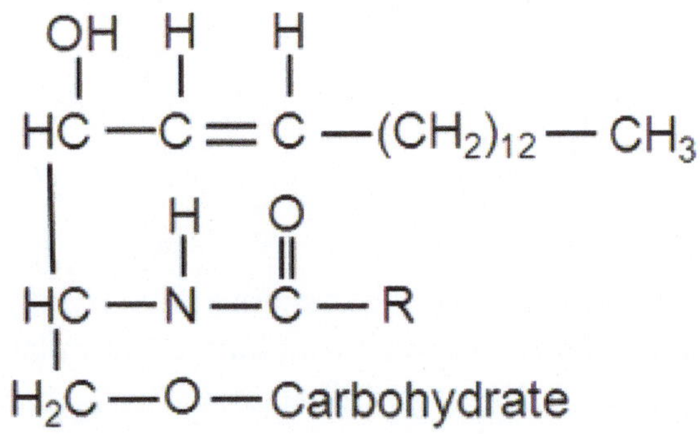

Glycerol Ethers

- These also called plasmalogens.
- Structurally they contain either an O-alkyl (-O-CH2-) or O-alkenyl ether (-O-CH=CH-) species at C-1 (Sn1) of glycerol. A basic O-alkenyl ether species is shown in the adjoining picture :

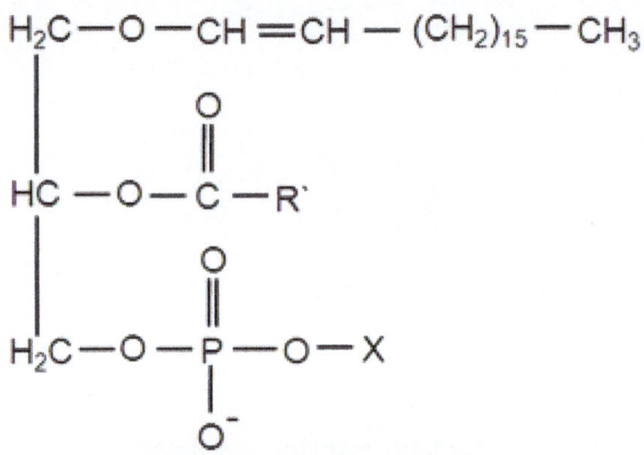

Basic structure of typical plasmalogens

- One of the physiologically important alkyl ether plasmalogens is platelet activating factor (PAF) which is a choline plasmalogen in which the C-2 (sn2) position of glycerol is esterified with an acetyl group instead of a long chain fatty acid. PAF mediates hypersensitivity and acute inflammatory reactions.

Terpenoids and Sterols

- These are very distinct group of lipids composed of the monomer repeating units called "isoprenoid units". Steroids, carotenoids, rubber and terpenes fall in this class of lipids.

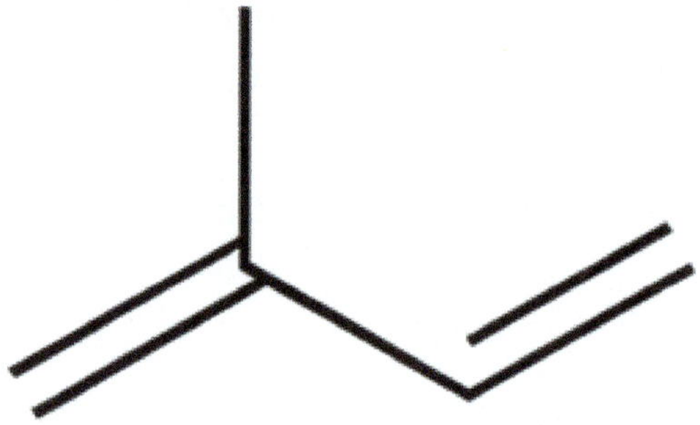

An isoprene unit

- Terpenes are the chief constituents of the essential oils, balsams, resins, waxes, and rubber. Plant terpenoids are known for their aromatic qualities. They play important part in traditional herbal remedies and are under study for antibacterial, and pharmaceutical applications.
- The steroids and sterols in animals are biologically produced from terpenoid precursors.
- Structure of β - carotene and cholesterol, few among important compound of this class are shown below :

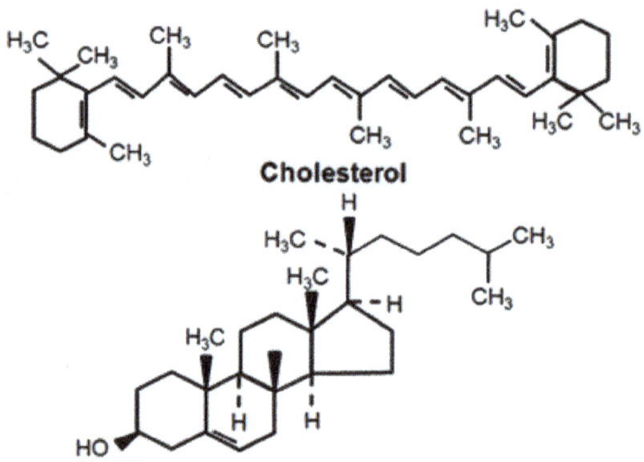

β – carotene

Waxes

These are class of lipids found as protective coating on fruits and leaves or secreted by insects. Chemically these are complex mixture of long chain alkanes and derivatives of secondary alcohol and ketones.

Lipoproteins

- A lipoprotein is a complex assembly containing both proteins and lipids bonded covalently or non-covalently to the proteins. Enzymes, transporters, structural proteins, antigens, adhesins and toxins are lipoproteins. For eg: HDL (high density lipoprotein), LDL (low density lipoprotein). The lipoprotein particle is composed of an outer shell of phospholipid, which makes it water soluble; a core of fats and a surface apoprotein molecule that allows recognition by the tissues and uptake of the particle.
- The general structure of a lipoprotein is given below :

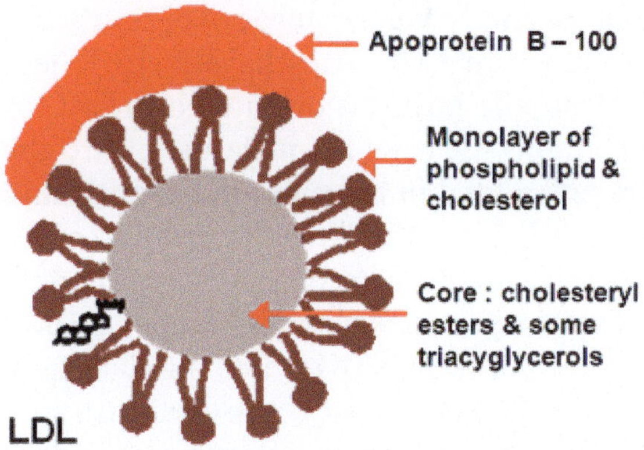

- The lipoprotein particle is composed of an outer shell of phospholipid, which makes it water soluble; a core of fats and a surface apoprotein molecule that allows recognition by the tissues and uptake of the particle.

Classification and Functions of Lipoproteins

The major function of lipoprotein particles is to transport lipids (fats) (such as triacylglycerol) around the body in the blood. Lipoproteins may be classified on the basis of their density. They may be categorised as :

- Chylomicron(lowest in density) which carry triglycerides from the intestine to the adipose tissue.
- VLDL (very low density lipoprotein); carry (newly ynthesised) triacylglycerol from the liver to adipose tissue.
- IDL(intermediate density lipoprotein); internediate between LDL and VLDL.
- LDL(low density lipoprotein); carry cholesterol from the liver to cell/tissues of the body.
- HDL(high density lipoprotein); carry cholesterol from the body tissues back to the liver.

Lipoprotein

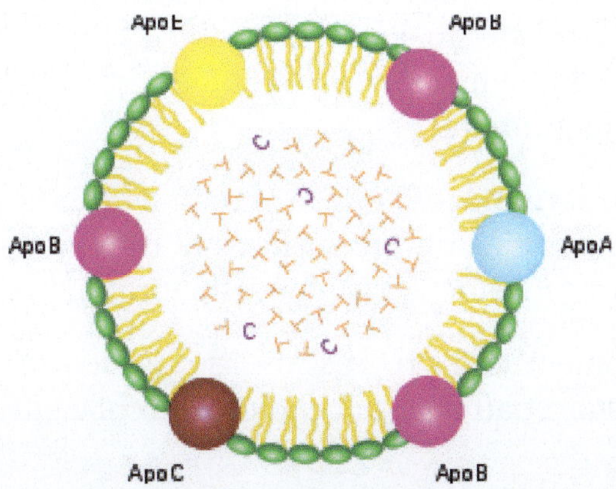

Lipoprotein structure (chylomicron)
ApoA, ApoB, ApoC, ApoE (apolipoproteins); T (triacylglycerol); C (cholesterol); green (phospholipids)

A lipoprotein is a biochemical assembly whose purpose is to transport hydrophobic lipid (a.k.a. fat) molecules in water, as in blood or ECF. They have a single-layer phospholipid and cholesterol outer shell, with the hydrophilic portions oriented outward toward the surrounding water and lipophilic portions of each molecule oriented inwards toward the lipids molecules within the particles. Apolipoproteins are embedded in the membrane, both stabilising the complex and giving it functional identity determining its fate. Thus the complex serves to emulsify the fats. Many enzymes, transporters, structural proteins, antigens, adhesins, and toxins are lipoproteins. Examples include the plasma lipoprotein particles classified under HDL, LDL, IDL, VLDL and ULDL (commonly called chylomicron) lipoproteins, which enable fats to be carried in all extracellular water including the blood stream (an example of emulsification), subgroups of which are primary drivers/modulators of atherosclerosis, the transmembrane proteins of the mitochondrion and the chloroplast, and bacterial lipoproteins.

Scope

Transmembrane Lipoproteins

The lipids are often an essential part of the complex, even if they seem to have no catalytic activity by themselves. To isolate transmembrane lipoproteins from their associated biological membranes, detergents are often needed.

Plasma Lipoprotein Particles

Because fats are insoluble in water, these cannot be transported in blood on its own. Instead, they are attached to hydrophilic proteins that function as transport vehicles. The role of lipoprotein particles is to transport triacylglycerols (a.k.a. triglycerides) and cholesterol in the blood between all the tissues of the body. The most common being the liver and the adipocytes of adipose tissue. Particles are synthesized in the small intestine and the liver, but interestingly not in the adipocytes.

All cells use and rely on fats and cholesterol as building-blocks to create the multiple membranes that cells use both to control internal water content and internal water-soluble elements and to organize their internal structure and protein enzymatic systems.

The lipoprotein particles have hydrophilic groups of phospholipids, cholesterol, and apoproteins directed outward. Such characteristics make them soluble in the salt water-based blood pool. Triglyceride-fats and cholesteryl esters are carried internally, shielded from the water by the phospholipid monolayer and the apoproteins.

The interaction of the proteins forming the surface of the particles (with enzymes in the blood; with each other; and with specific proteins on the surfaces of cells) determines whether triglycerides and cholesterol will be added to or removed from the lipoprotein transport particles.

Regarding atheroma development and progression as opposed to regression, the key issue has always been cholesterol transport patterns, not cholesterol concentration itself.

Function

The handling of lipoprotein particles in the body is referred to as *lipoprotein particle metabolism*.

It is divided into two pathways, exogenous and endogenous, depending in large part on whether the lipoprotein particles in question are composed chiefly of dietary (exogenous) lipids or whether they originated in the liver (endogenous), through de novo synthesis of triacylglycerols.

The hepatocytes are the main platform for the handling of triacylglycerols and cholesterol; the liver can also store certain amounts of glycogen and triacylglycerols. While adipocytes are the main storage cells for triacylglycerols, they do not produce any lipoproteins.

Exogenous Pathway

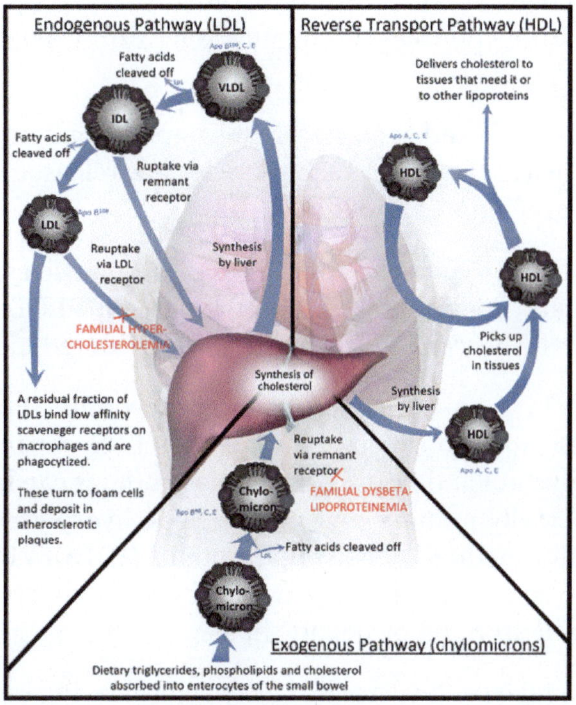

Simplified flowchart showing the essentials of lipoprotein metabolism.

Bile emulsifies fats contained in the chyme, then pancreatic lipase cleaves triacylglycerol molecules into two fatty acids and one 2-monoacylglycerol. Enterocytes readily absorb these small molecules from the chymus. Inside of the enterocytes, fatty acids and monoacylglycerides are transformed again into triacylglycerides. Then these lipids (i.e. triacylglycerols, phospholipids, cholesterol, and cholesteryl esters) are assembled with apolipoprotein B-48 into *nascent chylomicrons*. These particles are then secreted into the lacteals in a process that depends heavily on apolipoprotein B-48. As they circulate through the lymphatic vessels, nascent chylomicrons bypass the liver circulation and are drained via the thoracic duct into the bloodstream.

In the blood stream, *nascent chylomicron particles* interact with HDL particles resulting in HDL donation of apolipoprotein C-II and apolipoprotein E to the nascent chylomicron. The chylomicron at this stage is then considered mature. Via apolipoprotein C-II, mature chylomicrons activate lipoprotein lipase (LPL), an enzyme on endothelial cells lining the blood vessels. LPL catalyzes the hydrolysis of triacylglycerol (glycerol covalently joined to three fatty acids) that ultimately releases glycerol and fatty acids from the chylomicrons. Glycerol and fatty acids can then be absorbed in peripheral tissues, especially adipose and muscle, for energy and storage.

The hydrolyzed chylomicrons are now called *chylomicron remnants*. The chylomicron remnants continue circulating the bloodstream until they interact via apolipoprotein E with chylomicron remnant receptors, found chiefly in the liver. This interaction causes the endocytosis of the chylomicron remnants, which are subsequently hydrolyzed within lysosomes. Lysosomal hydrolysis releases glycerol and fatty acids into the cell, which can be used for energy or stored for later use.

Endogenous Pathway

The liver is the central platform for the handling of lipids: it is able to store glycerols and fats in its cells, the hepatocytes. Hepatocytes are also able to create triacylglycerols via de novo synthesis. They also produce the bile from cholesterol. The intestines are responsible for absorbing cholesterol. They transfer it over into the blood stream.

In the hepatocytes, triacylglycerols and cholesteryl esters are assembled with apolipoprotein B-100 to form *nascent VLDL particles*. Nascent VLDL particles are released into the bloodstream via a process that depends upon apolipoprotein B-100.

In the blood stream, *nascent VLDL particles* bump with HDL particles; as a result, HDL particles donate apolipoprotein C-II and apolipoprotein E to the nascent VLDL particle; Once loaded with apolipoproteins C-II and E, the nascent VLDL particle is considered mature.

Again, like chylomicrons, VLDL particles circulate and encounter lipoprotein lipase (LPL) expressed on endothelial cells. Apolipoprotein C-II activates LPL, causing hydrolysis of the VLDL particle and the release of glycerol and fatty acids. These products can be absorbed from the blood by peripheral tissues, principally adipose and muscle. The hydrolyzed VLDL particles are now called *VLDL remnants* or intermediate-density lipoproteins (IDLs). VLDL remnants can circulate and, via an interaction between apolipoprotein E and the remnant receptor, be absorbed by the liver, or they can be further hydrolyzed by hepatic lipase.

Hydrolysis by hepatic lipase releases glycerol and fatty acids, leaving behind *IDL remnants*, called low-density lipoproteins (LDL), which contain a relatively high cholesterol content. LDL circulates and is absorbed by the liver and peripheral cells. Binding of LDL to its target tissue occurs through an interaction between the LDL receptor and apolipoprotein B-100 on the LDL particle. Absorption occurs through endocytosis, and the internalized LDL particles are hydrolyzed within lysosomes, releasing lipids, chiefly cholesterol.

Classification

By Density

Lipoproteins may be classified as follows, listed from larger and less dense to smaller and denser. Lipoproteins are larger and less dense when the fat to protein ratio is increased. They are classified on the basis of electrophoresis, ultracentrifugation and Nuclear Magnetic Resonance spectroscopy via the Vantera Analyzer.

- Chylomicrons carry triglycerides (fat) from the intestines to the liver, to skeletal muscle, and to adipose tissue.

- Very-low-density lipoproteins (VLDL) carry (newly synthesised) triglycerides from the liver to adipose tissue.
- Intermediate-density lipoproteins (IDL) are intermediate between VLDL and LDL. They are not usually detectable in the blood when fasting.
- Low-density lipoproteins (LDL) carry 3,000 to 6,000 fat molecules (phospholipids, cholesterol, triglycerides, etc.) around the body. LDL particles are sometimes referred to as "bad" lipoprotein because concentrations, dose related, correlate with atherosclerosis progression.
    - large buoyant LDL (lb LDL) particles
    - small dense LDL (sd LDL) particles
    - Lipoprotein(a) is a lipoprotein particle of a certain phenotype
- High-density lipoproteins (HDL) collect fat molecules (phospholipids, cholesterol, triglycerides, etc.) from the body's cells/tissues, and take it back to the liver. HDLs are sometimes referred to as "good" lipoprotein because higher concentrations correlate with low rates of atherosclerosis progression and/or regression.

For young healthy research subjects, ~70 kg (154 lb), this data represents averages across individuals studied:

Density (g/mL)	**Class**	Diameter (nm)	% protein	% cholesterol	% phospholipid	% triacylglycerol & cholesterol ester
>1.063	HDL	5–15	33	30	29	4
1.019–1.063	LDL	18–28	25	50	21	8
1.006–1.019	IDL	25–50	18	29	22	31
0.95–1.006	VLDL	30–80	10	22	18	50
<0.95	Chylomicrons	100-1000	<2	8	7	84

However, this data is not reliable for any one individual or for the general clinical population.

Alpha and Beta

It is also possible to classify lipoproteins as "alpha" and "beta", according to the classification of proteins in serum protein electrophoresis. This terminology is sometimes used in describing lipid disorders such as abetalipoproteinemia.

Studies

Atherosclerosis is the leading cause of coronary artery disease, which is the leading cause of mortality in the world. Since the 1980s, many studies have examined possible *correlations* between the incidence of the disease and plasma lipoprotein particle concentrations in the blood. Hypotheses exist for possible *causations*. Studies have shown correlation between atherosclerosis and concentrations of particles. Further studies looked for correlations between nutrition and concentration of the distinguishable lipoprotein particles, e.g. whether the ratio of dietary fat raises or lowers levels of LDL particles in the blood. Studies have shown that different phenotypes do exist regarding the amount of particles and reaction to diet composition.

6

Principles of Nucleic Acids Structure

Nucleic acids store DNA and RNA in them, which are the basic building-blocks of life. DNA is responsible for heredity. The structure of every protein is due to the information programmed in the nucleotide sequence of DNA. Nucleic acid regulates protein synthesis, and replicates itself during cell division. This section has been carefully written to provide an easy understanding of the varied facets of nucleic acids.

Nucleic Acids

- Nucleic acids are the most important biomolecules of the cells forming very basis of central dogma of life.
- Most importantly, the nucleic acids : deoxyribonucleic acid (DNA) and ribonucleic acid (RNA), are the molecular repositories of genetic informations. DNA is the master molecule responsible for heredity. The ability to store and transmit genetic information from one generation to the next is a fundamental condition for life.
- The structure of every protein, and ultimately of every biomolecule and cellular component, is a product of information programmed into the nucleotide sequence of a cell's DNA. This way nucleic acid governs the functions and physiology of the cells.
- Nucleotides are monomeric unit or building blocks of nucleic acids.

Functions of Nucleic Acid

- The concept of Central Dogma explains the functions of nucleic acid in nutshell as below :

DNA

Reverse transcription ↑↓ *Transcription*

m-RNA

↓ *Translation*

Proteins

- DNA is very basis of life. It is the master molecule responsible for hereditary and genetic material of the cell carrying all the in formations.
- It is able to replicates it self during cell division and the process called replication.
- It synthesises a complementary messenger RNA which is responsible for carrying the information for protein synthesis.

- It also regulates protein synthesis.

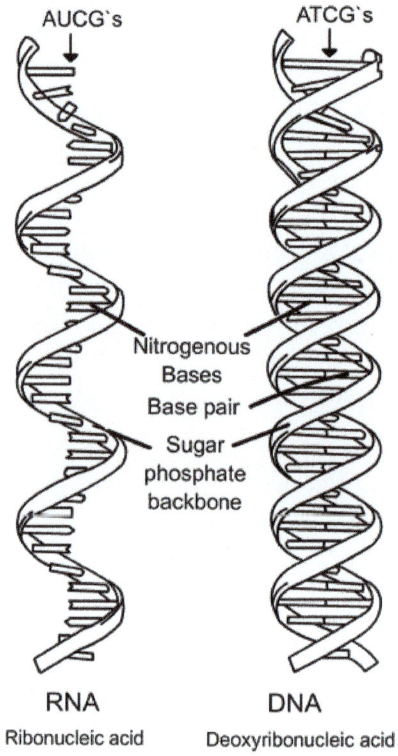

DNA Content of Organisms

- The overall DNA contained in one cell is called a genome.
- Prokaryotic cells contain one molecule of DNA, while human somatic cells contain 46 molecules of DNA per nucleus.
- The somatic cells of eukaryotes commonly have two sets of genes derived from both parents, and these cells are called diploids.
- Cells that have only one set of genes are called haploids. For example, most prokaryotes are haploids.
- DNA content greatly varies among organisms. Generally, DNA content per cell is larger in eukaryotes than in prokaryotes.
- Human somatic cells contain approximately 1,000 times as much DNA as those of E. coli (per haploid). For diploid cells, the amount per cell is 6 pg.
- Although it may generally be seen that among eukaryotes, the higher an organism, the more DNA content, there can be variations among organisms of the same group.
- For example, among vertebrates, such variations may be seen in fish and amphibians, with some species having more DNA content than humans.
- Some higher plant species also have more DNA than humans.

Amount of DNA Per Haploid in Several Organisms

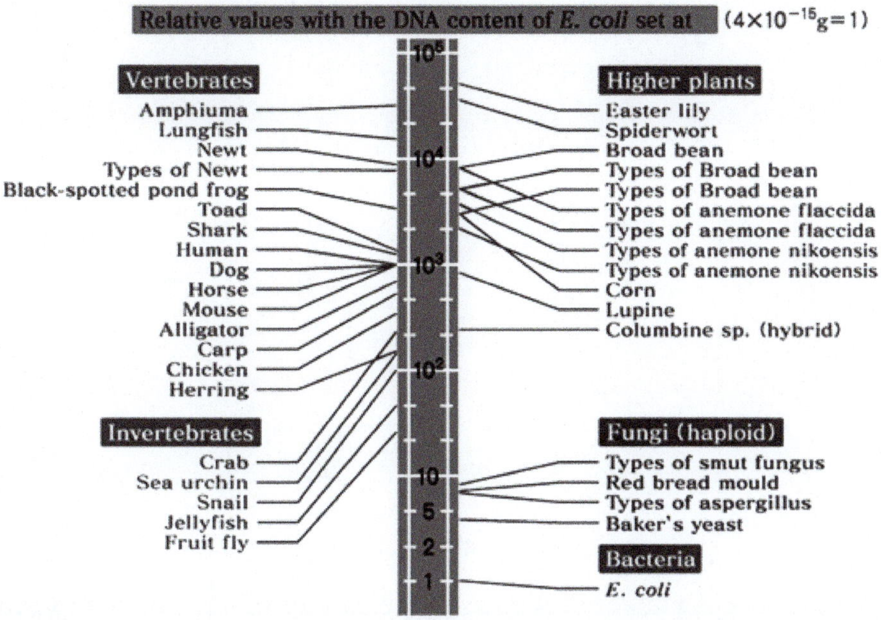

Nucleic Acid Structure

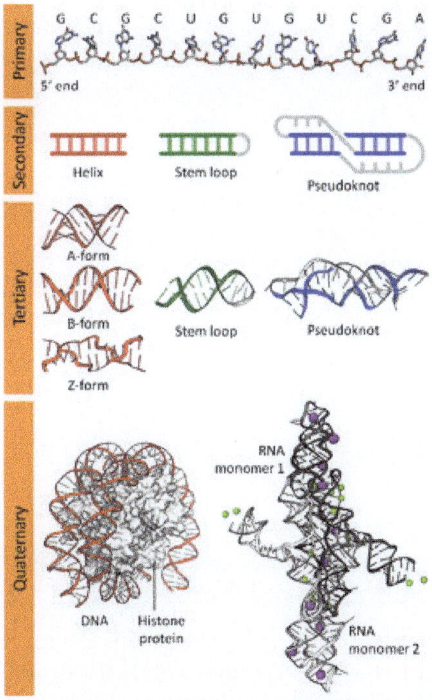

Interactive image of nucleic acid structure (primary, secondary, tertiary, and quaternary) using DNA helices and examples from the VS ribozyme and telomerase and nucleosome. (PDB: ADNA, 1BNA, 4OCB, 4R4V, 1YMO, 1EQZ)

Nucleic acid structure refers to the structure of nucleic acids such as DNA and RNA. Chemically speaking, DNA and RNA are very similar. Nucleic acid structure is often divided into four different levels: primary, secondary, tertiary and quaternary.

Primary Structure

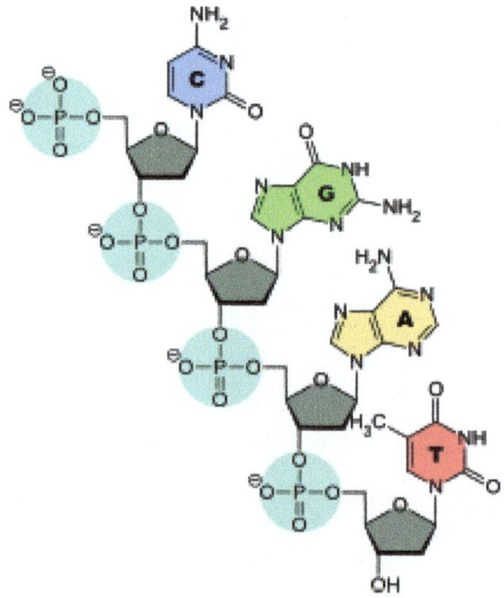

Chemical structure of DNA

Primary structure consists of a linear sequence of nucleotides that are linked together by phosphodiester bonds. It is this linear sequence of nucleotides that make up the Primary structure of DNA or RNA. Nucleotides consist of 3 components:

1. Nitrogenous base
    1. Adenine
    2. Guanine
    3. Cytosine
    4. Thymine (present in DNA only)
    5. Uracil (present in RNA only)
2. 5-carbon sugar which is called deoxyribose (found in DNA) and ribose (found in RNA).
3. One or more phosphate groups.

The nitrogen bases adenine and guanine are purine in structure and form a glycosidic bond between their 9 nitrogen and the 1' -OH group of the deoxyribose. Cytosine, thymine and uracil are pyrimidines, hence the glycosidic bonds forms between their 1 nitrogen and the 1' -OH of the deoxyribose. For both the purine and pyrimidine bases, the phosphate group forms a bond with the deoxyribose sugar through an ester bond between one of its negatively charged oxygen groups and the 5' -OH of the sugar. The polarity in DNA and RNA is derived from the oxygen and nitrogen atoms in the backbone. Nucleic acids are formed when nucleotides come together through phosphodiester linkages between the 5' and 3' carbon atoms. A Nucleic acid sequence is the order of nucleotides within a DNA (GACT) or RNA (GACU) molecule that is determined by a series of letters. Sequences are presented from the 5' to 3' end and determine the covalent structure of the entire molecule. Sequences can be complementary to another sequence in that the base on each position is complementary as well as in the reverse order. An example of a complementary sequence to AGCT is TCGA. DNA is double-stranded containing both a sense strand and an antisense strand. Therefore, the complementary sequence will be to the sense strand.

Nucleic acid design can be used to create nucleic acid complexes with complicated secondary structures such as this four-arm junction. These four strands associate into this structure because it maximizes the number of correct base pairs, with A's matched to T's and C's matched to G's. Image from Mao, 2004.

Complexes with Alkali Metal Ions

There are three potential metal binding groups on nucleic acids: phosphate, sugar and base moieties. Solid-state structure of complexes with alkali metal ions have been reviewed.

Secondary Structure

Secondary structure is the set of interactions between bases, i.e., which parts of strands are bound to each other. In DNA double helix, the two strands of DNA are held together by hydrogen bonds. The nucleotides on one strand base pairs with the nucleotide on the other strand. The secondary structure is responsible for the shape that the nucleic acid assumes. The bases in the DNA are classified as purines and pyrimidines. The purines are adenine and guanine. Purines consist of a double ring structure, a six membered and a five membered ring containing nitrogen. The pyrimidines are cytosine and thymine. It has a single ringed structure, a six membered ring containing nitrogen. A purine base always pairs with a pyrimidine base (guanine (G) pairs with cytosine (C) and adenine (A) pairs with thymine (T) or uracil (U)). DNA's secondary structure is predominantly determined by base-pairing of the two polynucleotide strands wrapped around each other to form a double helix. Although the two strands are aligned by hydrogen bonds in base pairs, the stronger forces holding the two strands together are stacking interactions between the bases. These stacking interactions are stabilized by Van der Waals forces and hydrophobic interactions, and show a large amount of local structural variability. There are also two grooves in the double helix, which are called major groove and minor groove based on their relative size.

The secondary structure of RNA consists of a single polynucleotide. Base pairing in RNA occurs when RNA folds between complementarity regions. Both single- and double-stranded regions are often found in RNA molecules. The antiparallel strands form a helical shape. The four basic elements in the secondary structure of RNA are helices, loops, bulges, and junctions. Stem-loop or hairpin loop is the most common element of RNA secondary structure. Stem-loop is formed when the RNA chains fold back on themselves to form a double helical tract called the stem, the unpaired nucleotides forms single stranded region called the loop. Secondary structure of RNA can be predicted by experimental data on the secondary structure elements, helices, loops and bulges. Bulges and internal loops are formed by separation of the double helical tract on either one strand (bulge) or on both strands (internal loops) by unpaired nucleotides. A tetraloop is a four-base pairs hairpin RNA structure. There are three common families of tetraloop in ribosomal RNA: UNCG, GNRA, and CUUG (N is one of the four nucleotides and R is a purine).UNCG is the most stable tetraloop. Pseudoknot is a RNA secondary structure first identified in turnip yellow mosaic virus. Pseudoknots are formed when nucleotides from the hairpin loop pairs with a single stranded region outside of the hairpin to form a helical segment. H-type fold pseudoknots are best characterized. In H-type fold, nucleotides in the hairpin loop pairs with the bases outside the hairpin stem forming second stem and loop. This causes formation of pseudoknots with two stems and two loops. Pseudoknots are functional elements in RNA structure having diverse function and found in most classes of RNA. DotKnot-PW method is used for comparative pseudoknots prediction. The main points in the DotKnot-PW method is scoring the similarities found in stems, secondary elements and H-type pseudoknots.

Tertiary Structure

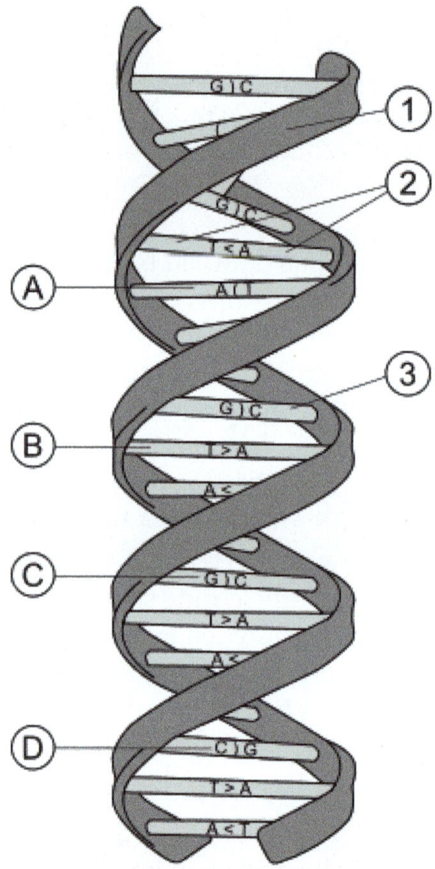

DNA structure and bases

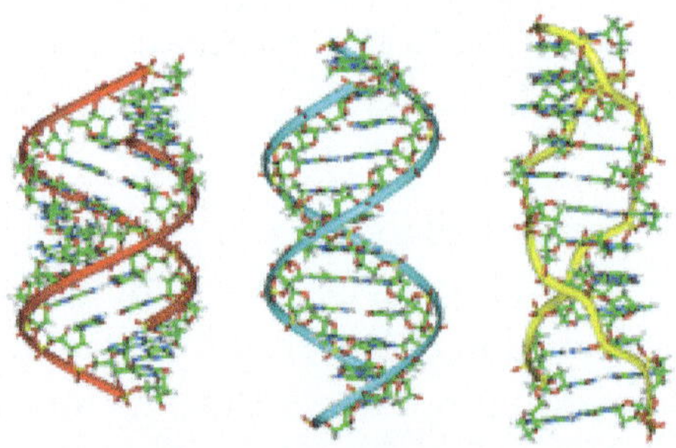

A-B-Z-DNA Side View

Tertiary structure refers to the locations of the atoms in three-dimensional space, taking into consideration geometrical and steric constraints. It is a higher order than the secondary structure, in which large-scale folding in a linear polymer occurs and the entire chain is folded into a specific

3-dimensional shape. There are 4 areas in which the structural forms of DNA can differ.

1. Handedness – right or left
2. Length of the helix turn
3. Number of base pairs per turn
4. Difference in size between the major and minor grooves

The tertiary arrangement of DNA's double helix in space includes B-DNA, A-DNA and Z-DNA.

B-DNA is the most common form of DNA in vivo and is a more narrow, elongated helix than A-DNA. Its wide major groove makes it more accessible to proteins. On the other hand, it has a narrow minor groove. B-DNA's favored conformations occur at high water concentrations; the hydration of the minor groove appears to favor B-DNA. B-DNA base pairs are nearly perpendicular to the helix axis. The sugar pucker which determines the shape of the a-helix, whether the helix will exist in the A-form or in the B-form, occurs at the C2'-endo.

A-DNA, is a form of the DNA duplex observed under dehydrating conditions. It is shorter and wider than B-DNA. RNA adopts this double helical form, and RNA-DNA duplexes are mostly A-form, but B-form RNA-DNA duplexes have been observed. In localized single strand dinucleotide contexts, RNA can also adopt the B-form without pairing to DNA. A-DNA has a deep, narrow major groove which does not make it easily accessible to proteins. On the other hand, its wide, shallow minor groove makes it accessible to proteins but with lower information content than the major groove. Its favored conformation is at low water concentrations. A-DNAs base pairs are tilted relative to the helix axis, and are displaced from the axis. The sugar pucker occurs at the C3'-endo and in RNA 2'-OH inhibits C2'-endo conformation. Long considered little more than a laboratory artifice, A-DNA is now known to have several biological functions.

Z-DNA is a relatively rare left-handed double-helix. Given the proper sequence and superhelical tension, it can be formed in vivo but its function is unclear. It has a more narrow, more elongated helix than A or B. Z-DNA's major groove is not really a groove, and it has a narrow minor groove. The most favored conformation occurs when there are high salt concentrations. There are some base substitutions but they require an alternating purine-pyrimidine sequence. The N2-amino of G H-bonds to 5' PO, which explains the slow exchange of protons and the need for the G purine. Z-DNA base pairs are nearly perpendicular to the helix axis. Z-DNA does not contain single base-pairs but rather a GpC repeat with P-P distances varying for GpC and CpG. On the GpC stack there is good base overlap, whereas on the CpG stack there is less overlap. Z-DNA's zigzag backbone is due to the C sugar conformation compensating for G glycosidic bond conformation. The conformation of G is syn, C2'-endo and for C it is anti, C3'-endo.

A linear DNA molecule having free ends can rotate, to adjust to changes of various dynamic processes in the cell, by changing how many times the two chains of its double helix twist around each other. Some DNA molecules are circular and are topologically constrained. More recently circular RNA was described as well to be a natural pervasive class of nucleic acids, expressed in many organisms.

A covalently closed, circular DNA (also known as cccDNA) is topologically constrained as the number of times the chains coiled around one other cannot change. This cccDNA can be supercoiled, which is the tertiary structure of DNA. Supercoiling is characterized by the linking number, twist and writhe. The linking number (Lk) for circular DNA is defined as the number of times one strand would have to pass through the other strand to completely separate the two strands. The linking number for circular DNA can only be changed by breaking of a covalent bond in one of the two strands. Always an integer, the linking number of a cccDNA is the sum of two components: twists (Tw) and writhes (Wr).

$$Lk = Tw + Wr$$

Twists are the number of times the two strands of DNA are twisted around each other. Writhes are number of times the DNA helix crosses over itself. DNA in cells is negatively supercoiled and has the tendency to unwind. Hence the separation of strands is easier in negatively supercoiled DNA than in relaxed DNA. The two components of supercoiled DNA are solenoid and plectonemic. The plectonemic supercoil is found in prokaryotes, while the solenoidal supercoiling is mostly seen in eukaryotes.

Quaternary Structure

The quaternary structure of nucleic acids is similar to that of protein quaternary structure. Although some of the concepts are not exactly the same, the quaternary structure refers to a higher-level of organization of nucleic acids. Moreover, it refers to interactions of the nucleic acids with other molecules. The most commonly seen form of higher-level organization of nucleic acids is seen in the form of chromatin which leads to its interactions with the small proteins histones. Also, the quaternary structure refers to the interactions between separate RNA units in the ribosome or spliceosome.

Nitrogenous Bases

- Purines and Pyrimidines are two types of bases which occur in nucleic acids.

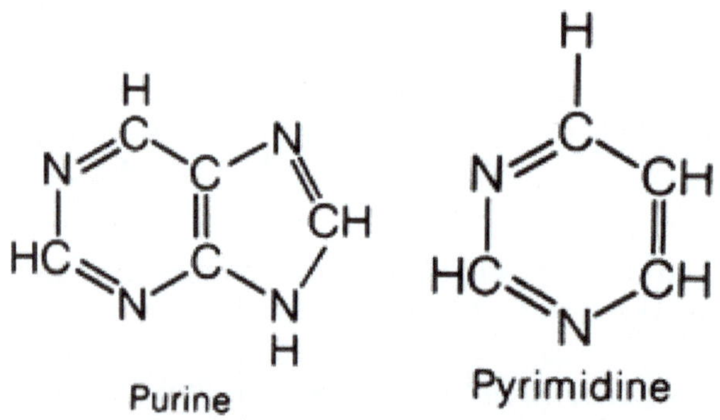

- There are five major bases found in cells. The derivatives of purine are called adenine and guanine, while those of pyrimidine are called thymine, cytosine and uracil.
- The common abbreviations used for these five bases are, A, G, T, C and U. DNA contains A, G, C and T, whereas RNA contains A, G, C and U bases.

Structure of Nitrogenous Bases

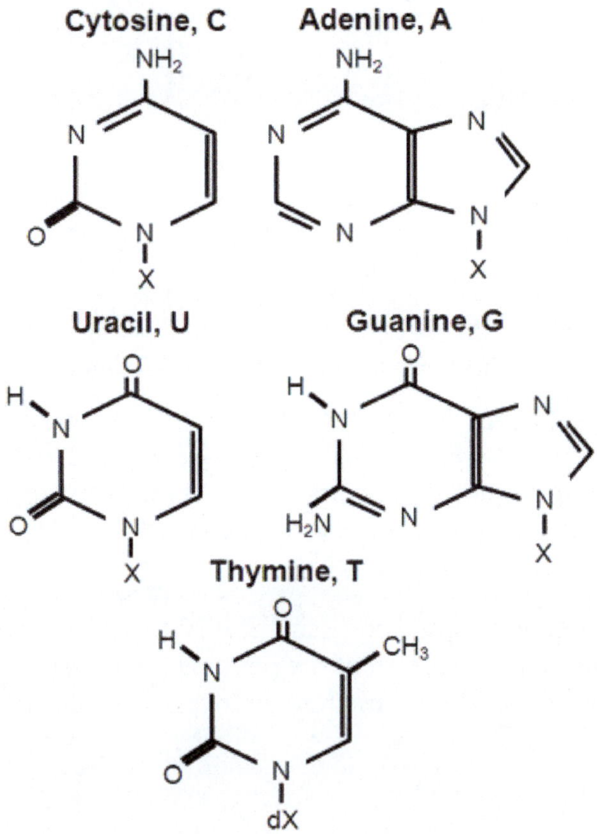

- Sugar : Ribose and 2-deoxy ribose are the two sugars found in RNA and DNA respectively. The recurring deoxyribonucleotide units of DNA contain 2- deoxy-D-ribose, and the ribonucleotide units of RNA contain D-ribose. In nucleotides, these are found in their furanose (closed five-membered ring) form.

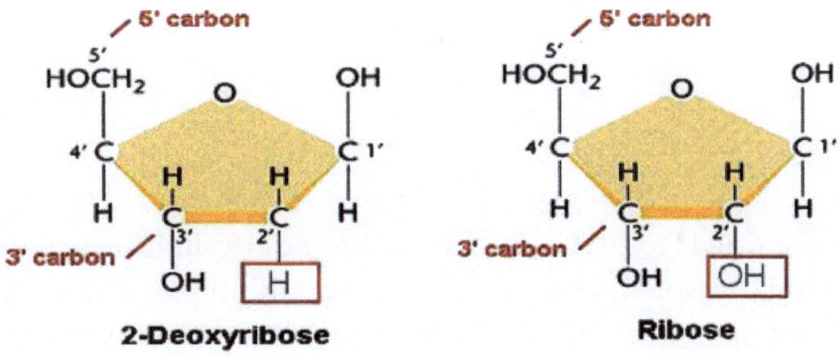

- Phosphoric acid gets attached to C-5 OH group of the sugar.
- The base and sugar are called "nucleoside". The bond between them is called the beta-glycosidic linkage. The position of attachment is shown below. Examples of nucleosides include cytidine, uridine, adenosine, guanosine, thymidine.

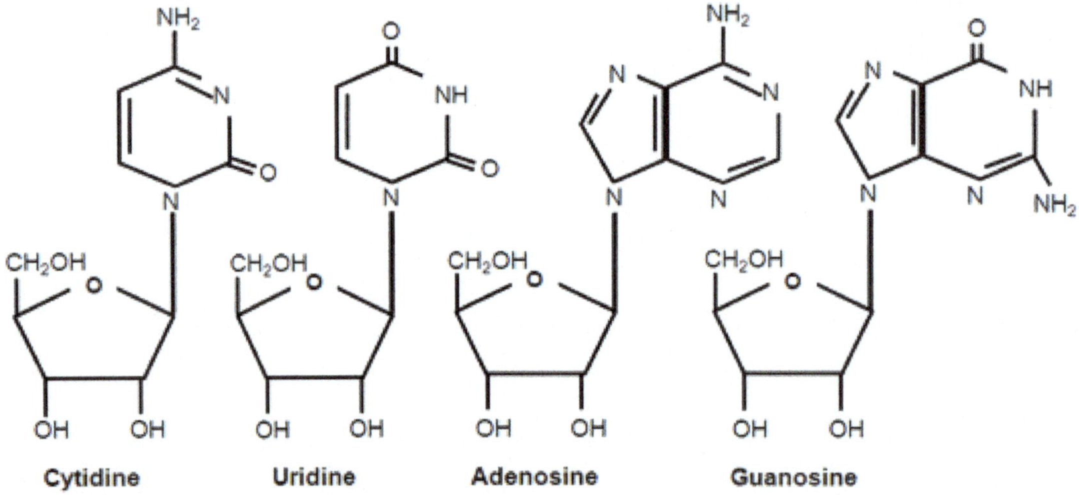

- Nucleotides are composed of three components namely nitrogenous bases, pentose sugar and phosphate group. The phosphate is attached to 5' CH_2OH group of sugar part of nucleoside. A typical nucleotide is shown below :
- The base of a nucleotide is joined covalently (at N-1 of pyrimidines and N-9 of purines) in an N—glycosyl bond to the 1 carbon of the pentose, and the phosphate is esterified to the 5 carbon. The N-glycosyl bond is formed by removal of a water molecule.

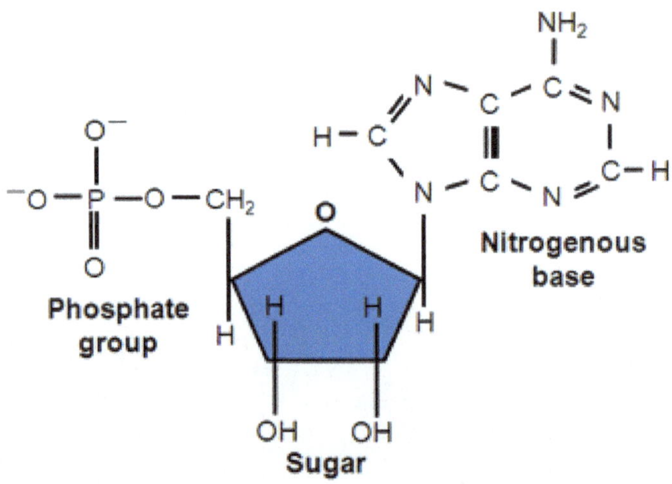

- The base of a nucleotide is joined covalently (at N-1 of pyrimidines and N-9 of purines) in an N—glycosyl bond to the 1 carbon of the pentose, and the phosphate is esterified to the 5 carbon. The N-glycosyl bond is formed by removal of a water molecule.

Polynucleotides

- Polynucleotides are formed by joining of nucleotides by phosphodiester linkages.
- The bond formation takes place between the alcohol of a 5'-phosphate of one nucleotide and the 3'-hydroxyl of the next, resulting into a phosphodiester bond.

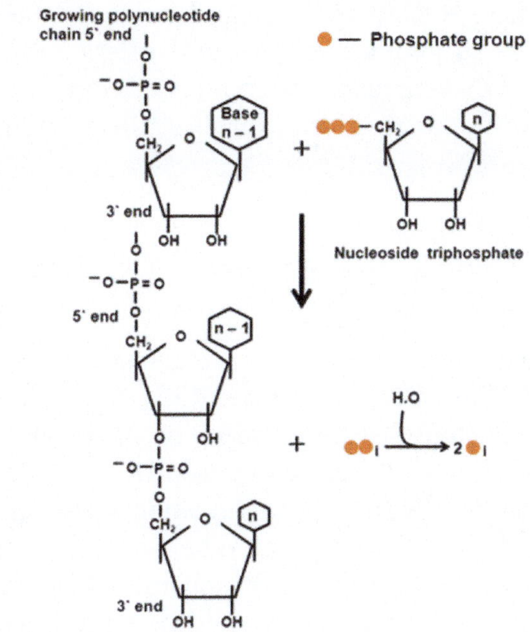

- DNA and RNA are polynucleotides.

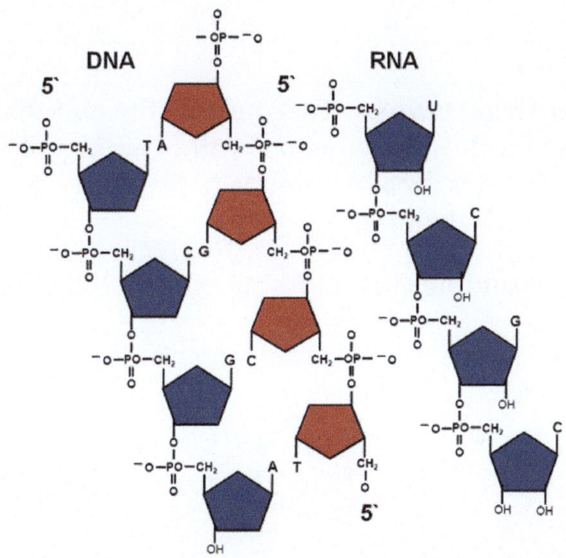

- In DNA and RNA the nucleotides are arranged in linear way and proceeds in the 5' ----> 3' direction. A common representation of ploynucleotide for example can be seen as below : 5'pApTpGpC OH3'

Purine

A purine is a heterocyclic aromatic organic compound that consists of a pyrimidine ring fused to an imidazole ring. Purine gives its name to the wider class of molecules, purines, which include substituted purines and their tautomers, are the most widely occurring nitrogen-containing heterocycle in nature.

Purines are found in high concentration in meat and meat products, especially internal organs such as liver and kidney. In general, plant-based diets are low in purines. Examples of high-purine sources include: sweetbreads, anchovies, sardines, liver, beef kidneys, brains, meat extracts (e.g., Oxo, Bovril), herring, mackerel, scallops, game meats, beer (from the yeast) and gravy.

A moderate amount of purine is also contained in beef, pork, poultry, fish and seafood, asparagus, cauliflower, spinach, mushrooms, green peas, lentils, dried peas, beans, oatmeal, wheat bran, wheat germ, and haws.

Purines and pyrimidines make up the two groups of nitrogenous bases, including the two groups of nucleotide bases. Two of the four deoxyribonucleotides (deoxyadenosine and deoxyguanosine) and two of the four ribonucleotides (adenosine, or AMP, and guanosine, or GMP), the respective building blocks of DNA and RNA, are purines. In order to form DNA and RNA, both purines and pyrimidines are needed by the cell in approximately equal quantities. Both purine and pyrimidine are self-inhibiting and activating. When purines are formed, they inhibit the enzymes required for more purine formation. This self-inhibition occurs as they also activate the enzymes needed for pyrimidine formation. Pyrimidine simultaneously self-inhibits and activates purine in similar manner. Because of this, there is nearly an equal amount of both substances in the cell at all times.

Notable Purines

There are many naturally occurring purines. They include the nucleobases adenine (2) and guanine (3). In DNA, these bases form hydrogen bonds with their complementary pyrimidines, thymine and cytosine, respectively. This is called complementary base pairing. In RNA, the complement of adenine is uracil instead of thymine.

Other notable purines are hypoxanthine (4), xanthine (5), theobromine (6), caffeine (7), uric acid (8) and isoguanine (9).

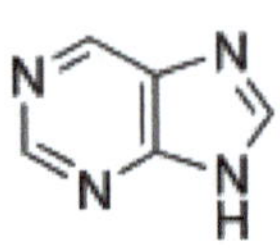
purine
1

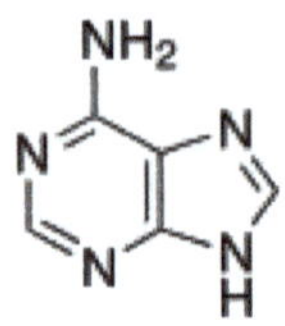
adenine
2

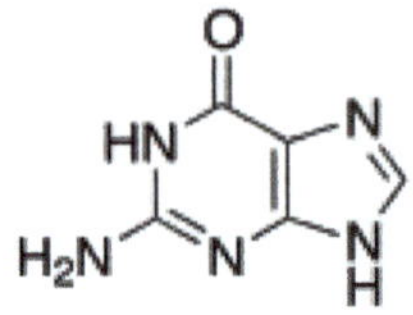
guanine
3

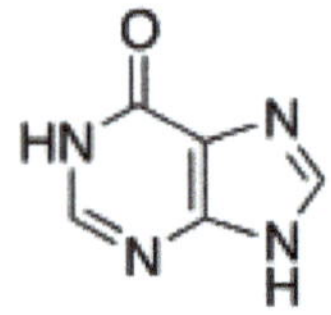
hypoxanthine
4

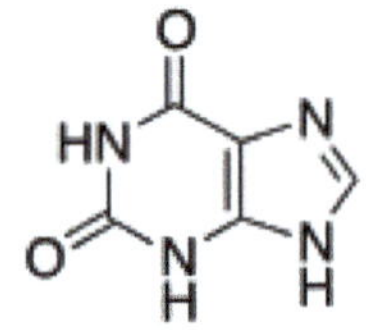
xanthine
5

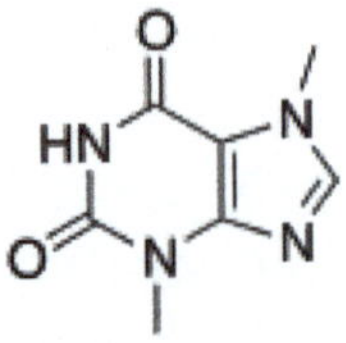
theobromine
6

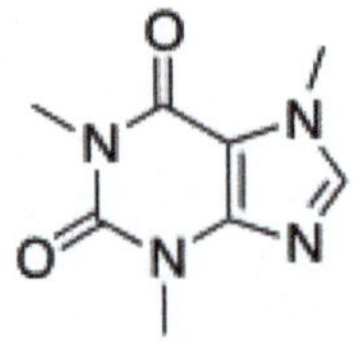
caffeine
7

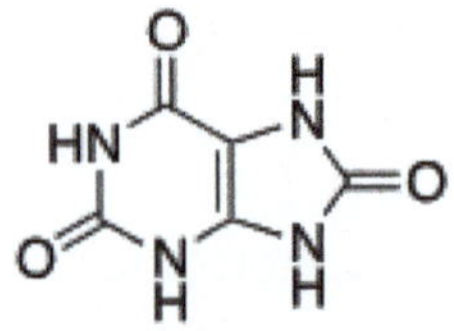
uric acid
8

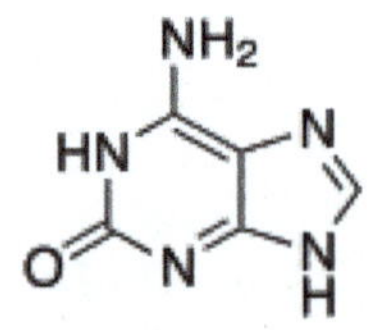
isoguanine
9

Functions

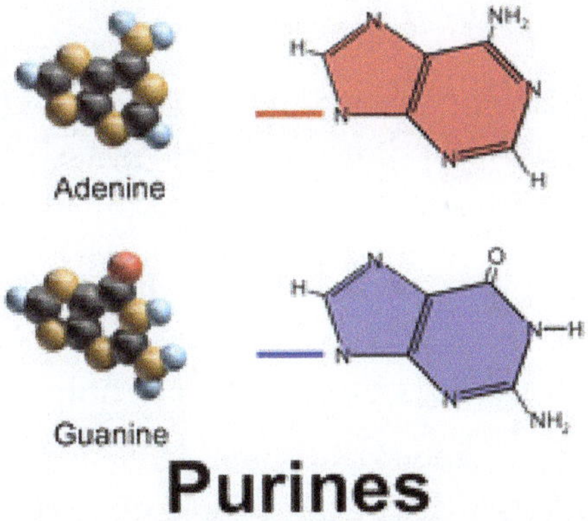

The main purine-derived nucleobases.

Aside from the crucial roles of purines (adenine and guanine) in DNA and RNA, purines are also significant components in a number of other important biomolecules, such as ATP, GTP, cyclic AMP, NADH, and coenzyme A. Purine (1) itself, has not been found in nature, but it can be produced by organic synthesis.

They may also function directly as neurotransmitters, acting upon purinergic receptors. Adenosine activates adenosine receptors.

History

The word *purine* (*pure urine*) was coined by the German chemist Emil Fischer in 1884. He synthesized it for the first time in 1898. The starting material for the reaction sequence was uric acid (8), which had been isolated from kidney stones by Carl Wilhelm Scheele in 1776. Uric acid (8) was reacted with PCl_5 to give 2,6,8-trichloropurine (10), which was converted with HI and PH_4I to give 2,6-diiodopurine (11). The product was reduced to purine (1) using zinc dust.

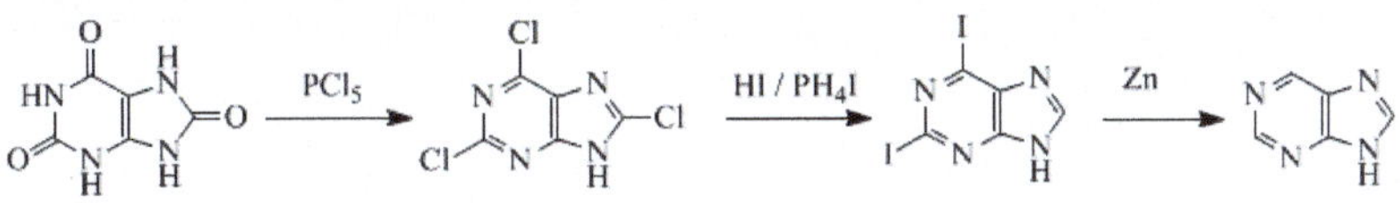

8 uric acid   10   11   1 purine

Metabolism

Many organisms have metabolic pathways to synthesize and break down purines.

Purines are biologically synthesized as nucleosides (bases attached to ribose).

Accumulation of modified purine nucleotides is defective to various cellular processes, especially those involving DNA and RNA. To be viable, organisms possess a number of (deoxy)purine phosphohydrolases, which hydrolyze these purine derivatives removing them from the active NTP and dNTP pools. Deamination of purine bases can result in accumulation of such nucleotides as ITP, dITP, XTP and dXTP.

Defects in enzymes that control purine production and breakdown can severely alter a cell's DNA sequences, which may explain why people who carry certain genetic variants of purine metabolic enzymes have a higher risk for some types of cancer.

Higher levels of meat and seafood consumption are associated with an increased risk of gout, whereas a higher level of consumption of dairy products is associated with a decreased risk. Moderate intake of purine-rich vegetables or protein is not associated with an increased risk of gout.

Laboratory Synthesis

In addition to *in vivo* synthesis of purines in purine metabolism, purine can also be created artificially.

Purine (1) is obtained in good yield when formamide is heated in an open vessel at 170 °C for 28 hours.

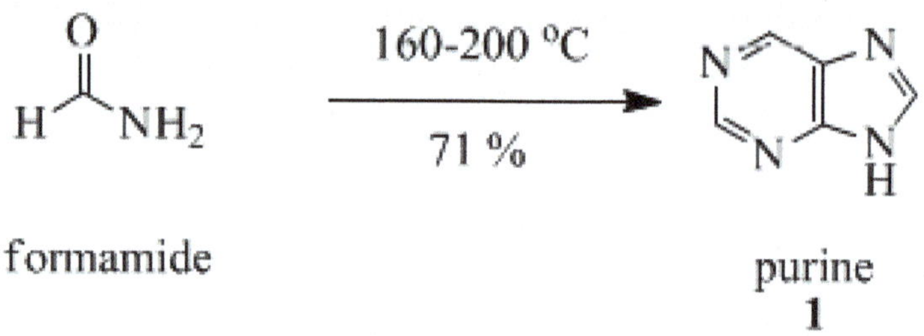

This remarkable reaction and others like it have been discussed in the context of the origin of life.

Oro, Orgel and co-workers have shown that four molecules of HCN tetramerize to form diaminomaleodinitrile (12), which can be converted into almost all naturally-occurring purines. For example, five molecules of HCN condense in an exothermic reaction to make adenine, especially in the presence of ammonia.

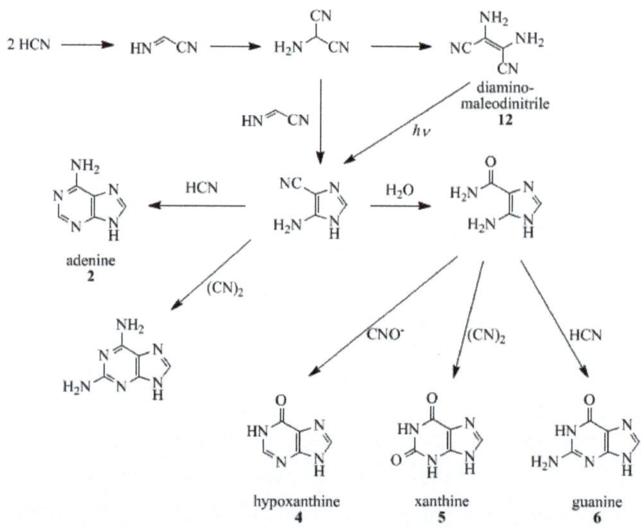

The Traube purine synthesis (1900) is a classic reaction (named after Wilhelm Traube) between an amine-substituted pyrimidine and formic acid.

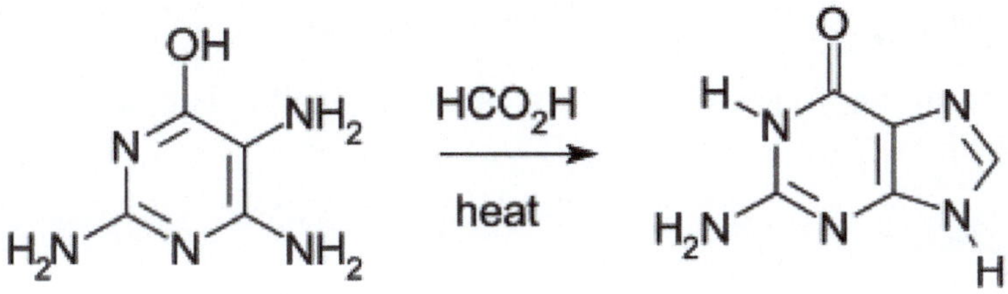

Pyrimidine

Pyrimidine is an aromatic heterocyclic organic compound similar to pyridine. One of the three diazines (six-membered heterocyclics with two nitrogen atoms in the ring), it has the nitrogen atoms at positions 1 and 3 in the ring. The other diazines are pyrazine (nitrogen atoms at the 1 and 4 positions) and pyridazine (nitrogen atoms at the 1 and 2 positions). In nucleic acids, three types of nucleobases are pyrimidine derivatives: cytosine (C), thymine (T), and uracil (U).

Occurrence and History

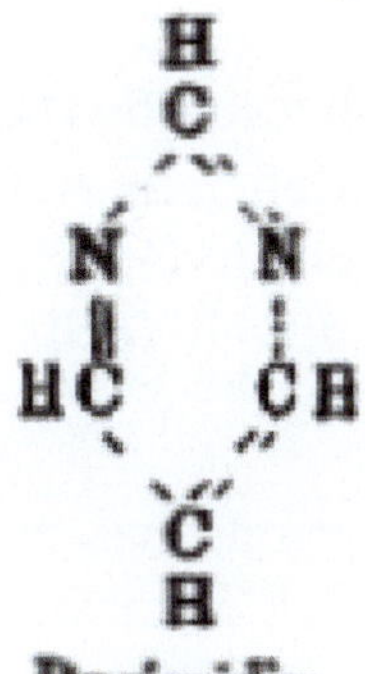

Pinner's 1885 structure for pyrimidine

The pyrimidine ring system has wide occurrence in nature as substituted and ring fused compounds and derivatives, including the nucleotides, thiamine (vitamin B1) and alloxan. It is also found in many synthetic compounds such as barbiturates and the HIV drug, zidovudine. Although pyrimidine derivatives such as uric acid and alloxan were known in the early 19th century, a laboratory synthesis of a pyrimidine was not carried out until 1879, when Grimaux reported the preparation of barbituric acid from Ivy urea and malonic acid in the presence of phosphorus oxychloride. The systematic study of pyrimidines began in 1884 with Pinner, who synthesized derivatives by condensing ethyl acetoacetate with amidines. Pinner first proposed the name "pyrimidin" in 1885. The parent compound was first prepared by Gabriel and Colman in 1900, by conversion of barbituric acid to 2,4,6-trichloropyrimidine followed by reduction using zinc dust in hot water.

Nomenclature

The nomenclature of pyrimidines is straightforward. However, like other heterocyclics, tautomeric hydroxyl groups yield complications since they exist primarily in the cyclic amide form. For example, 2-hydroxypyrimidine is more properly named 2-pyrimidone. A partial list of trivial names of various pyrimidines exists.

Physical Properties

Physical properties are shown in the data box. A more extensive discussion, including spectra, can be found in Brown *et al.*

Chemical Properties

Per the classification by Albert six-membered heterocycles can be described as π-deficient. Substitution by electronegative groups or additional nitrogen atoms in the ring significantly increase the π-deficiency. These effects also decrease the basicity.

Like pyridines, in pyrimidines the π-electron density is decreased to an even greater extent. Therefore, electrophilic aromatic substitution is more difficult while nucleophilic aromatic substitution is facilitated. An example of the last reaction type is the displacement of the amino group in 2-aminopyrimidine by chlorine and its reverse.

Electron lone pair availability (basicity) is decreased compared to pyridine. Compared to pyridine, *N*-alkylation and *N*-oxidation are more difficult. The pK_a value for protonated pyrimidine is 1.23 compared to 5.30 for pyridine. Protonation and other electrophilic additions will occur at only one nitrogen due to further deactivation by the second nitrogen. The 2-, 4-, and 6- positions on the pyrimidine ring are electron deficient analogous to those in pyridine and nitro- and dinitrobenzene. The 5-position is less electron deficient and substitutents there are quite stable. However, electrophilic substitution is relatively facile at the 5-position, including nitration and halogenation.

Reduction in resonance stabilization of pyrimidines may lead to addition and ring cleavage reactions rather than substitutions. One such manifestation is observed in the Dimroth rearrangement.

Pyrimidine is also found in meteorites, but scientists still do not know its origin. Pyrimidine also photolytically decomposes into uracil under ultraviolet light.

Synthesis

As is often the case with parent heterocyclic ring systems, the synthesis of pyrimidine is not that common and is usually performed by removing functional groups from derivatives. Primary syntheses in quantity involving formamide have been reported.

As a class, pyrimidines are typically synthesized by the principal synthesis involving cyclization of β-dicarbonyl compounds with N–C–N compounds. Reaction of the former with amidines to give 2-substituted pyrimidines, with urea to give 2-pyrimidinones, and guanidines to give 2-aminopyrimidines are typical.

Pyrimidines can be prepared via the Biginelli reaction. Many other methods rely on condensation of carbonyls with diamines for instance the synthesis of 2-thio-6-methyluracil from thiourea and ethyl acetoacetate or the synthesis of 4-methylpyrimidine with 4,4-dimethoxy-2-butanone and formamide.

A novel method is by reaction of *N*-vinyl and *N*-aryl amides with carbonitriles under electrophilic activation of the amide with 2-chloro-pyridine and trifluoromethanesulfonic anhydride:

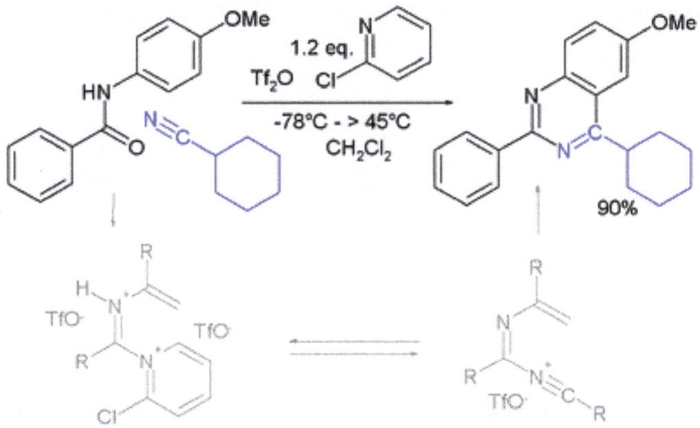

Reactions

Because of the decreased basicity compared to pyridine, electrophilic substitution of pyrimidine is less facile. Protonation or alkylation typically takes place at only one of the ring nitrogen atoms. Mono-*N*-oxidation occurs by reaction with peracids.

Electrophilic *C*-substitution of pyrimidine occurs at the 5-position, the least electron-deficient. Nitration, nitrosation, azo coupling, halogenation, sulfonation, formylation, hydroxymethylation, and aminomethylation have been observed with substituted pyrimidines.

Nucleophilic *C*-substitution should be facilitated at the 2-, 4-, and 6-positions but there are only a few examples. Amination and hydroxylation has been observed for substituted pyrimidines. Reactions with Grignard or alkyllithium reagents yield 4-alkyl- or 4-aryl pyrimidine after aromatization.

Free radical attack has been observed for pyrimidine and photochemical reactions have been observed for substituted pyrimidines. Pyrimidine can be hydrogenated to give tetrahydropyrimidine.

Nucleotides

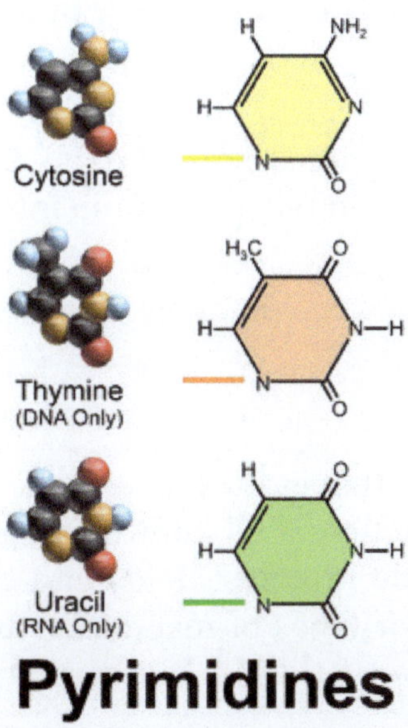

The pyrimidine nitrogen bases found in DNA and RNA.

Three nucleobases found in nucleic acids, cytosine (C), thymine (T), and uracil (U), are pyrimidine derivatives:

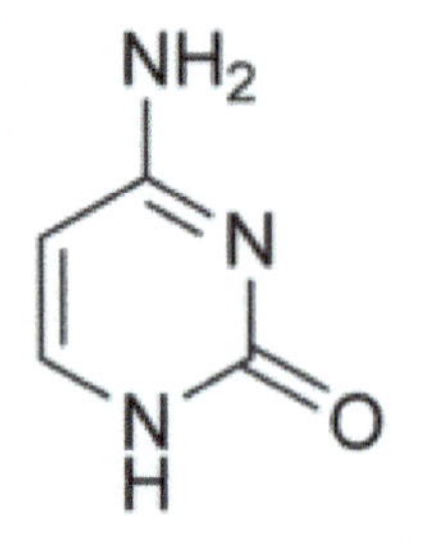	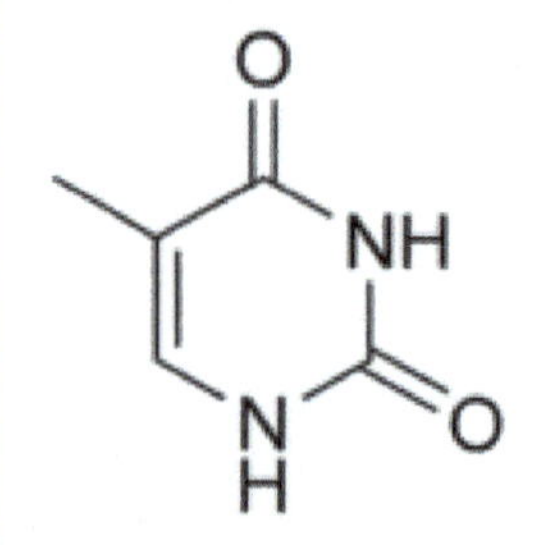	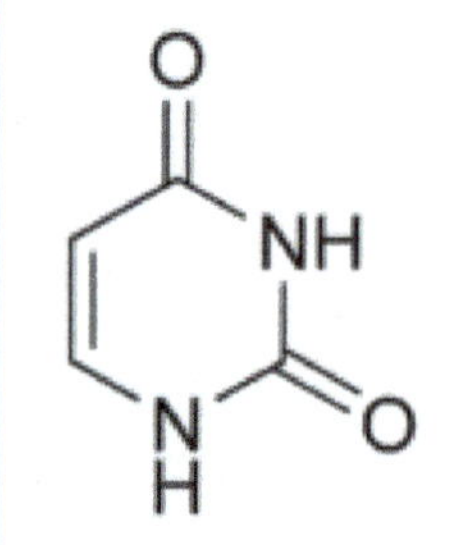
Cytosine (C)	Thymine (T)	Uracil (U)

In DNA and RNA, these bases form hydrogen bonds with their complementary purines. Thus, in DNA, the purines adenine (A) and guanine (G) pair up with the pyrimidines thymine (T) and cytosine (C), respectively.

In RNA, the complement of adenine (A) is uracil (U) instead of thymine (T), so the pairs that form are adenine:uracil and guanine:cytosine.

Very rarely, thymine can appear in RNA, or uracil in DNA, but when the other three major pyrimidine bases are represented, some minor pyrimidine bases can also occur in nucleic acids. These minor pyrimidines are usually methylated versions of major ones and are postulated to have regulatory functions.

These hydrogen bonding modes are for classical Watson–Crick base pairing. Other hydrogen bonding modes ("wobble pairings") are available in both DNA and RNA, although the additional 2′-hydroxyl group of RNA expands the configurations, through which RNA can form hydrogen bonds.

Theoretical Aspects

In March 2015, NASA Ames scientists reported that, for the first time, complex DNA and RNA organic compounds of life, including uracil, cytosine and thymine, have been formed in the laboratory under outer space conditions, using starting chemicals, such as pyrimidine, found in meteorites. Pyrimidine, like polycyclic aromatic hydrocarbons (PAHs), the most carbon-rich chemical found in the universe, may have been formed in red giants or in interstellar dust and gas clouds.

DNA

Deoxyribonucleic acid is a molecule that carries the genetic instructions used in the growth, development, functioning and reproduction of all known living organisms and many viruses. DNA and RNA are nucleic acids; alongside proteins, lipids and complex carbohydrates (polysaccharides), they are one of the four major types of macromolecules that are essential for all known forms of life. Most DNA molecules consist of two biopolymer strands coiled around each other to form a double helix.

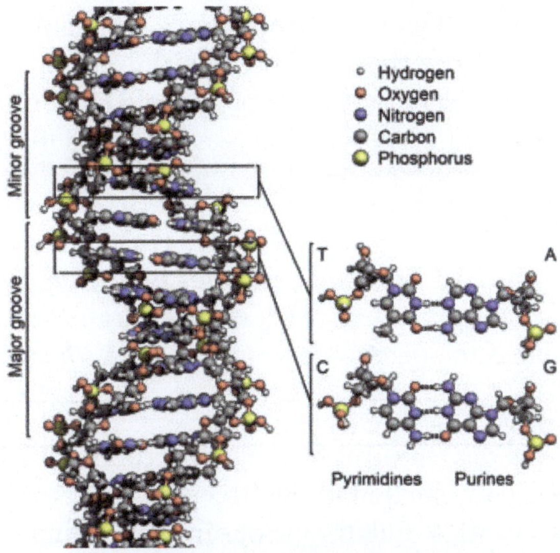

The structure of the DNA double helix. The atoms in the structure are colour-coded by element and the detailed structure of two base pairs are shown in the bottom right.

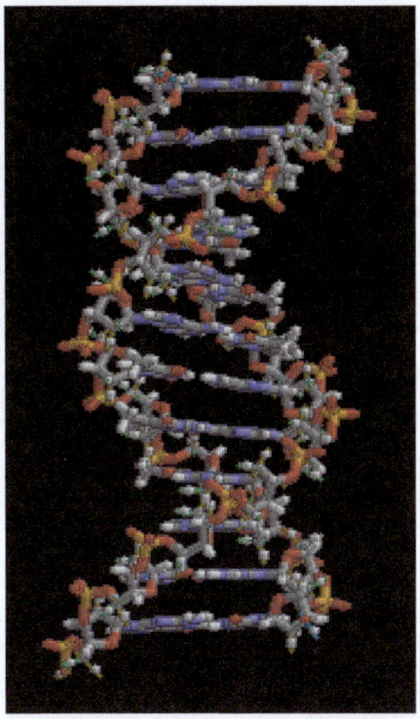

The structure of part of a DNA double helix

The two DNA strands are termed polynucleotides since they are composed of simpler monomer units called nucleotides. Each nucleotide is composed of one of four nitrogen-containing nucleobases—cytosine (C), guanine (G), adenine (A), or thymine (T)—a sugar called deoxyribose, and a phosphate group. The nucleotides are joined to one another in a chain by covalent bonds between the sugar of one nucleotide and the phosphate of the next, resulting in an alternating sugar-phosphate backbone. The nitrogenous bases of the two separate polynucleotide strands are bound together, according to base pairing rules (A with T, and C with G), with hydrogen bonds to make double-stranded DNA. The total amount of related DNA base pairs on Earth is estimated at 5.0×10^{37} and weighs 50 billion tonnes. In comparison the total mass of the biosphere has been estimated to be as much as 4 trillion tons of carbon (TtC).

DNA stores biological information. The DNA backbone is resistant to cleavage, and both strands of the double-stranded structure store the same biological information. This information is replicated as and when the two strands separate. A large part of DNA (more than 98% for humans) is non-coding, meaning that these sections do not serve as patterns for protein sequences.

The two strands of DNA run in opposite directions to each other and are thus antiparallel. Attached to each sugar is one of four types of nucleobases (informally, *bases*). It is the sequence of these four nucleobases along the backbone that encodes biological information. RNA strands are created using DNA strands as a template in a process called transcription. Under the genetic code, these RNA strands are translated to specify the sequence of amino acids within proteins in a process called translation.

Within eukaryotic cells DNA is organized into long structures called chromosomes. During cell division these chromosomes are duplicated in the process of DNA replication, providing each cell its own complete set of chromosomes. Eukaryotic organisms (animals, plants, fungi, and protists) store most of their DNA inside the cell nucleus and some of their DNA in organelles, such as mitochondria or chloroplasts. In contrast prokaryotes (bacteria and archaea) store their DNA only in the cytoplasm. Within the eukaryotic chromosomes, chromatin proteins such as histones compact and organize DNA. These compact structures guide the interactions between DNA and other proteins, helping control which parts of the DNA are transcribed.

DNA was first isolated by Friedrich Miescher in 1869. Its molecular structure was identified by James Watson and Francis Crick in 1953, whose model-building efforts were guided by X-ray diffraction data acquired by Raymond Gosling, who was a post-graduate student of Rosalind Franklin. DNA is used by researchers as a molecular tool to explore physical laws and theories, such as the ergodic theorem and the theory of elasticity. The unique material properties of DNA have made it an attractive molecule for material scientists and engineers interested in micro- and nano-fabrication. Among notable advances in this field are DNA origami and DNA-based hybrid materials.

Properties

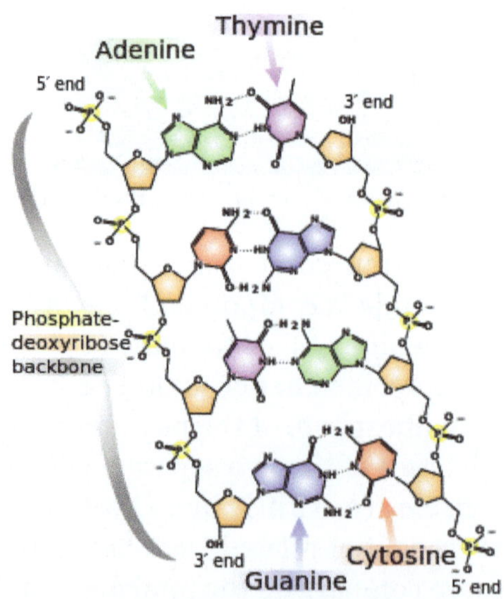

Chemical structure of DNA; hydrogen bonds shown as dotted lines

DNA is a long polymer made from repeating units called nucleotides. The structure of DNA is dynamic along its length, being capable of coiling into tight loops, and other shapes. In all species it is composed of two helical chains, bound to each other by hydrogen bonds. Both chains are coiled round the same axis, and have the same pitch of 34 ångströms (3.4 nanometres). The pair of chains has a radius of 10 ångströms (1.0 nanometre). According to another study, when measured in a different solution, the DNA chain measured 22 to 26 ångströms wide (2.2 to 2.6 nanometres), and one nucleotide unit measured 3.3 Å (0.33 nm) long. Although each individual nucleotide repeating unit is very small, DNA polymers can be very large molecules containing millions to hundreds of millions of nucleotides. For instance, the DNA in the largest human chromosome, chromosome number 1, consists of approximately 220 million base pairs and would be 85 mm long if straightened.

In living organisms, DNA does not usually exist as a single molecule, but instead as a pair of molecules that are held tightly together. These two long strands entwine like vines, in the shape of a double helix. The nucleotide contains both a segment of the backbone of the molecule (which holds the chain together) and a nucleobase (which interacts with the other DNA strand in the helix). A nucleobase linked to a sugar is called a nucleoside and a base linked to a sugar and one or more phosphate groups is called a nucleotide. A polymer comprising multiple linked nucleotides (as in DNA) is called a polynucleotide.

The backbone of the DNA strand is made from alternating phosphate and sugar residues. The sugar in DNA is 2-deoxyribose, which is a pentose (five-carbon) sugar. The sugars are joined together by phosphate groups that form phosphodiester bonds between the third and fifth carbon atoms of adjacent sugar rings. These asymmetric bonds mean a strand of DNA has a direction. In a double helix, the direction of the nucleotides in one strand is opposite to their direction in the other strand: the strands are *antiparallel*. The asymmetric ends of DNA strands are said to have a directionality of *five prime* (5′) and *three prime* (3′), with the 5′ end having a terminal phosphate group and the 3′ end a terminal hydroxyl group. One major difference between DNA and RNA is the sugar, with the 2-deoxyribose in DNA being replaced by the alternative pentose sugar ribose in RNA.

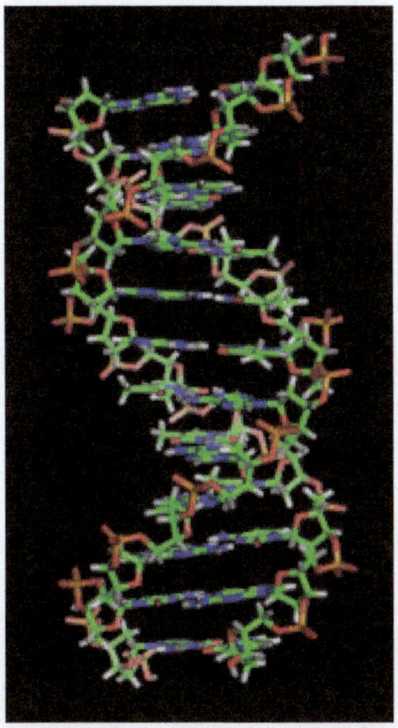

A section of DNA. The bases lie horizontally between the two spiraling strands. (animated version).

The DNA double helix is stabilized primarily by two forces: hydrogen bonds between nucleotides and base-stacking interactions among aromatic nucleobases. In the aqueous environment of the cell, the conjugated π bonds of nucleotide bases align perpendicular to the axis of the DNA molecule, minimizing their interaction with the solvation shell. The four bases found in DNA are adenine (A), cytosine (C), guanine (G) and thymine (T). These four bases are attached to the sugar-phosphate to form the complete nucleotide, as shown for adenosine monophosphate. Adenine pairs with thymine and guanine pairs with cytosine. It was represented by A-T base pairs and G-C base pairs.

Nucleobase Classification

The nucleobases are classified into two types: the purines, A and G, being fused five- and six-membered heterocyclic compounds, and the pyrimidines, the six-membered rings C and T. A fifth pyrimidine nucleobase, uracil (U), usually takes the place of thymine in RNA and differs from thymine by lacking a methyl group on its ring. In addition to RNA and DNA, many artificial nucleic acid analogues have been created to study the properties of nucleic acids, or for use in biotechnology.

Uracil is not usually found in DNA, occurring only as a breakdown product of cytosine. However, in several bacteriophages, *Bacillus subtilis* bacteriophages PBS1 and PBS2 and *Yersinia* bacteriophage piR1-37, thymine has been replaced by uracil. Another phage - Staphylococcal phage S6 - has been identified with a genome where thymine has been replaced by uracil.

Base J (beta-d-glucopyranosyloxymethyluracil), a modified form of uracil, is also found in several organisms: the flagellates *Diplonema* and *Euglena*, and all the kinetoplastid genera. Biosynthesis of J occurs in two steps: in the first step, a specific thymidine in DNA is converted into hydroxymethyldeoxyuridine; in the second, HOMedU is glycosylated to form J. Proteins that bind specifically to this base have been identified. These proteins appear to be distant relatives of the Tet1 oncogene that is involved in the pathogenesis of acute myeloid leukemia. J appears to act as a termination signal for RNA polymerase II.

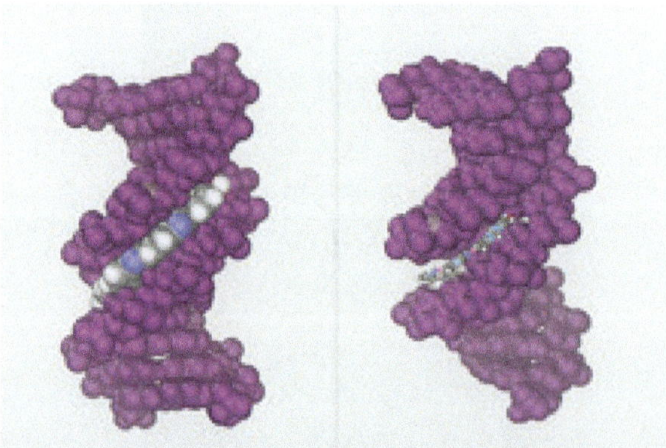

DNA major and minor grooves. The latter is a binding site for the Hoechst stain dye 33258.

Grooves

Twin helical strands form the DNA backbone. Another double helix may be found tracing the spaces, or grooves, between the strands. These voids are adjacent to the base pairs and may provide a binding site. As the strands are not symmetrically located with respect to each other, the grooves are unequally sized. One groove, the major groove, is 22 Å wide and the other, the minor groove, is 12 Å wide. The width of the major groove means that the edges of the bases are more accessible in the major groove than in the minor groove. As a result, proteins such as transcription factors that can bind to specific sequences in double-stranded DNA usually make contact with the sides of the bases exposed in the major groove. This situation varies in unusual conformations of DNA within the cell, but the major and minor grooves are always named to reflect the differences in size that would be seen if the DNA is twisted back into the ordinary B form.

Base Pairing

In a DNA double helix, each type of nucleobase on one strand bonds with just one type of nucleobase on the other strand. This is called complementary base pairing. Here, purines form hydrogen bonds to pyrimidines, with adenine bonding only to thymine in two hydrogen bonds, and cytosine bonding only to guanine in three hydrogen bonds. This arrangement of two nucleotides binding together across the double helix is called a base pair. As hydrogen bonds are not covalent, they can be broken and rejoined relatively easily. The two strands of DNA in a double helix can thus be pulled apart like a zipper, either by a mechanical force or high temperature. As a result of this base pair complementarity, all the information in the double-stranded sequence of a DNA helix is duplicated on each strand, which is vital in DNA replication. This reversible and specific interaction between complementary base pairs is critical for all the functions of DNA in living organisms.

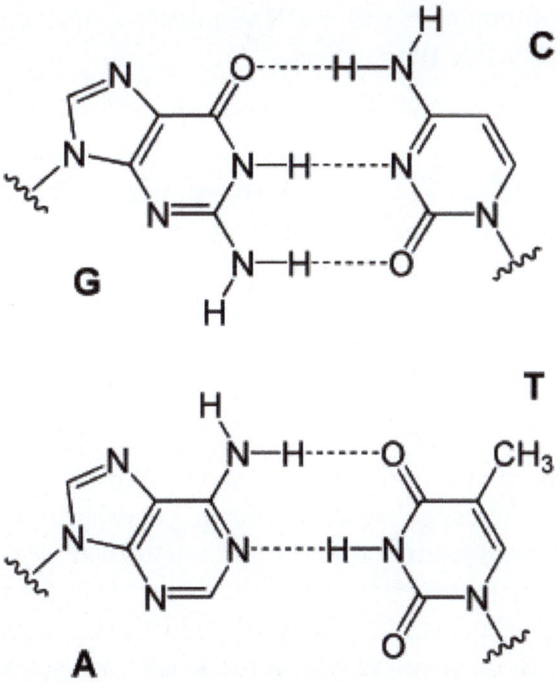

Top, a **GC** base pair with three hydrogen bonds. Bottom, an AT base pair with two hydrogen bonds. Non-covalent hydrogen bonds between the pairs are shown as dashed lines.

The two types of base pairs form different numbers of hydrogen bonds, AT forming two hydrogen bonds, and GC forming three hydrogen bonds. DNA with high GC-content is more stable than DNA with low GC-content.

As noted above, most DNA molecules are actually two polymer strands, bound together in a helical fashion by noncovalent bonds; this double stranded structure (dsDNA) is maintained largely by the intrastrand base stacking interactions, which are strongest for G,C stacks. The two strands can come apart – a process known as melting – to form two single-stranded DNA molecules (ssDNA) molecules. Melting occurs at high temperature, low salt and high pH (low pH also melts DNA, but since DNA is unstable due to acid depurination, low pH is rarely used).

The stability of the dsDNA form depends not only on the GC-content (% G,C basepairs) but also on sequence (since stacking is sequence specific) and also length (longer molecules are more stable). The stability can be measured in various ways; a common way is the "melting temperature", which is the temperature at which 50% of the ds molecules are converted to ss molecules; melting temperature is dependent on ionic strength and the concentration of DNA. As a result, it is both the percentage of GC base pairs and the overall length of a DNA double helix that determines the strength of the association between the two strands of DNA. Long DNA helices with a high GC-content have stronger-interacting strands, while short helices with high AT content have weaker-interacting strands. In biology, parts of the DNA double helix that need to separate easily, such as the TATAAT Pribnow box in some promoters, tend to have a high AT content, making the strands easier to pull apart.

In the laboratory, the strength of this interaction can be measured by finding the temperature necessary to break the hydrogen bonds, their melting temperature (also called T_m value). When all the base pairs in a DNA double helix melt, the strands separate and exist in solution as two entirely independent molecules. These single-stranded DNA molecules have no single common shape, but some conformations are more stable than others.

Sense and Antisense

A DNA sequence is called "sense" if its sequence is the same as that of a messenger RNA copy that is translated into protein. The sequence on the opposite strand is called the "antisense" sequence. Both sense and antisense sequences can exist on different parts of the same strand of DNA (i.e. both strands can contain both sense and antisense sequences). In both prokaryotes and eukaryotes, antisense RNA sequences are produced, but the functions of these RNAs are not entirely clear. One proposal is that antisense RNAs are involved in regulating gene expression through RNA-RNA base pairing.

A few DNA sequences in prokaryotes and eukaryotes, and more in plasmids and viruses, blur the distinction between sense and antisense strands by having overlapping genes. In these cases, some DNA sequences do double duty, encoding one protein when read along one strand, and a second protein when read in the opposite direction along the other strand. In bacteria, this overlap may be involved in the regulation of gene transcription, while in viruses, overlapping genes increase the amount of information that can be encoded within the small viral genome.

Supercoiling

DNA can be twisted like a rope in a process called DNA supercoiling. With DNA in its "relaxed" state, a strand usually circles the axis of the double helix once every 10.4 base pairs, but if the DNA is twisted the strands become more tightly or more loosely wound. If the DNA is twisted in the direction of the helix, this is positive supercoiling, and the bases are held more tightly together. If they are twisted in the opposite direction, this is negative supercoiling, and the bases come apart more easily. In nature, most DNA has slight negative supercoiling that is introduced by enzymes called topoisomerases. These enzymes are also needed to relieve the twisting stresses introduced into DNA strands during processes such as transcription and DNA replication.

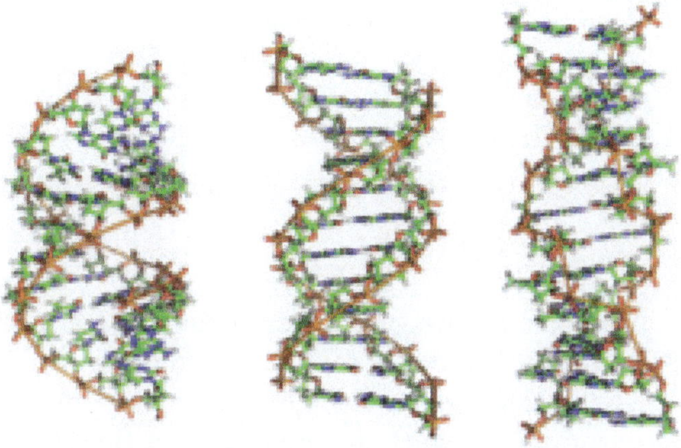

From left to right, the structures of A, B and Z DNA

Alternative DNA Structures

DNA exists in many possible conformations that include A-DNA, B-DNA, and Z-DNA forms, although, only B-DNA and Z-DNA have been directly observed in functional organisms. The conformation that DNA adopts depends on the hydration level, DNA sequence, the amount and direction of supercoiling, chemical modifications of the bases, the type and concentration of metal ions, and the presence of polyamines in solution.

The first published reports of A-DNA X-ray diffraction patterns—and also B-DNA—used analyses based on Patterson transforms that provided only a limited amount of structural information for oriented fibers of DNA. An alternative analysis was then proposed by Wilkins *et al.*, in 1953, for the *in vivo* B-DNA X-ray diffraction-scattering patterns of highly hydrated DNA fibers in terms of squares of Bessel functions. In the same journal, James Watson and Francis Crick presented their molecular modeling analysis of the DNA X-ray diffraction patterns to suggest that the structure was a double-helix.

Although the *B-DNA form* is most common under the conditions found in cells, it is not a well-defined conformation but a family of related DNA conformations that occur at the high hydration levels present in living cells. Their corresponding X-ray diffraction and scattering patterns are characteristic of molecular paracrystals with a significant degree of disorder.

Compared to B-DNA, the A-DNA form is a wider right-handed spiral, with a shallow, wide minor groove and a narrower, deeper major groove. The A form occurs under non-physiological conditions in partly dehydrated samples of DNA, while in the cell it may be produced in hybrid pairings of DNA and RNA strands, and in enzyme-DNA complexes. Segments of DNA where the bases have been chemically modified by methylation may undergo a larger change in conformation and adopt the Z form. Here, the strands turn about the helical axis in a left-handed spiral, the opposite of the more common B form. These unusual structures can be recognized by specific Z-DNA binding proteins and may be involved in the regulation of transcription.

Alternative DNA Chemistry

For many years exobiologists have proposed the existence of a shadow biosphere, a postulated microbial biosphere of Earth that uses radically different biochemical and molecular processes than currently known life. One of the proposals was the existence of lifeforms that use arsenic instead of phosphorus in DNA. A report in 2010 of the possibility in the bacterium GFAJ-1, was announced, though the research was disputed, and evidence suggests the bacterium actively prevents the incorporation of arsenic into the DNA backbone and other biomolecules.

Quadruplex Structures

At the ends of the linear chromosomes are specialized regions of DNA called telomeres. The main function of these regions is to allow the cell to replicate chromosome ends using the enzyme telomerase, as the enzymes that normally replicate DNA cannot copy the extreme 3′ ends of chromosomes. These specialized chromosome caps also help protect the DNA ends, and stop the DNA repair systems in the cell from treating them as damage to be corrected. In human cells, telomeres are usually lengths of single-stranded DNA containing several thousand repeats of a simple TTAGGG sequence.

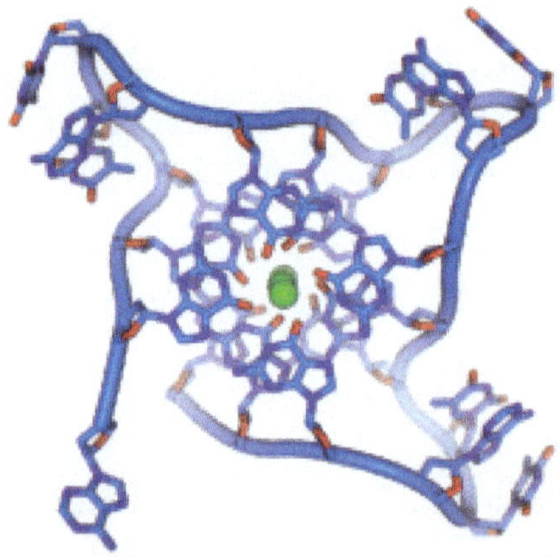

DNA quadruplex formed by telomere repeats. The looped conformation of the DNA backbone is very different from the typical DNA helix. The green spheres in the center represent potassium ions.

These guanine-rich sequences may stabilize chromosome ends by forming structures of stacked sets of four-base units, rather than the usual base pairs found in other DNA molecules. Here, four guanine bases form a flat plate and these flat four-base units then stack on top of each other, to form a stable G-quadruplex structure. These structures are stabilized by hydrogen bonding between the edges of the bases and chelation of a metal ion in the centre of each four-base unit. Other structures can also be formed, with the central set of four bases coming from either a single strand folded around the bases, or several different parallel strands, each contributing one base to the central structure.

In addition to these stacked structures, telomeres also form large loop structures called telomere loops, or T-loops. Here, the single-stranded DNA curls around in a long circle stabilized by telomere-binding proteins. At the very end of the T-loop, the single-stranded telomere DNA is held onto a region of double-stranded DNA by the telomere strand disrupting the double-helical DNA and base pairing to one of the two strands. This triple-stranded structure is called a displacement loop or D-loop.

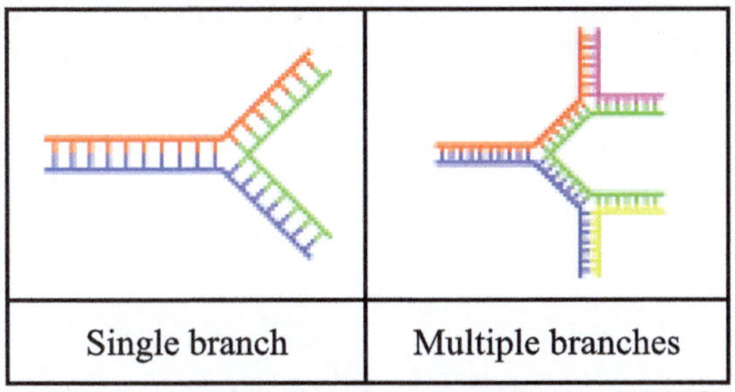

Branched DNA can form networks containing multiple branches.

Branched DNA

In DNA, fraying occurs when non-complementary regions exist at the end of an otherwise complementary double-strand of DNA. However, branched DNA can occur if a third strand of DNA is introduced and contains adjoining regions able to hybridize with the frayed regions of the pre-existing double-strand. Although the simplest example of branched DNA involves only three strands of DNA, complexes involving additional strands and multiple branches are also possible. Branched DNA can be used in nanotechnology to construct geometric shapes.

Chemical Modifications and Altered DNA Packaging

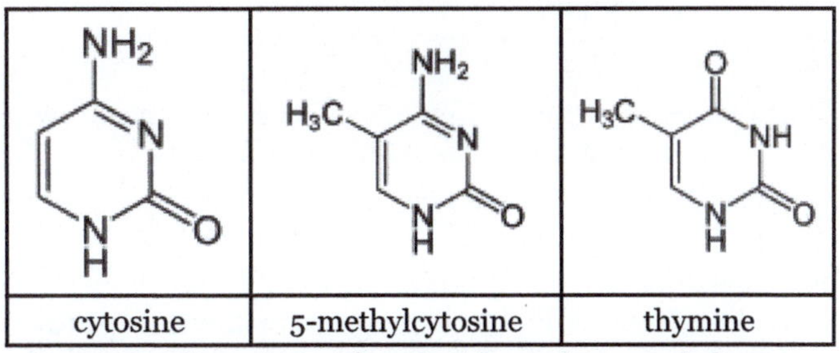

Structure of cytosine with and without the 5-methyl group. Deamination converts 5-methylcytosine into thymine.

Base Modifications and DNA Packaging

The expression of genes is influenced by how the DNA is packaged in chromosomes, in a structure called chromatin. Base modifications can be involved in packaging, with regions that have low or no gene expression usually containing high levels of methylation of cytosine bases. DNA packaging and its influence on gene expression can also occur by covalent modifications of the histone protein core around which DNA is wrapped in the chromatin structure or else by remodeling carried out by chromatin remodeling complexes. There is, further, crosstalk between DNA methylation and histone modification, so they can coordinately affect chromatin and gene expression.

For one example, cytosine methylation produces 5-methylcytosine, which is important for X-inactivation of chromosomes. The average level of methylation varies between organisms – the worm *Caenorhabditis elegans* lacks cytosine methylation, while vertebrates have higher levels, with up to 1% of their DNA containing 5-methylcytosine. Despite the importance of 5-methylcytosine, it can deaminate to leave a thymine base, so methylated cytosines are particularly prone to mutations. Other base modifications include adenine methylation in bacteria, the presence of 5-hydroxymethylcytosine in the brain, and the glycosylation of uracil to produce the "J-base" in kinetoplastids.

Damage

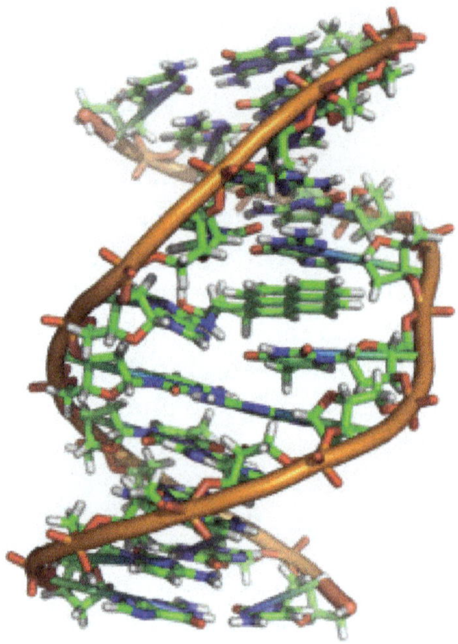

A covalent adduct between a metabolically activated form of benzo[*a*]pyrene, the major mutagen in tobacco smoke, and DNA

DNA can be damaged by many sorts of mutagens, which change the DNA sequence. Mutagens include oxidizing agents, alkylating agents and also high-energy electromagnetic radiation such as ultraviolet light and X-rays. The type of DNA damage produced depends on the type of mutagen. For example, UV light can damage DNA by producing thymine dimers, which are cross-links between pyrimidine bases. On the other hand, oxidants such as free radicals or hydrogen peroxide produce multiple forms of damage, including base modifications, particularly of guanosine, and double-strand breaks. A typical human cell contains about 150,000 bases that have suffered oxidative damage. Of these oxidative lesions, the most dangerous are double-strand breaks, as these are difficult to repair and can produce point mutations, insertions, deletions from the DNA sequence, and chromosomal translocations. These mutations can cause cancer. Because of inherent limits in the DNA repair mechanisms, if humans lived long enough, they would all eventually develop cancer. DNA damages that are naturally occurring, due to normal cellular processes that produce reactive oxygen species, the hydrolytic activities of cellular water, etc., also occur frequently. Although most of these damages are repaired, in any cell some DNA damage may remain despite the action of repair processes. These remaining DNA damages accumulate with age in mammalian postmitotic tissues. This accumulation appears to be an important underlying cause of aging.

Many mutagens fit into the space between two adjacent base pairs, this is called *intercalation*. Most intercalators are aromatic and planar molecules; examples include ethidium bromide, acridines, daunomycin, and doxorubicin. For an intercalator to fit between base pairs, the bases must separate, distorting the DNA strands by unwinding of the double helix. This inhibits both transcription and DNA replication, causing toxicity and mutations. As a result, DNA intercalators may be carcinogens, and in the case of thalidomide, a teratogen. Others such as benzo[*a*]pyrene diol epoxide and aflatoxin form DNA adducts that induce errors in replication. Nevertheless, due to their ability to inhibit DNA transcription and replication, other similar toxins are also used in chemotherapy to inhibit rapidly growing cancer cells.

Biological Functions

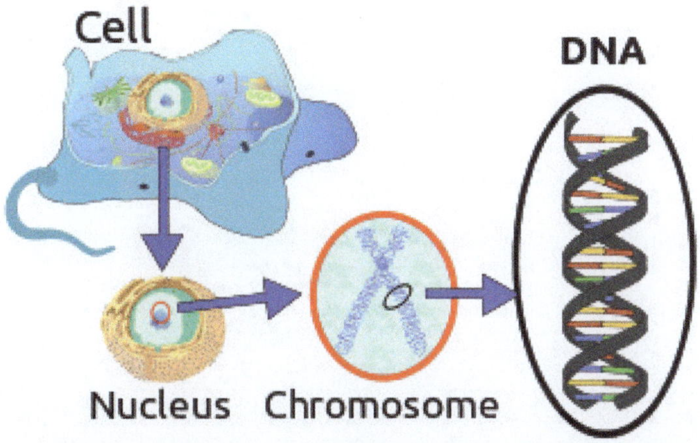

Location of eukaryote nuclear DNA within the chromosomes.

DNA usually occurs as linear chromosomes in eukaryotes, and circular chromosomes in prokaryotes. The set of chromosomes in a cell makes up its genome; the human genome has approximately 3 billion base pairs of DNA arranged into 46 chromosomes. The information carried by DNA is held in the sequence of pieces of DNA called genes. Transmission of genetic information in genes is achieved via complementary base pairing. For example, in transcription, when a cell uses the information in a gene, the DNA sequence is copied into a complementary RNA sequence through the attraction between the DNA and the correct RNA nucleotides. Usually, this RNA copy is then used to make a matching protein sequence in a process called translation, which depends on the same interaction between RNA nucleotides. In alternative fashion, a cell may simply copy its genetic information in a process called DNA replication. The details of these functions are covered in other articles; here the focus is on the interactions between DNA and other molecules that mediate the function of the genome.

Genes and Genomes

Genomic DNA is tightly and orderly packed in the process called DNA condensation, to fit the small available volumes of the cell. In eukaryotes, DNA is located in the cell nucleus, with small amounts in mitochondria and chloroplasts. In prokaryotes, the DNA is held within an irregularly shaped body in the cytoplasm called the nucleoid. The genetic information in a genome is held within genes, and the complete set of this information in an organism is called its genotype. A gene is a unit of heredity and is a region of DNA that influences a particular characteristic in an organism. Genes contain an open reading frame that can be transcribed, and regulatory sequences such as promoters and enhancers, which control transcription of the open reading frame.

In many species, only a small fraction of the total sequence of the genome encodes protein. For example, only about 1.5% of the human genome consists of protein-coding exons, with over 50% of human DNA consisting of non-coding repetitive sequences. The reasons for the presence of so much noncoding DNA in eukaryotic genomes and the extraordinary differences in genome size, or *C-value*, among species, represent a long-standing puzzle known as the "C-value enigma". However, some DNA sequences that do not code protein may still encode functional non-coding RNA molecules, which are involved in the regulation of gene expression.

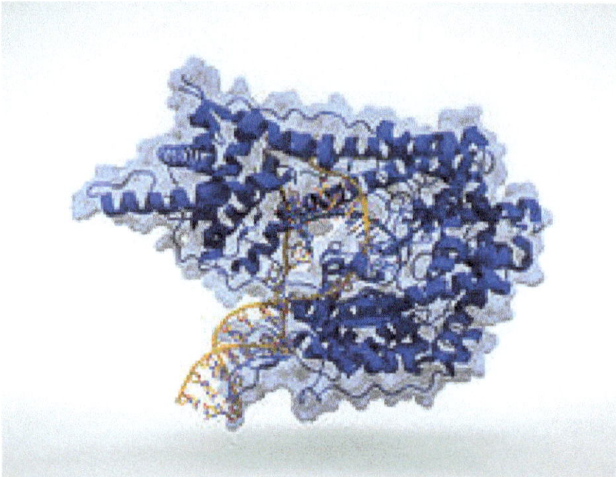

T7 RNA polymerase (blue) producing an mRNA (green) from a DNA template (orange).

Some noncoding DNA sequences play structural roles in chromosomes. Telomeres and centromeres typically contain few genes but are important for the function and stability of chromosomes. An abundant form of noncoding DNA in humans are pseudogenes, which are copies of genes that have been disabled by mutation. These sequences are usually just molecular fossils, although they can occasionally serve as raw genetic material for the creation of new genes through the process of gene duplication and divergence.

Transcription and Translation

A gene is a sequence of DNA that contains genetic information and can influence the phenotype of an organism. Within a gene, the sequence of bases along a DNA strand defines a messenger RNA sequence, which then defines one or more protein sequences. The relationship between the nucleotide sequences of genes and the amino-acid sequences of proteins is determined by the rules of translation, known collectively as the genetic code. The genetic code consists of three-letter 'words' called *codons* formed from a sequence of three nucleotides (e.g. ACT, CAG, TTT).

In transcription, the codons of a gene are copied into messenger RNA by RNA polymerase. This RNA copy is then decoded by a ribosome that reads the RNA sequence by base-pairing the messenger RNA to transfer RNA, which carries amino acids. Since there are 4 bases in 3-letter combinations, there are 64 possible codons (4^3 combinations). These encode the twenty standard amino acids, giving most amino acids more than one possible codon. There are also three 'stop' or 'nonsense' codons signifying the end of the coding region; these are the TAA, TGA, and TAG codons.

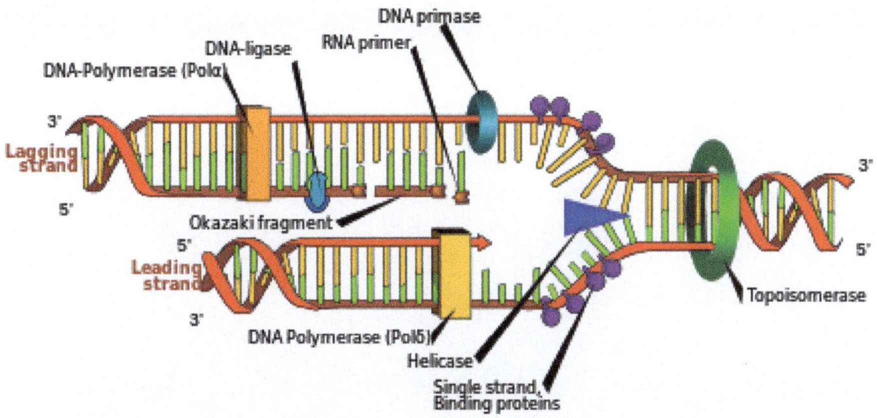

DNA replication. The double helix is unwound by a helicase and topoisomerase. Next, one DNA polymerase produces the leading strand copy. Another DNA polymerase binds to the lagging strand. This enzyme makes discontinuous segments (called Okazaki fragments) before DNA ligase joins them together.

Replication

Cell division is essential for an organism to grow, but, when a cell divides, it must replicate the DNA in its genome so that the two daughter cells have the same genetic information as their parent. The double-stranded structure of DNA provides a simple mechanism for DNA replication. Here, the two strands are separated and then each strand's complementary DNA sequence is recreated by an enzyme called DNA polymerase. This enzyme makes the complementary strand by finding the correct base through complementary base pairing and bonding it onto the original strand. As DNA polymerases can only extend a DNA strand in a 5′ to 3′ direction, different mechanisms are used to copy the antiparallel strands of the double helix. In this way, the base on the old strand dictates which base appears on the new strand, and the cell ends up with a perfect copy of its DNA.

Extracellular Nucleic Acids

Naked extracellular DNA (eDNA), most of it released by cell death, is nearly ubiquitous in the environment. Its concentration in soil may be as high as 2 μg/L, and its concentration in natural aquatic environments may be as high at 88 μg/L. Various possible functions have been proposed for eDNA: it may be involved in horizontal gene transfer; it may provide nutrients; and it may act as a buffer to recruit or titrate ions or antibiotics. Extracellular DNA acts as a functional extracellular matrix component in the biofilms of several bacterial species. It may act as a recognition factor to regulate the attachment and dispersal of specific cell types in the biofilm; it may contribute to biofilm formation; and it may contribute to the biofilm's physical strength and resistance to biological stress.

Cell-free fetal DNA is found in the blood of the mother, and can be sequenced to determine a great deal of information about the developing fetus.

Interactions with Proteins

All the functions of DNA depend on interactions with proteins. These protein interactions can be non-specific, or the protein can bind specifically to a single DNA sequence. Enzymes can also bind to DNA and of these, the polymerases that copy the DNA base sequence in transcription and DNA replication are particularly important.

DNA-binding Proteins

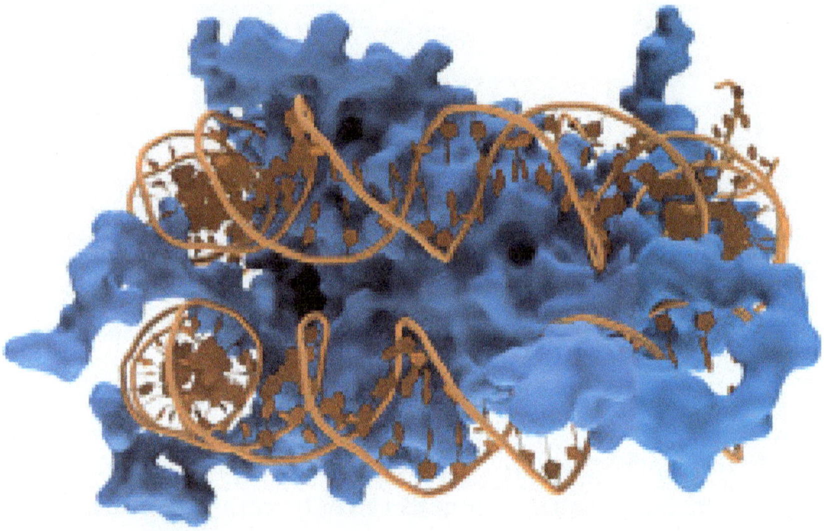

Interaction of DNA (in orange) with histones (in blue). These proteins' basic amino acids bind to the acidic phosphate groups on DNA.

Structural proteins that bind DNA are well-understood examples of non-specific DNA-protein interactions. Within chromosomes, DNA is held in complexes with structural proteins. These proteins organize the DNA into a compact structure called chromatin. In eukaryotes, this structure involves DNA binding to a complex of small basic proteins called histones, while in prokaryotes multiple types of proteins are involved. The histones form a disk-shaped complex called a nucleosome, which contains two complete turns of double-stranded DNA wrapped around its surface. These non-specific interactions are formed through basic residues in the histones, making ionic bonds to the acidic sugar-phosphate backbone of the DNA, and are thus largely independent of the base sequence. Chemical modifications of these basic amino acid residues include methylation, phosphorylation, and acetylation. These chemical changes alter the strength of the interaction between the DNA and the histones, making the DNA more or less accessible to transcription factors and changing the rate of transcription. Other non-specific DNA-binding proteins in chromatin include the high-mobility group proteins, which bind to bent or distorted DNA. These proteins are important in bending arrays of nucleosomes and arranging them into the larger structures that make up chromosomes.

A distinct group of DNA-binding proteins is the DNA-binding proteins that specifically bind single-stranded DNA. In humans, replication protein A is the best-understood member of this family and is used in processes where the double helix is separated, including DNA replication, recombination, and DNA repair. These binding proteins seem to stabilize single-stranded DNA and protect it from forming stem-loops or being degraded by nucleases.

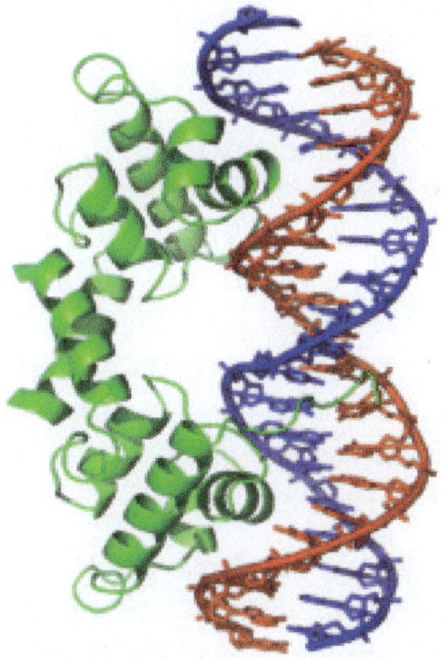

The lambda repressor helix-turn-helix transcription factor bound to its DNA target

In contrast, other proteins have evolved to bind to particular DNA sequences. The most intensively studied of these are the various transcription factors, which are proteins that regulate transcription. Each transcription factor binds to one particular set of DNA sequences and activates or inhibits the transcription of genes that have these sequences close to their promoters. The transcription factors do this in two ways. Firstly, they can bind the RNA polymerase responsible for transcription, either directly or through other mediator proteins; this locates the polymerase at the promoter and allows it to begin transcription. Alternatively, transcription factors can bind enzymes that modify the histones at the promoter. This changes the accessibility of the DNA template to the polymerase.

As these DNA targets can occur throughout an organism's genome, changes in the activity of one type of transcription factor can affect thousands of genes. Consequently, these proteins are often the targets of the signal transduction processes that control responses to environmental changes or cellular differentiation and development. The specificity of these transcription factors' interactions with DNA come from the proteins making multiple contacts to the edges of the DNA bases, allowing them to "read" the DNA sequence. Most of these base-interactions are made in the major groove, where the bases are most accessible.

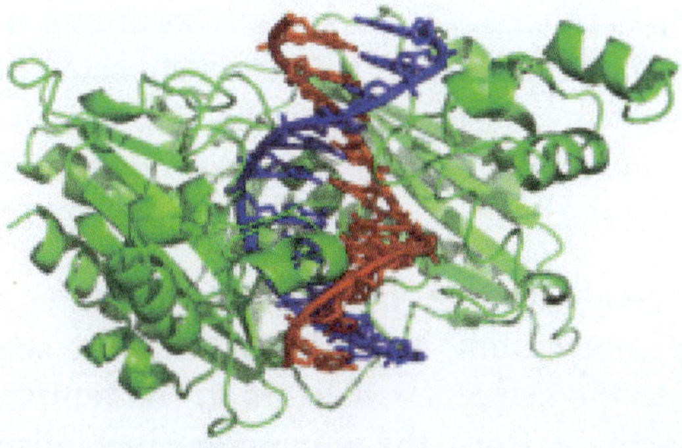

The restriction enzyme EcoRV (green) in a complex with its substrate DNA

DNA-modifying Enzymes

Nucleases and Ligases

Nucleases are enzymes that cut DNA strands by catalyzing the hydrolysis of the phosphodiester bonds. Nucleases that hydrolyse nucleotides from the ends of DNA strands are called exonucleases, while endonucleases cut within strands. The most frequently used nucleases in molecular biology are the restriction endonucleases, which cut DNA at specific sequences. For instance, the EcoRV enzyme shown to the left recognizes the 6-base sequence 5′-GATATC-3′ and makes a cut at the horizontal line. In nature, these enzymes protect bacteria against phage infection by digesting the phage DNA when it enters the bacterial cell, acting as part of the restriction modification system. In technology, these sequence-specific nucleases are used in molecular cloning and DNA fingerprinting.

Enzymes called DNA ligases can rejoin cut or broken DNA strands. Ligases are particularly important in lagging strand DNA replication, as they join together the short segments of DNA produced at the replication fork into a complete copy of the DNA template. They are also used in DNA repair and genetic recombination.

Topoisomerases and Helicases

Topoisomerases are enzymes with both nuclease and ligase activity. These proteins change the amount of supercoiling in DNA. Some of these enzymes work by cutting the DNA helix and allowing one section to rotate, thereby reducing its level of supercoiling; the enzyme then seals the DNA break. Other types of these enzymes are capable of cutting one DNA helix and then passing a second strand of DNA through this break, before rejoining the helix. Topoisomerases are required for many processes involving DNA, such as DNA replication and transcription.

Helicases are proteins that are a type of molecular motor. They use the chemical energy in nucleoside triphosphates, predominantly adenosine triphosphate (ATP), to break hydrogen bonds between bases and unwind the DNA double helix into single strands. These enzymes are essential for most processes where enzymes need to access the DNA bases.

Polymerases

Polymerases are enzymes that synthesize polynucleotide chains from nucleoside triphosphates. The sequence of their products is created based on existing polynucleotide chains—which are called *templates*. These enzymes function by repeatedly adding a nucleotide to the 3′ hydroxyl group at the end of the growing polynucleotide chain. As a consequence, all polymerases work in a 5′ to 3′ direction. In the active site of these enzymes, the incoming nucleoside triphosphate base-pairs to the template: this allows polymerases to accurately synthesize the complementary strand of their template. Polymerases are classified according to the type of template that they use.

In DNA replication, DNA-dependent DNA polymerases make copies of DNA polynucleotide chains. To preserve biological information, it is essential that the sequence of bases in each copy are precisely complementary to the sequence of bases in the template strand. Many DNA polymerases have a proofreading activity. Here, the polymerase recognizes the occasional mistakes in the synthesis reaction by the lack of base pairing between the mismatched nucleotides. If a mismatch is detected, a 3′ to 5′ exonuclease activity is activated and the incorrect base removed. In most organisms, DNA polymerases function in a large complex called the replisome that contains multiple accessory subunits, such as the DNA clamp or helicases.

RNA-dependent DNA polymerases are a specialized class of polymerases that copy the sequence of an RNA strand into DNA. They include reverse transcriptase, which is a viral enzyme involved in the infection of cells by retroviruses, and telomerase, which is required for the replication of telomeres. Telomerase is an unusual polymerase because it contains its own RNA template as part of its structure.

Transcription is carried out by a DNA-dependent RNA polymerase that copies the sequence of a DNA strand into RNA. To begin transcribing a gene, the RNA polymerase binds to a sequence of DNA called a promoter and separates the DNA strands. It then copies the gene sequence into a messenger RNA transcript until it reaches a region of DNA called the terminator, where it halts and detaches from the DNA. As with human DNA-dependent DNA polymerases, RNA polymerase II, the enzyme that transcribes most of the genes in the human genome, operates as part of a large protein complex with multiple regulatory and accessory subunits.

Genetic Recombination

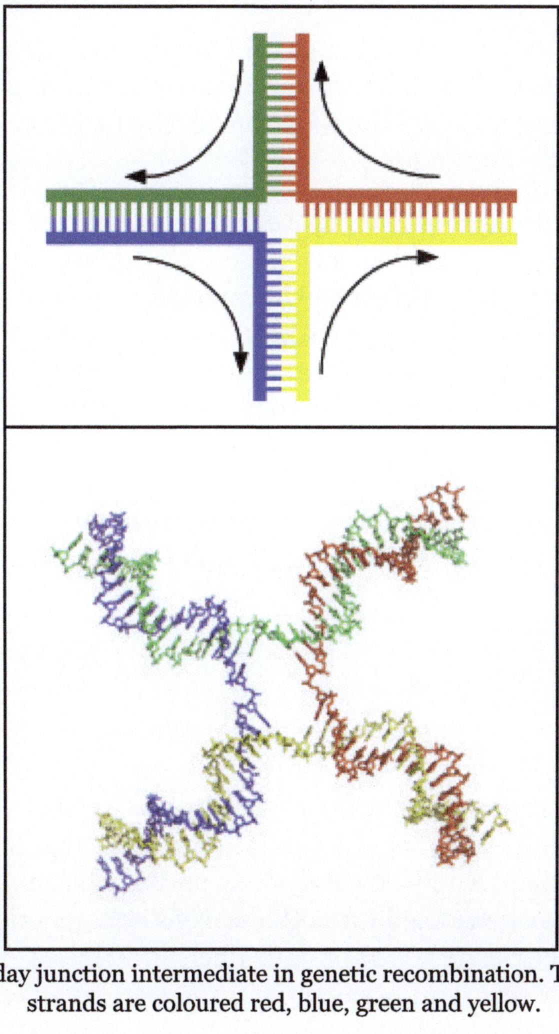

Structure of the Holliday junction intermediate in genetic recombination. The four separate DNA strands are coloured red, blue, green and yellow.

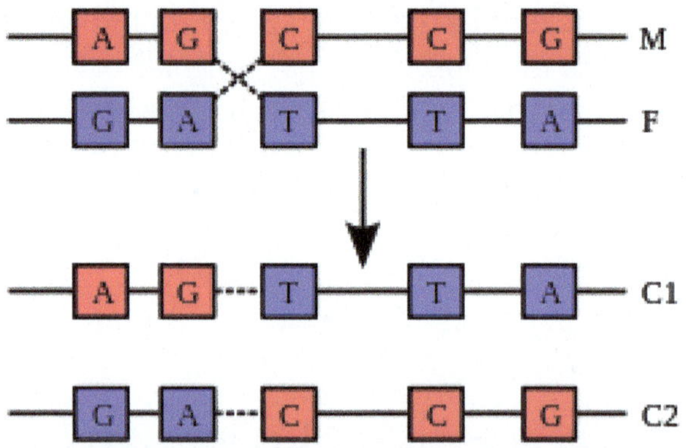

Recombination involves the breaking and rejoining of two chromosomes (M and F) to produce two rearranged chromosomes (C1 and C2).

A DNA helix usually does not interact with other segments of DNA, and in human cells, the different chromosomes even occupy separate areas in the nucleus called "chromosome territories". This physical separation of different chromosomes is important for the ability of DNA to function as a stable repository for information, as one of the few times chromosomes interact is in chromosomal crossover which occurs during sexual reproduction, when genetic recombination occurs. Chromosomal crossover is when two DNA helices break, swap a section and then rejoin.

Recombination allows chromosomes to exchange genetic information and produces new combinations of genes, which increases the efficiency of natural selection and can be important in the rapid evolution of new proteins. Genetic recombination can also be involved in DNA repair, particularly in the cell's response to double-strand breaks.

The most common form of chromosomal crossover is homologous recombination, where the two chromosomes involved share very similar sequences. Non-homologous recombination can be damaging to cells, as it can produce chromosomal translocations and genetic abnormalities. The recombination reaction is catalyzed by enzymes known as recombinases, such as RAD51. The first step in recombination is a double-stranded break caused by either an endonuclease or damage to the DNA. A series of steps catalyzed in part by the recombinase then leads to joining of the two helices by at least one Holliday junction, in which a segment of a single strand in each helix is annealed to the complementary strand in the other helix. The Holliday junction is a tetrahedral junction structure that can be moved along the pair of chromosomes, swapping one strand for another. The recombination reaction is then halted by cleavage of the junction and re-ligation of the released DNA.

Evolution

DNA contains the genetic information that allows all modern living things to function, grow and reproduce. However, it is unclear how long in the 4-billion-year history of life DNA has performed this function, as it has been proposed that the earliest forms of life may have used RNA as their genetic material. RNA may have acted as the central part of early cell metabolism as it can both transmit genetic information and carry out catalysis as part of ribozymes. This ancient RNA world where nucleic acid would have been used for both catalysis and genetics may have influenced the evolution of the current genetic code based on four nucleotide bases. This would occur, since the number of different bases in such an organism is a trade-off between a small number of bases increasing replication accuracy and a large number of bases increasing the catalytic efficiency of ribozymes. However, there is no direct evidence of ancient genetic systems, as recovery of DNA from most fossils is impossible because DNA survives in the environment for less than one million years, and slowly degrades into short fragments in solution. Claims for older DNA have been made, most notably a report of the isolation of a viable bacterium from a salt crystal 250 million years old, but these claims are controversial.

Building blocks of DNA (adenine, guanine, and related organic molecules) may have been formed extraterrestrially in outer space. Complex DNA and RNA organic compounds of life, including uracil, cytosine, and thymine, have also been formed in the laboratory under conditions mimicking those found in outer space, using starting chemicals, such as pyrimidine, found in meteorites. Pyrimidine, like polycyclic aromatic hydrocarbons (PAHs), the most carbon-rich chemical found in the universe, may have been formed in red giants or in interstellar cosmic dust and gas clouds.

Uses in Technology

Genetic Engineering

Methods have been developed to purify DNA from organisms, such as phenol-chloroform extraction, and to manipulate it in the laboratory, such as restriction digests and the polymerase chain reaction. Modern biology and biochemistry make intensive use of these techniques in recombinant DNA technology. Recombinant DNA is a man-made DNA sequence that has been assembled from other DNA sequences. They can be transformed into organisms in the form of plasmids or in the appropriate format, by using a viral vector. The genetically modified organisms produced can be used to produce products such as recombinant proteins, used in medical research, or be grown in agriculture.

DNA Profiling

Forensic scientists can use DNA in blood, semen, skin, saliva or hair found at a crime scene to identify a matching DNA of an individual, such as a perpetrator. This process is formally termed DNA profiling, but may also be called "genetic fingerprinting". In DNA profiling, the lengths of variable sections of repetitive DNA, such as short tandem repeats and minisatellites, are compared between people. This method is usually an extremely reliable technique for identifying a matching DNA. However, identification can be complicated if the scene is contaminated with DNA from several people. DNA profiling was developed in 1984 by British geneticist Sir Alec Jeffreys, and first used in forensic science to convict Colin Pitchfork in the 1988 Enderby murders case.

The development of forensic science and the ability to now obtain genetic matching on minute samples of blood, skin, saliva, or hair has led to re-examining many cases. Evidence can now be uncovered that was scientifically impossible at the time of the original examination. Combined with the removal of the double jeopardy law in some places, this can allow cases to be reopened where prior trials have failed to produce sufficient evidence to convince a jury. People charged with serious crimes may be required to provide a sample of DNA for matching purposes. The most obvious defense to DNA matches obtained forensically is to claim that cross-contamination of evidence has occurred. This has resulted in meticulous strict handling procedures with new cases of serious crime. DNA profiling is also used successfully to positively identify victims of mass casualty incidents, bodies or body parts in serious accidents, and individual victims in mass war graves, via matching to family members.

DNA profiling is also used in DNA paternity testing to determine if someone is the biological parent or grandparent of a child with the probability of parentage is typically 99.99% when the alleged parent is biologically related to the child. Normal DNA sequencing methods happen after birth, but there are new methods to test paternity while a mother is still pregnant.

DNA Enzymes or Catalytic DNA

Deoxyribozymes, also called DNAzymes or catalytic DNA, are first discovered in 1994. They are mostly single stranded DNA sequences isolated from a large pool of random DNA sequences through a combinatorial approach called in vitro selection or systematic evolution of ligands by exponential enrichment (SELEX). DNAzymes catalyze variety of chemical reactions including RNA-DNA cleavage, RNA-DNA ligation, amino acids phosphorylation-dephosphorylation, carbon-carbon bond formation, and etc. DNAzymes can enhance catalytic rate of chemical reactions up to 100,000,000,000-fold over the uncatalyzed reaction. The most extensively studied class of DNAzymes is RNA-cleaving types which have been used to detect different metal ions and designing therapeutic agents. Several metal-specific DNAzymes have been reported including the GR-5 DNAzyme (lead-specific), the CA1-3 DNAzymes (copper-specific), the 39E DNAzyme (uranyl-specific) and the NaA43 DNAzyme (sodium-specific). The NaA43 DNAzyme, which is reported to be more than 10,000-fold selective for sodium over other metal ions, was used to make a real-time sodium sensor in living cells.

Bioinformatics

Bioinformatics involves the development of techniques to store, data mine, search and manipulate biological data, including DNA nucleic acid sequence data. These have led to widely applied advances in computer science, especially string searching algorithms, machine learning, and database theory. String searching or matching algorithms, which find an occurrence of a sequence of letters inside a larger sequence of letters, were developed to search for specific sequences of nucleotides. The DNA sequence may be aligned with other DNA sequences to identify homologous sequences and locate the specific mutations that make them distinct. These techniques, especially multiple sequence alignment, are used in studying phylogenetic relationships and protein function. Data sets representing entire genomes' worth of DNA sequences, such as those produced by the Human Genome Project, are difficult to use without the annotations that identify the locations of genes and regulatory elements on each chromosome. Regions of DNA sequence that have the characteristic patterns associated with protein- or RNA-coding genes can be identified by gene finding algorithms, which allow researchers to predict the presence of particular gene products and their possible functions in an organism even before they have been isolated experimentally. Entire genomes may also be compared, which can shed light on the evolutionary history of particular organism and permit the examination of complex evolutionary events.

DNA Nanotechnology

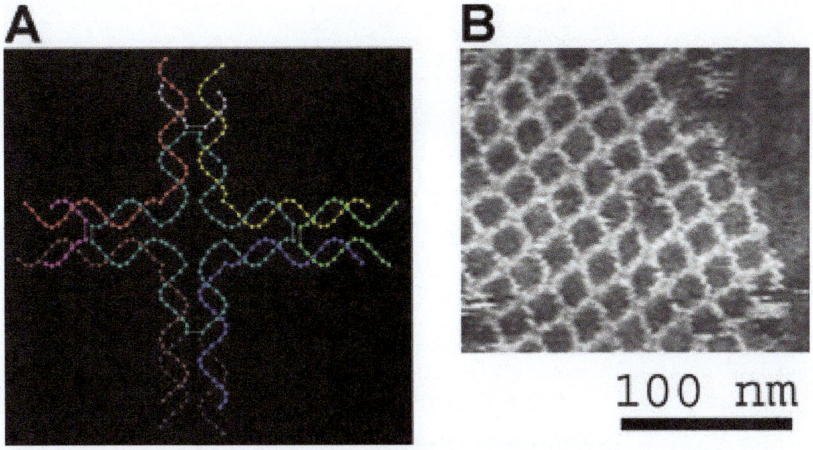

The DNA structure at left (schematic shown) will self-assemble into the structure visualized by atomic force microscopy at right. DNA nanotechnology is the field that seeks to design nanoscale structures using the molecular recognition properties of DNA molecules. Image from Strong, 2004.

DNA nanotechnology uses the unique molecular recognition properties of DNA and other nucleic acids to create self-assembling branched DNA complexes with useful properties. DNA is thus used as a structural material rather than as a carrier of biological information. This has led to the creation of two-dimensional periodic lattices (both tile-based and using the *DNA origami* method) and three-dimensional structures in the shapes of polyhedra. Nanomechanical devices and algorithmic self-assembly have also been demonstrated, and these DNA structures have been used to template the arrangement of other molecules such as gold nanoparticles and streptavidin proteins.

History and Anthropology

Because DNA collects mutations over time, which are then inherited, it contains historical information, and, by comparing DNA sequences, geneticists can infer the evolutionary history of organisms, their phylogeny. This field of phylogenetics is a powerful tool in evolutionary biology. If DNA sequences within a species are compared, population geneticists can learn the history of particular populations. This can be used in studies ranging from ecological genetics to anthropology; For example, DNA evidence is being used to try to identify the Ten Lost Tribes of Israel.

Information Storage

In a paper published in *Nature* in January 2013, scientists from the European Bioinformatics Institute and Agilent Technologies proposed a mechanism to use DNA's ability to code information as a means of digital data storage. The group was able to encode 739 kilobytes of data into DNA code, synthesize the actual DNA, then sequence the DNA and decode the information back to its original form, with a reported 100% accuracy. The encoded information consisted of text files and audio files. A prior experiment was published in August 2012. It was conducted by researchers at Harvard University, where the text of a 54,000-word book was encoded in DNA.

Moreover, in living cells, the storage can be turned active by enzymes. Light-gated protein domains fused to DNA processing enzymes are suitable for that task *in vitro*. Fluorescent exonucleases can transmit the output according to the nucleotide they have read.

History of DNA Research

James Watson and Francis Crick (right), co-originators of the double-helix model, with Maclyn McCarty (left).

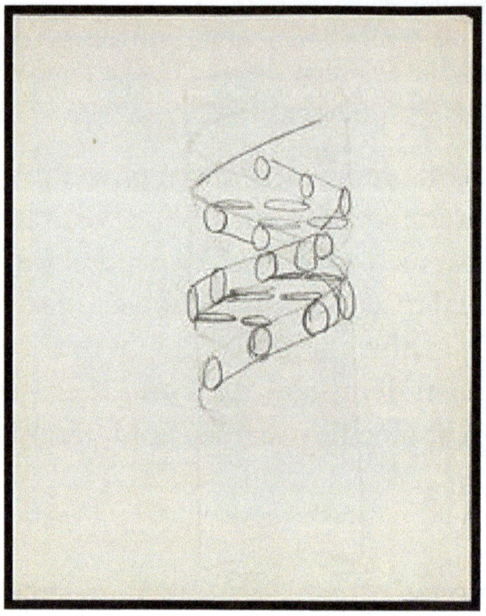

Pencil sketch of the DNA double helix by Francis Crick in 1953

DNA was first isolated by the Swiss physician Friedrich Miescher who, in 1869, discovered a microscopic substance in the pus of discarded surgical bandages. As it resided in the nuclei of cells, he called it "nuclein". In 1878, Albrecht Kossel isolated the non-protein component of "nuclein", nucleic acid, and later isolated its five primary nucleobases. In 1919, Phoebus Levene identified the base, sugar, and phosphate nucleotide unit. Levene suggested that DNA consisted of a string of nucleotide units linked together through the phosphate groups. Levene thought the chain was short and the bases repeated in a fixed order. In 1937, William Astbury produced the first X-ray diffraction patterns that showed that DNA had a regular structure.

In 1927, Nikolai Koltsov proposed that inherited traits would be inherited via a "giant hereditary molecule" made up of "two mirror strands that would replicate in a semi-conservative fashion using each strand as a template". In 1928, Frederick Griffith in his experiment discovered that traits of the "smooth" form of *Pneumococcus* could be transferred to the "rough" form of the same bacteria by mixing killed "smooth" bacteria with the live "rough" form. This system provided the first clear suggestion that DNA carries genetic information—the Avery–MacLeod–McCarty experiment—when Oswald Avery, along with coworkers Colin MacLeod and Maclyn McCarty, identified DNA as the transforming principle in 1943. DNA's role in heredity was confirmed in 1952 when

Alfred Hershey and Martha Chase in the Hershey–Chase experiment showed that DNA is the genetic material of the T2 phage.

In 1953, James Watson and Francis Crick suggested what is now accepted as the first correct double-helix model of DNA structure in the journal *Nature*. Their double-helix, molecular model of DNA was then based on one X-ray diffraction image taken by Rosalind Franklin and Raymond Gosling, and the information that the DNA bases are paired.

Experimental evidence supporting the Watson and Crick model was published in a series of five articles in the same issue of *Nature*. Of these, Franklin and Gosling's paper was the first publication of their own X-ray diffraction data and original analysis method that partly supported the Watson and Crick model; this issue also contained an article on DNA structure by Maurice Wilkins and two of his colleagues, whose analysis and *in vivo* B-DNA X-ray patterns also supported the presence *in vivo* of the double-helical DNA configurations as proposed by Crick and Watson for their double-helix molecular model of DNA in the prior two pages of *Nature*. In 1962, after Franklin's death, Watson, Crick, and Wilkins jointly received the Nobel Prize in Physiology or Medicine. Nobel Prizes are awarded only to living recipients. A debate continues about who should receive credit for the discovery.

In an influential presentation in 1957, Crick laid out the central dogma of molecular biology, which foretold the relationship between DNA, RNA, and proteins, and articulated the "adaptor hypothesis". Final confirmation of the replication mechanism that was implied by the double-helical structure followed in 1958 through the Meselson–Stahl experiment. Further work by Crick and coworkers showed that the genetic code was based on non-overlapping triplets of bases, called codons, allowing Har Gobind Khorana, Robert W. Holley, and Marshall Warren Nirenberg to decipher the genetic code. These findings represent the birth of molecular biology.

RNA

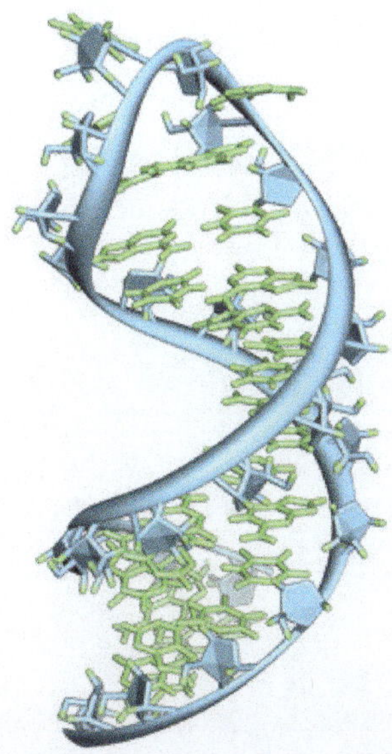

A hairpin loop from a pre-mRNA. Highlighted are the nucleobases (green) and the ribose-phosphate backbone (blue). Note that this is a single strand of RNA that folds back upon itself.

Ribonucleic acid (RNA) is a polymeric molecule essential in various biological roles in coding, decoding, regulation, and expression of genes. RNA and DNA are nucleic acids, and, along with proteins and carbohydrates, constitute the four major macromolecules essential for all known forms of life. Like DNA, RNA is assembled as a chain of nucleotides, but unlike DNA it is more often found in nature as a single-strand folded onto itself, rather than a paired double-strand. Cellular organisms use messenger RNA (*mRNA*) to convey genetic information (using the letters G, U, A, and C to denote the nitrogenous bases guanine, uracil, adenine, and cytosine) that directs synthesis of specific proteins. Many viruses encode their genetic information using an RNA genome.

Some RNA molecules play an active role within cells by catalyzing biological reactions, controlling gene expression, or sensing and communicating responses to cellular signals. One of these active processes is protein synthesis, a universal function where RNA molecules direct the assembly of proteins on ribosomes. This process uses transfer RNA (*tRNA*) molecules to deliver amino acids to the ribosome, where ribosomal RNA (*rRNA*) then links amino acids together to form proteins.

Comparison with DNA

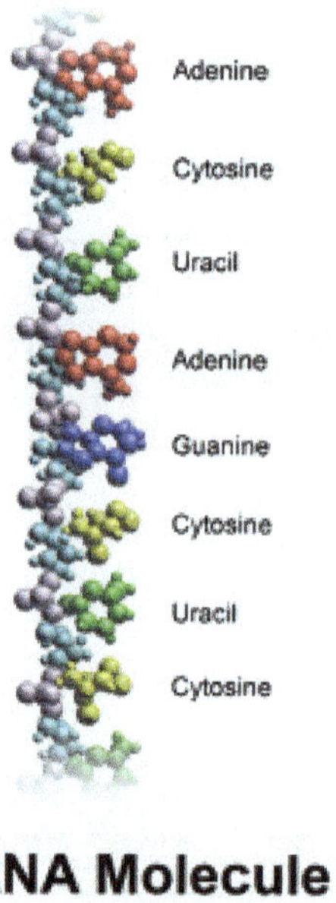

Bases in an RNA molecule.

Three-dimensional representation of the 50S ribosomal subunit. Ribosomal RNA is in ochre, proteins in blue. The active site is a small segment of rRNA, indicated in red.

The chemical structure of RNA is very similar to that of DNA, but differs in three main ways:

- Unlike double-stranded DNA, RNA is a single-stranded molecule in many of its biological roles and has a much shorter chain of nucleotides. However, RNA can, by complementary base pairing, form intrastrand (i.e., single-strand) double helixes, as in tRNA.
- While DNA contains *deoxyribose*, RNA contains *ribose* (in deoxyribose there is no hydroxyl group attached to the pentose ring in the 2' position). These hydroxyl groups make RNA less stable than DNA because it is more prone to hydrolysis.
- The complementary base to adenine in DNA is thymine, whereas in RNA, it is uracil, which is an unmethylated form of thymine.

Like DNA, most biologically active RNAs, including mRNA, tRNA, rRNA, snRNAs, and other non-coding RNAs, contain self-complementary sequences that allow parts of the RNA to fold and pair with itself to form double helices. Analysis of these RNAs has revealed that they are highly structured. Unlike DNA, their structures do not consist of long double helices, but rather collections of short helices packed together into structures akin to proteins. In this fashion, RNAs can achieve chemical catalysis (like enzymes). For instance, determination of the structure of the ribosome—an enzyme that catalyzes peptide bond formation—revealed that its active site is composed entirely of RNA.

Structure

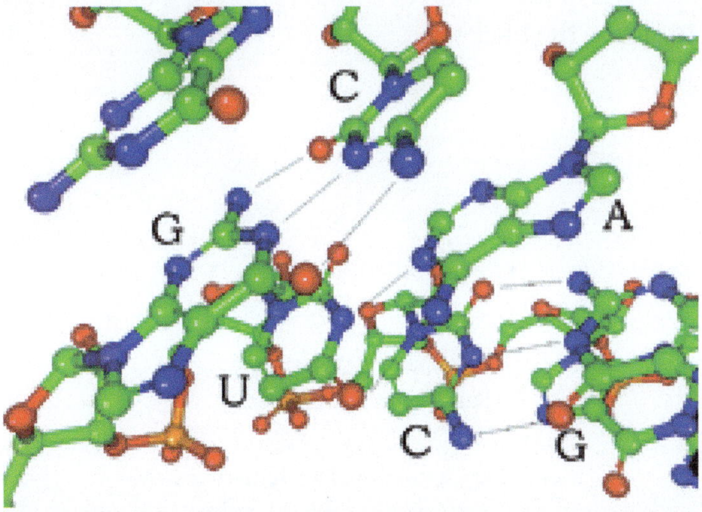

Watson-Crick base pairs in a siRNA (hydrogen atoms are not shown)

Each nucleotide in RNA contains a ribose sugar, with carbons numbered 1' through 5'. A base is attached to the 1' position, in general, adenine (A), cytosine (C), guanine (G), or uracil (U). Adenine and guanine are purines, cytosine and uracil are pyrimidines. A phosphate group is attached to the 3' position of one ribose and the 5' position of the next. The phosphate groups have a negative charge each, making RNA a charged molecule (polyanion). The bases form hydrogen bonds between cytosine and guanine, between adenine and uracil and between guanine and uracil. However, other interactions are possible, such as a group of adenine bases binding to each other in a bulge, or the GNRA tetraloop that has a guanine–adenine base-pair.

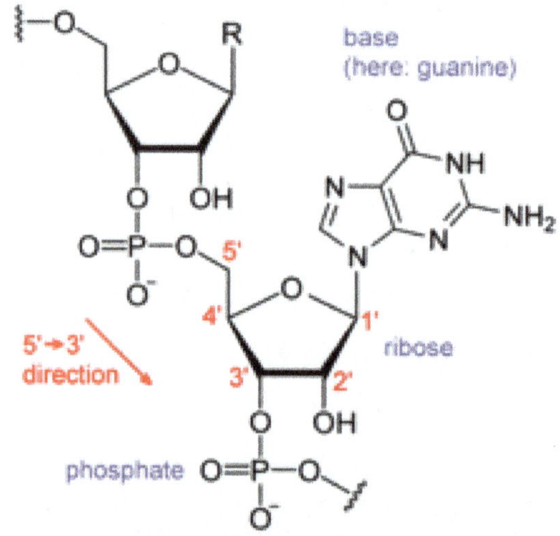

Chemical structure of RNA

An important structural feature of RNA that distinguishes it from DNA is the presence of a hydroxyl group at the 2' position of the ribose sugar. The presence of this functional group causes the helix to mostly adopt the A-form geometry, although in single strand dinucleotide contexts, RNA can rarely also adopt the B-form most commonly observed in DNA. The A-form geometry results in a very deep and narrow major groove and a shallow and wide minor groove. A second consequence of the presence of the 2'-hydroxyl group is that in conformationally flexible regions of an RNA molecule (that is, not involved in formation of a double helix), it can chemically attack the adjacent phosphodiester bond to cleave the backbone.

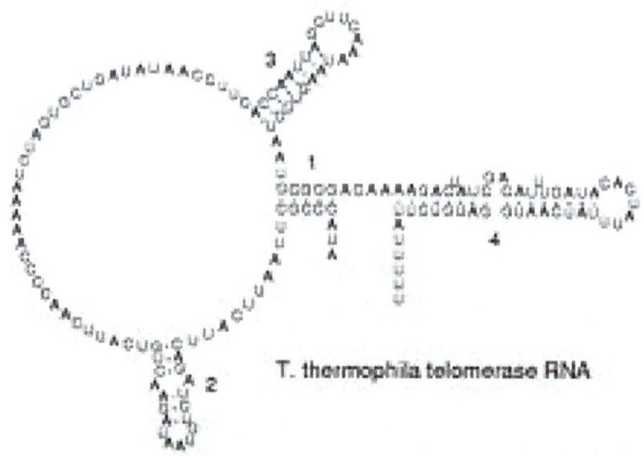

Secondary structure of a telomerase RNA.

RNA is transcribed with only four bases (adenine, cytosine, guanine and uracil), but these bases and attached sugars can be modified in numerous ways as the RNAs mature. Pseudouridine (Ψ), in which the linkage between uracil and ribose is changed from a C–N bond to a C–C bond, and ribothymidine (T) are found in various places (the most notable ones being in the TΨC loop of tRNA). Another notable modified base is hypoxanthine, a deaminated adenine base whose nucleoside is called inosine (I). Inosine plays a key role in the wobble hypothesis of the genetic code.

There are more than 100 other naturally occurring modified nucleosides, The greatest structural diversity of modifications can be found in tRNA, while pseudouridine and nucleosides with

2'-O-methylribose often present in rRNA are the most common. The specific roles of many of these modifications in RNA are not fully understood. However, it is notable that, in ribosomal RNA, many of the post-transcriptional modifications occur in highly functional regions, such as the peptidyl transferase center and the subunit interface, implying that they are important for normal function.

The functional form of single-stranded RNA molecules, just like proteins, frequently requires a specific tertiary structure. The scaffold for this structure is provided by secondary structural elements that are hydrogen bonds within the molecule. This leads to several recognizable "domains" of secondary structure like hairpin loops, bulges, and internal loops. Since RNA is charged, metal ions such as Mg^{2+} are needed to stabilise many secondary and tertiary structures.

The naturally occurring enantiomer of RNA is D-RNA composed of D-ribonucleotides. All chirality centers are located in the D-ribose. By the use of L-ribose or rather L-ribonucleotides, L-RNA can be synthesized. L-RNA is much more stable against degradation by RNase.

Like other structured biopolymers such as proteins, one can define topology of a folded RNA molecule. This is often done based on arrangement of intra-chain contacts within a folded RNA, termed as circuit topology.

Synthesis

Synthesis of RNA is usually catalyzed by an enzyme—RNA polymerase—using DNA as a template, a process known as transcription. Initiation of transcription begins with the binding of the enzyme to a promoter sequence in the DNA (usually found "upstream" of a gene). The DNA double helix is unwound by the helicase activity of the enzyme. The enzyme then progresses along the template strand in the 3' to 5' direction, synthesizing a complementary RNA molecule with elongation occurring in the 5' to 3' direction. The DNA sequence also dictates where termination of RNA synthesis will occur.

Primary transcript RNAs are often modified by enzymes after transcription. For example, a poly(A) tail and a 5' cap are added to eukaryotic pre-mRNA and introns are removed by the spliceosome.

There are also a number of RNA-dependent RNA polymerases that use RNA as their template for synthesis of a new strand of RNA. For instance, a number of RNA viruses (such as poliovirus) use this type of enzyme to replicate their genetic material. Also, RNA-dependent RNA polymerase is part of the RNA interference pathway in many organisms.

Types of RNA

Overview

Messenger RNA (mRNA) is the RNA that carries information from DNA to the ribosome, the sites of protein synthesis (translation) in the cell. The coding sequence of the mRNA determines the amino acid sequence in the protein that is produced. However, many RNAs do not code for protein (about 97% of the transcriptional output is non-protein-coding in eukaryotes).

Structure of a hammerhead ribozyme, a ribozyme that cuts RNA

These so-called non-coding RNAs ("ncRNA") can be encoded by their own genes (RNA genes), but can also derive from mRNA introns. The most prominent examples of non-coding RNAs are transfer RNA (tRNA) and ribosomal RNA (rRNA), both of which are involved in the process of translation. There are also non-coding RNAs involved in gene regulation, RNA processing and other roles. Certain RNAs are able to catalyse chemical reactions such as cutting and ligating other RNA molecules, and the catalysis of peptide bond formation in the ribosome; these are known as ribozymes.

In Length

According to the length of RNA chain, RNA includes small RNA and long RNA. Usually, small RNAs are shorter than 200 nt in length, and long RNAs are greater than 200 nt long. Long RNAs, also called large RNAs, mainly include long non-coding RNA (lncRNA) and mRNA. Small RNAs mainly include 5.8S ribosomal RNA (rRNA), 5S rRNA, transfer RNA (tRNA), microRNA (miRNA), small interfering RNA (siRNA), small nucleolar RNA (snoRNAs), Piwi-interacting RNA (piRNA), tRNA-derived small RNA (tsRNA) and small rDNA-derived RNA (srRNA).

In Translation

Messenger RNA (mRNA) carries information about a protein sequence to the ribosomes, the protein synthesis factories in the cell. It is coded so that every three nucleotides (a codon) corresponds to one amino acid. In eukaryotic cells, once precursor mRNA (pre-mRNA) has been transcribed from DNA, it is processed to mature mRNA. This removes its introns—non-coding sections of the pre-mRNA. The mRNA is then exported from the nucleus to the cytoplasm, where it is bound to ribosomes and translated into its corresponding protein form with the help of tRNA. In prokaryotic cells, which do not have nucleus and cytoplasm compartments, mRNA can bind to ribosomes while it is being transcribed from DNA. After a certain amount of time the message degrades into its component nucleotides with the assistance of ribonucleases.

Transfer RNA (tRNA) is a small RNA chain of about 80 nucleotides that transfers a specific amino acid to a growing polypeptide chain at the ribosomal site of protein synthesis during translation. It has sites for amino acid attachment and an anticodon region for codon recognition that binds to a specific sequence on the messenger RNA chain through hydrogen bonding.

Ribosomal RNA (rRNA) is the catalytic component of the ribosomes. Eukaryotic ribosomes contain four different rRNA molecules: 18S, 5.8S, 28S and 5S rRNA. Three of the rRNA molecules are synthesized in the nucleolus, and one is synthesized elsewhere. In the cytoplasm, ribosomal RNA and protein combine to form a nucleoprotein called a ribosome. The ribosome binds mRNA and carries out protein synthesis. Several ribosomes may be attached to a single mRNA at any time. Nearly all the RNA found in a typical eukaryotic cell is rRNA.

Transfer-messenger RNA (tmRNA) is found in many bacteria and plastids. It tags proteins encoded by mRNAs that lack stop codons for degradation and prevents the ribosome from stalling.

Regulatory RNAs

Several types of RNA can downregulate gene expression by being complementary to a part of an mRNA or a gene's DNA. MicroRNAs (miRNA; 21-22 nt) are found in eukaryotes and act through RNA interference (RNAi), where an effector complex of miRNA and enzymes can cleave complementary mRNA, block the mRNA from being translated, or accelerate its degradation.

While small interfering RNAs (siRNA; 20-25 nt) are often produced by breakdown of viral RNA, there are also endogenous sources of siRNAs. siRNAs act through RNA interference in a fashion similar to miRNAs. Some miRNAs and siRNAs can cause genes they target to be methylated, thereby decreasing or increasing transcription of those genes. Animals have Piwi-interacting RNAs (piRNA; 29-30 nt) that are active in germline cells and are thought to be a defense against transposons and play a role in gametogenesis.

Many prokaryotes have CRISPR RNAs, a regulatory system similar to RNA interference. Antisense RNAs are widespread; most downregulate a gene, but a few are activators of transcription. One way antisense RNA can act is by binding to an mRNA, forming double-stranded RNA that is enzymatically degraded. There are many long noncoding RNAs that regulate genes in eukaryotes, one such RNA is Xist, which coats one X chromosome in female mammals and inactivates it.

An mRNA may contain regulatory elements itself, such as riboswitches, in the 5' untranslated region or 3' untranslated region; these cis-regulatory elements regulate the activity of that mRNA. The untranslated regions can also contain elements that regulate other genes.

In RNA Processing

Many RNAs are involved in modifying other RNAs. Introns are spliced out of pre-mRNA by spliceosomes, which contain several small nuclear RNAs (snRNA), or the introns can be ribozymes that are spliced by themselves. RNA can also be altered by having its nucleotides modified to nucleotides other than A, C, G and U. In eukaryotes, modifications of RNA nucleotides are in general directed by small nucleolar RNAs (snoRNA; 60-300 nt), found in the nucleolus and cajal bodies. snoRNAs associate with enzymes and guide them to a spot on an RNA by basepairing to that RNA. These enzymes then perform the nucleotide modification. rRNAs and tRNAs are extensively modified, but snRNAs and mRNAs can also be the target of base modification. RNA can also be methylated.

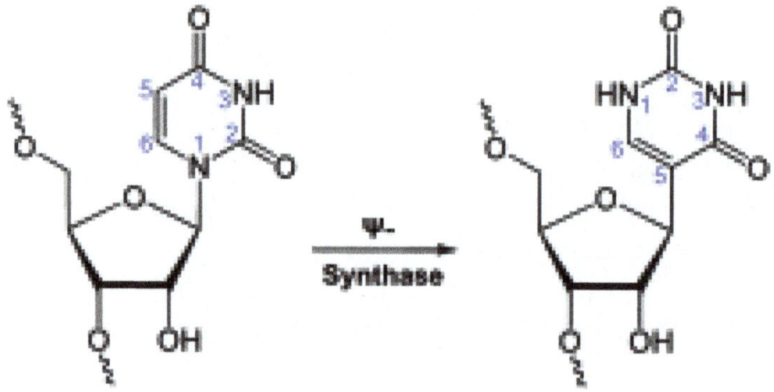

Uridine to pseudouridine is a common RNA modification.

RNA genomes

Like DNA, RNA can carry genetic information. RNA viruses have genomes composed of RNA that encodes a number of proteins. The viral genome is replicated by some of those proteins, while other proteins protect the genome as the virus particle moves to a new host cell. Viroids are another group of pathogens, but they consist only of RNA, do not encode any protein and are replicated by a host plant cell's polymerase.

In Reverse Transcription

Reverse transcribing viruses replicate their genomes by reverse transcribing DNA copies from their RNA; these DNA copies are then transcribed to new RNA. Retrotransposons also spread by copying DNA and RNA from one another, and telomerase contains an RNA that is used as template for building the ends of eukaryotic chromosomes.

Double-stranded RNA

Double-stranded RNA (dsRNA) is RNA with two complementary strands, similar to the DNA found in all cells. dsRNA forms the genetic material of some viruses (double-stranded RNA viruses). Double-stranded RNA such as viral RNA or siRNA can trigger RNA interference in eukaryotes, as well as interferon response in vertebrates.

Circular RNA

Recently, it was shown that there is a single stranded covalently closed, *i.e.* circular form of RNA expressed throughout the animal and plant kingdom. circRNAs are thought to arise via a "back-splice" reaction where the spliceosome joins a downstream donor to an upstream acceptor splice site. So far the function of circRNAs is largely unknown, although for few examples a microRNA sponging activity has been demonstrated.

Key Discoveries in RNA Biology

Robert W. Holley, left, poses with his research team.

Research on RNA has led to many important biological discoveries and numerous Nobel Prizes. Nucleic acids were discovered in 1868 by Friedrich Miescher, who called the material 'nuclein' since it was found in the nucleus. It was later discovered that prokaryotic cells, which do not have a nucleus, also contain nucleic acids. The role of RNA in protein synthesis was suspected already in 1939. Severo Ochoa won the 1959 Nobel Prize in Medicine (shared with Arthur Kornberg) after he discovered an enzyme that can synthesize RNA in the laboratory. However, the enzyme discovered by Ochoa (polynucleotide phosphorylase) was later shown to be responsible for RNA degradation, not RNA synthesis. In 1956 Alex Rich and David Davies hybridized two separate strands of RNA to form the first crystal of RNA whose structure could be determined by X-ray crystallography.

The sequence of the 77 nucleotides of a yeast tRNA was found by Robert W. Holley in 1965, winning Holley the 1968 Nobel Prize in Medicine (shared with Har Gobind Khorana and Marshall Nirenberg). In 1967, Carl Woese hypothesized that RNA might be catalytic and suggested that the earliest forms of life (self-replicating molecules) could have relied on RNA both to carry genetic information and to catalyze biochemical reactions—an RNA world.

During the early 1970s, retroviruses and reverse transcriptase were discovered, showing for the first time that enzymes could copy RNA into DNA (the opposite of the usual route for transmission of genetic information). For this work, David Baltimore, Renato Dulbecco and Howard Temin were awarded a Nobel Prize in 1975. In 1976, Walter Fiers and his team determined the first complete nucleotide sequence of an RNA virus genome, that of bacteriophage MS2.

In 1977, introns and RNA splicing were discovered in both mammalian viruses and in cellular genes, resulting in a 1993 Nobel to Philip Sharp and Richard Roberts. Catalytic RNA molecules (ribozymes) were discovered in the early 1980s, leading to a 1989 Nobel award to Thomas Cech and Sidney Altman. In 1990, it was found in *Petunia* that introduced genes can silence similar genes of the plant's own, now known to be a result of RNA interference.

At about the same time, 22 nt long RNAs, now called microRNAs, were found to have a role in the development of *C. elegans*. Studies on RNA interference gleaned a Nobel Prize for Andrew Fire and Craig Mello in 2006, and another Nobel was awarded for studies on the transcription of RNA to Roger Kornberg in the same year. The discovery of gene regulatory RNAs has led to attempts to develop drugs made of RNA, such as siRNA, to silence genes.

Evolution

In March 2015, complex DNA and RNA nucleotides, including uracil, cytosine and thymine, were reportedly formed in the laboratory under outer space conditions, using starter chemicals, such as pyrimidine, an organic compound commonly found in meteorites. Pyrimidine, like polycyclic aromatic hydrocarbons (PAHs), is the most carbon-rich compounds found in the Universe and may have been formed in red giants or in interstellar dust and gas clouds.

References

- Sedova A, Banavali NK (2017). "Geometric Patterns for Neighboring Bases Near the Stacked State in Nucleic Acid Strands". Biochemistry. 56 (10): 1426–1443. doi:10.1021/acs.biochem.6b01101. PMID 28187685
- Mao C (December 2004). "The emergence of complexity: lessons from DNA". PLoS Biology. 2 (12): e431. doi:10.1371/journal.pbio.0020431. PMC 535573. PMID 15597116
- Krieger M, Scott MP, Matsudaira PT, Lodish HF, Darnell JE, Lawrence Z, Kaiser C, Berk A (2004). "Section 4.1: Structure of Nucleic Acids". Molecular cell biology. New York: W.H. Freeman and CO. ISBN 0-7167-4366-3
- Lythgoe, B.; Rayner, L. S. (1951). "Substitution Reactions of Pyrimidine and its 2- and 4-Phenyl Derivatives". Journal of the Chemical Society. 1951: 2323–2329. doi:10.1039/JR9510002323
- McGuigan, Hugh (1921). An Introduction To Chemical Pharmacology. P. Blakiston's Sons & Co. p. 283. Retrieved July 18, 2012
- Yamada, H.; Okamoto, T. (1972). "A One-step Synthesis of Purine Ring from Formamide". Chemical & Pharmaceutical Bulletin. 20 (3): 623. doi:10.1248/cpb.20.623
- Lu XJ, Shakked Z, Olson WK (July 2000). "A-form conformational motifs in ligand-bound DNA structures". Journal of Molecular Biology. 300 (4): 819–40. doi:10.1006/jmbi.2000.3690. PMID 10891271
- Marlaire, Ruth (3 March 2015). "NASA Ames reproduces the building blocks of life in laboratory" (Press release). NASA. Retrieved 5 March 2015
- Tarrago-Litvak L, Andréola ML, Nevinsky GA, Sarih-Cottin L, Litvak S (May 1994). "The reverse transcriptase of HIV-1: from enzymology to therapeutic intervention". FASEB Journal. 8 (8): 497–503. PMID 7514143
- Alberts B, Johnson A, Lewis J, Raff M, Roberts K, Walter P (2014). Molecular Biology of the Cell (6th ed.). Garland. p. Chapter 4: DNA, Chromosomes and Genomes. ISBN 9780815344322
- Lagoja, Irene M. (2007). "Pyrimidine as Constituent of Natural Biologically Active Compounds" (PDF). Chemistry and Biodiversity. 2 (1): 1–50. doi:10.1002/cbdv.200490173. PMID 17191918
- Nuwer R (18 July 2015). "Counting All the DNA on Earth". The New York Times. New York: The New York Times Company. ISSN 0362-4331. Retrieved 2015-07-18
- Mandelkern M, Elias JG, Eden D, Crothers DM (October 1981). "The dimensions of DNA in solution". Journal of Molecular Biology. 152 (1): 153–61. doi:10.1016/0022-2836(81)90099-1. PMID 7338906
- Alberts B, Johnson A, Lewis J, Raff M, Roberts K, Peter W (2002). Molecular Biology of the Cell (Fourth ed.). New York and London: Garland Science. ISBN 0-8153-3218-1. OCLC 145080076
- Munroe SH (November 2004). "Diversity of antisense regulation in eukaryotes: multiple mechanisms, emerging patterns". Journal of Cellular Biochemistry. 93 (4): 664–71. doi:10.1002/jcb.20252. PMID 15389973
- Abbreviations and Symbols for Nucleic Acids, Polynucleotides and their Constituents IUPAC-IUB Commission on Biochemical Nomenclature (CBN). Retrieved 3 January 2006

- Hermann T; Patel DJ (2000). "RNA bulges as architectural and recognition motifs". Structure. 8 (3): R47–R54. doi:10.1016/S0969-2126(00)00110-6. PMID 10745015
- Cooper GC; Hausman RE (2004). The Cell: A Molecular Approach (3rd ed.). Sinauer. pp. 261–76, 297, 339–44. ISBN 0-87893-214-3. OCLC 174924833
- Mattick, JS (2001). "Noncoding RNAs: the architects of eukaryotic complexity". EMBO Reports. 2 (11): 986–91. doi:10.1093/embo-reports/kve230. PMC 1084129. PMID 11713189
- Palmer J (2 December 2010). "Arsenic-loving bacteria may help in hunt for alien life". BBC News. Retrieved 2 December 2010
- Nudler E; Gottesman ME (2002). "Transcription termination and anti-termination in E. coli". Genes to Cells. 7 (8): 755–68. doi:10.1046/j.1365-2443.2002.00563.x. PMID 12167155

7

Understanding Central Dogma of Molecular Biology

According to central dogma, the hereditary information is stored in the DNA nucleus. The sequence in gene expression runs from DNA to RNA to protein. Therefore, the central dogma comprises of three important biomolecules, DNA, RNA, and protein. The chapter strategically encompasses and incorporates the major components and key concepts biomolecular engineering, providing a complete understanding.

Central Dogma

- First proposed by Francis Crick in 1958.
- According to this dogma, DNA holds the coded hereditary information in the nucleus. The sequence involved in the expression of hereditary characteristics runs from DNA to RNA to protein.

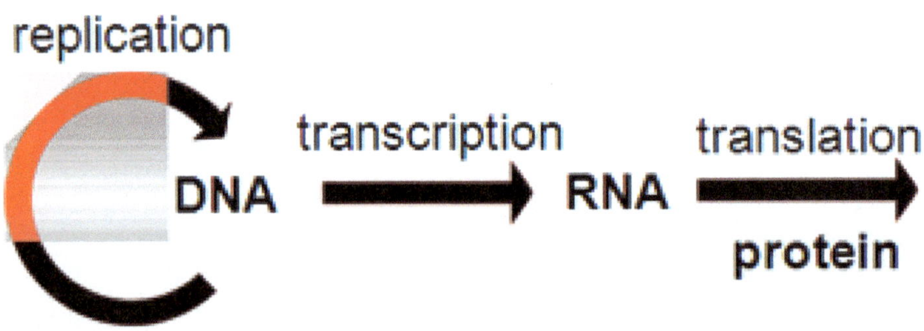

- Thus, the Central Dogma involves three important biomolecules , namely DNA, RNA, and proteins.
- It also involves three processes, namely DNA replication, transcription, and translation through protein synthesis.
- It forms the basis of storage, transmission, and expression of hereditary information.

The Modified Central Dogma

An important revolution in the Central Dogma came with the discovery of retroviruses, which transcribed RNA into DNA using an enzyme called reverse transcriptase. This lead to an exception to the Central Dogma which was later modified as shown below :

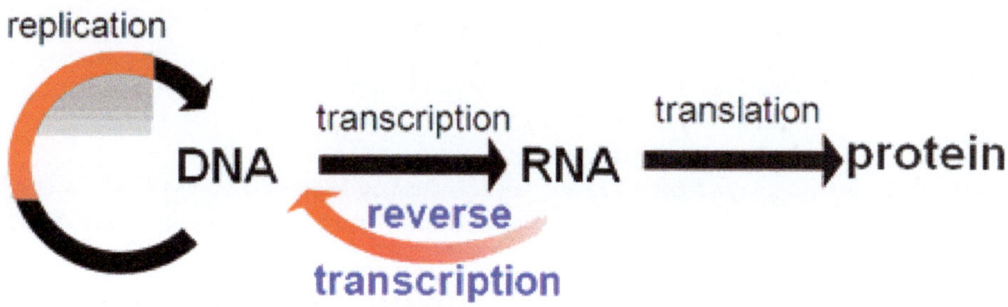

REPLICATION	Replication is the process for synthesis of daughter DNA from parental DNA : DNA dependent DNA synthesis
TRANSCRIPTION	Transcription is the process synthesis of RNA using DNA as the template: DNA dependent RNA synthesis
TRANSLATION	It is the process of protein synthesis using mRNA molecules as the template
REVERSE TRANSCRIPTION	Process of synthesis of DNA using RNA as the template: RNA dependent DNA synthesis

Enzymes

Enzymes Involved

Enzyme	Function
DNA protein.	Recognition of origin points on DNA.
DNA B protein or helicase.	Unwinding of DNA.
DNA C protein.	Assist DNA B binding.
DNA polymerase*	Elongation of the DNA strands, repair and proofreading etc.
DNA G protein or primase.	Synthesis of RNA primer.
Single-strand binding proteins (SSB).	Bind to single-strands.
DNA Topoisomerase II or gyrase.	Release super coil constraint.

DNA-dependent DNA Polymerase

- First discovered in 1958 by Arthur Kornberg; Later, DNA-pol II and DNA-pol III were identified in experiments using mutant *E.coli* cell line.

	Function
DNA polymerase I	• Proofreading (3'-5' exonuclease activity) • Repairing DNA damage • 5'-3' exonuclease activity • Ploymerization rate : 16-20 nucleotides/second • Processivity : 3-200 nucleotides added before polymerase dissociates
DNA polymerase II	• 3'-5' exonuclease activity for proofreading • Temporary functional when DNA-pol I and I III are not functional • Capable for doing synthesis on the damaged template • DNA repairing • Ploymerization rate: 40 nucleotides/second • Processivity: 1500 nucleotides added before polymerase dissociates
DNA polymerase III	• 3'-5' exonuclease activity for proofreading • Having the highest polymerization activity • Responsible for the elongation process • Ploymerization rate: 250-1000 nucleotides/second • Processivity: >500000 nucleotides added before polymerase dissociates

Structure of DNA Polymerase III

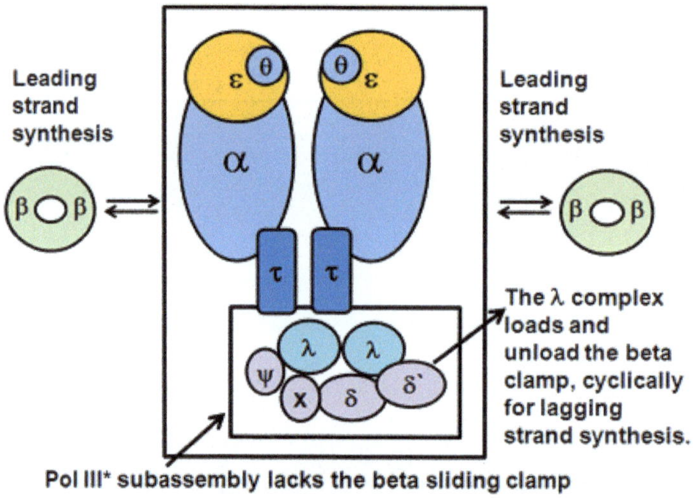

Klenow Fragment

The Klenow fragment is a large protein fragment produced when prokaryotic DNA polymerase I is cleaved by the protease subtilisin. Two fragments (one large and one small) are generated as a result of proteolysis :

- Large fragment is called Klenow fragment. It possess 5´ --> 3' DNA polymerization activity and 3´ --> 5´exonuclease activity for proofreading, but 5' --> 3' exonuclease activity is lost.
- Small fragment retains 5´- --> 3´ exonuclease activity but does not have the other two activities exhibited by the Klenow fragment (i.e. 5'- --> 3' polymerase activity, and 3'- --> 5' exonuclease activity).
- The Klenow fragment of DNA polymerase I is widely used in molecular biology.

DNA-polymerase of Eukaryotes

Eukaryotes DNA polymerases have different nomenclature and activities are differentially distributed. These are :

- DNA-pol α: initiate replication and synthesize primers.
- DNA-pol β: replication with low fidelity.
- DNA-pol γ: polymerization in mitochondria.
- DNA-pol δ: elongation.
- DNA-pol ε: proofreading and filling gap.

Primase

- Also called DNAG.
- Primers are short RNA fragments of a several nucleotides long. Primase is able to synthesize primers using free NTPs as the substrate and the ssDNA as the template.
- Primers provide free 3´-OH groups to react with the α - P atom of dNTP to form phosphoester bonds.
- Primase, DNAB, DNAC and an origin form a complex at the initiation phase.

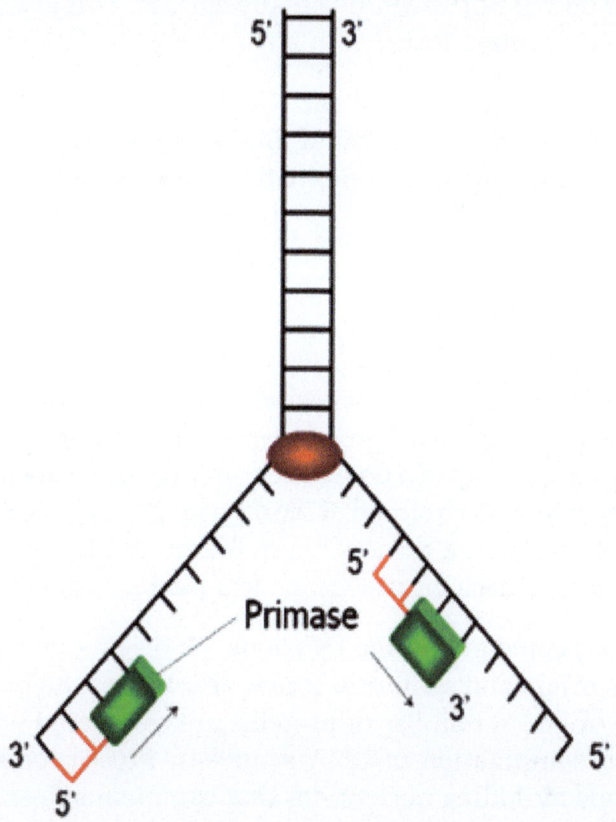

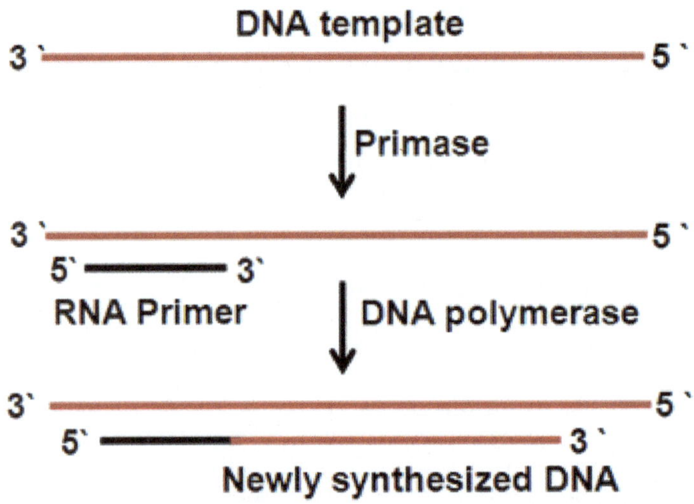

- Helicase : Also referred to as DNAB, it opens the double strand DNA near the origin of replication with the assistance of DNAA and DNAC; and at the expense of ATP.
- SSB protein : To avoid the recontact of the strands opened by helicase, single strand DNA binding protein attach to each single strand preventing the reformation of dsDNA. It also protects the vulnerable single stranded DNA against nucleases.
- Topoisomerase II or DNA gyrase : Topoisomerase II are functional in releasing the torsional strain generated because of unwinding of the dsDNA. This also prevents the formation of supercoil ahead of replication forks.
- DNA Ligase : Connects two adjacent ssDNA strands by joining the 3´-OH of one DNA strand to the 5´-P of another DNA strand. It also is responsible for sealing the nick in the process of replication, repairing, recombination, and splicing.

DNA Replication

In molecular biology, DNA replication is the biological process of producing two identical replicas of DNA from one original DNA molecule. This process occurs in all living organisms and is the basis for biological inheritance. DNA is made up of a double helix of two complementary strands. During replication, these strands are separated. Each strand of the original DNA molecule then serves as a template for the production of its counterpart, a process referred to as semiconservative replication. Cellular proofreading and error-checking mechanisms ensure near perfect fidelity for DNA replication.

In a cell, DNA replication begins at specific locations, or origins of replication, in the genome. Unwinding of DNA at the origin and synthesis of new strands results in replication forks growing bi-directionally from the origin. A number of proteins are associated with the replication fork to help in the initiation and continuation of DNA synthesis. Most prominently, DNA polymerase synthesizes the new strands by adding nucleotides that complement each (template) strand. DNA replication occurs during the S-stage of interphase.

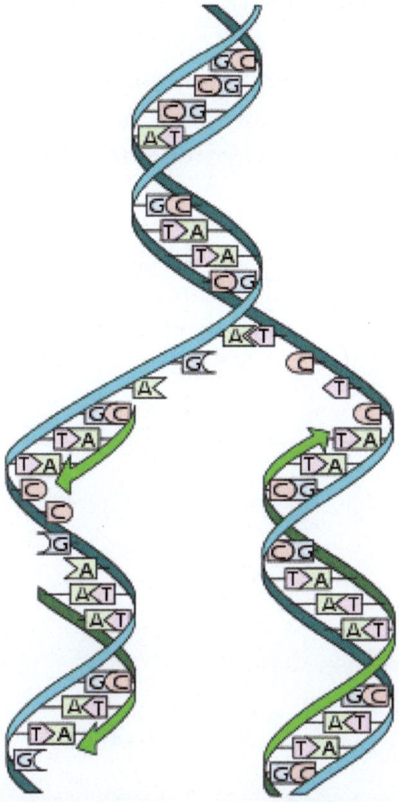

DNA replication: The double helix is un'zipped' and unwound, then each separated strand (turquoise) acts as a template for replicating a new partner strand (green). Nucleotides (bases) are matched to synthesize the new partner strands into two new double helices.

DNA replication can also be performed *in vitro* (artificially, outside a cell). DNA polymerases isolated from cells and artificial DNA primers can be used to initiate DNA synthesis at known sequences in a template DNA molecule. The polymerase chain reaction (PCR), a common laboratory technique, cyclically applies such artificial synthesis to amplify a specific target DNA fragment from a pool of DNA.

DNA Structures

DNA usually exists as a double-stranded structure, with both strands coiled together to form the characteristic double-helix. Each single strand of DNA is a chain of four types of nucleotides. Nucleotides in DNA contain a deoxyribose sugar, a phosphate, and a nucleobase. The four types of nucleotide correspond to the four nucleobases adenine, cytosine, guanine, and thymine, commonly abbreviated as A,C, G and T. Adenine and guanine are purine bases, while cytosine and thymine are pyrimidines. These nucleotides form phosphodiester bonds, creating the phosphate-deoxyribose backbone of the DNA double helix with the nuclei bases pointing inward (i.e., toward the opposing strand). Nucleotides (bases) are matched between strands through hydrogen bonds to form base pairs. Adenine pairs with thymine (two hydrogen bonds), and guanine pairs with cytosine (stronger: three hydrogen bonds).

DNA strands have a directionality, and the different ends of a single strand are called the "3' (three-prime) end" and the "5' (five-prime) end". By convention, if the base sequence of a single strand of

DNA is given, the left end of the sequence is the 5' end, while the right end of the sequence is the 3' end. The strands of the double helix are anti-parallel with one being 5' to 3', and the opposite strand 3' to 5'. These terms refer to the carbon atom in deoxyribose to which the next phosphate in the chain attaches. Directionality has consequences in DNA synthesis, because DNA polymerase can synthesize DNA in only one direction by adding nucleotides to the 3' end of a DNA strand.

The pairing of complementary bases in DNA (through hydrogen bonding) means that the information contained within each strand is redundant. Phosphodiester (intra-strand) bonds are stronger than hydrogen (inter-strand) bonds. This allows the strands to be separated from one another. The nucleotides on a single strand can therefore be used to reconstruct nucleotides on a newly synthesized partner strand.

DNA Polymerase

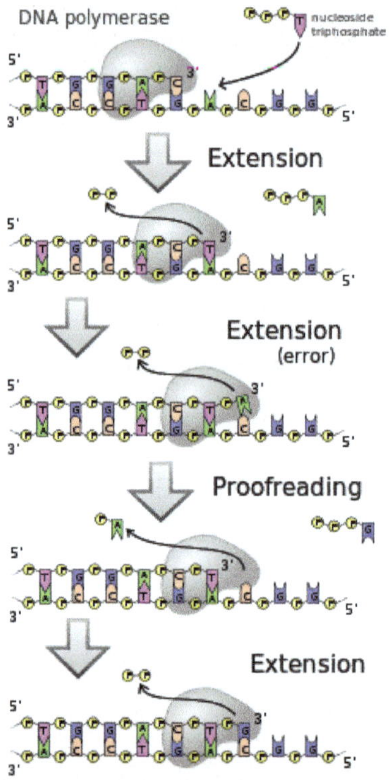

DNA polymerases adds nucleotides to the 3' end of a strand of DNA. If a mismatch is accidentally incorporated, the polymerase is inhibited from further extension. Proofreading removes the mismatched nucleotide and extension continues.

DNA polymerases are a family of enzymes that carry out all forms of DNA replication. DNA polymerases in general cannot initiate synthesis of new strands, but can only extend an existing DNA or RNA strand paired with a template strand. To begin synthesis, a short fragment of RNA, called a primer, must be created and paired with the template DNA strand.

DNA polymerase adds a new strand of DNA by extending the 3' end of an existing nucleotide chain, adding new nucleotides matched to the template strand one at a time via the creation of phosphodiester bonds. The energy for this process of DNA polymerization comes from hydrolysis of the high-energy phosphate (phosphoanhydride) bonds between the three phosphates attached to each unincorporated base. Free bases with their attached phosphate groups are called nucleotides; in particular, bases with three attached phosphate groups are called nucleoside triphosphates. When a nucleotide is being added to a growing DNA strand, the formation of a phosphodiester bond between the proximal phosphate of the nucleotide to the growing chain is accompanied by hydrolysis of a high-energy phosphate bond with release of the two distal phosphates as a pyrophosphate. Enzymatic hydrolysis of the resulting pyrophosphate into inorganic phosphate consumes a second high-energy phosphate bond and renders the reaction effectively irreversible.

In general, DNA polymerases are highly accurate, with an intrinsic error rate of less than one mistake for every 10^7 nucleotides added. In addition, some DNA polymerases also have proofreading ability; they can remove nucleotides from the end of a growing strand in order to correct mismatched bases. Finally, post-replication mismatch repair mechanisms monitor the DNA for errors, being capable of distinguishing mismatches in the newly synthesized DNA strand from the original strand sequence. Together, these three discrimination steps enable replication fidelity of less than one mistake for every 10^9 nucleotides added.

The rate of DNA replication in a living cell was first measured as the rate of phage T4 DNA elongation in phage-infected E. coli. During the period of exponential DNA increase at 37 °C, the rate was 749 nucleotides per second. The mutation rate per base pair per replication during phage T4 DNA synthesis is 1.7 per 10^8.

Replication Process

DNA replication, like all biological polymerization processes, proceeds in three enzymatically catalyzed and coordinated steps: initiation, elongation and termination.

Initiation

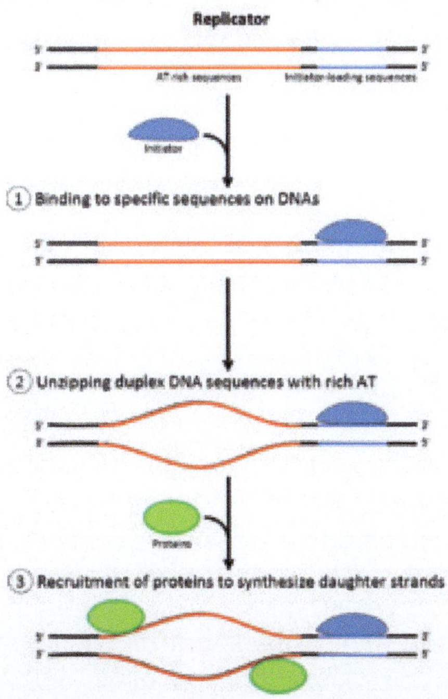

Role of initiators for initiation of DNA replication.

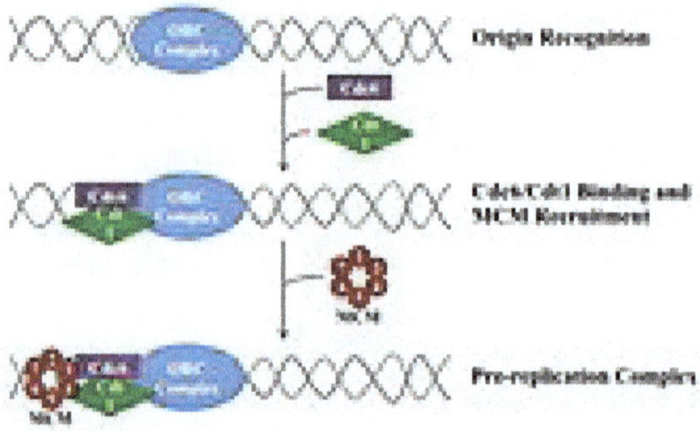

Formation of pre-replication complex.

For a cell to divide, it must first replicate its DNA. This process is initiated at particular points in the DNA, known as "origins", which are targeted by initiator proteins. In *E. coli* this protein is DnaA; in yeast, this is the origin recognition complex. Sequences used by initiator proteins tend to be "AT-rich" (rich in adenine and thymine bases), because A-T base pairs have two hydrogen bonds (rather than the three formed in a C-G pair) and thus are easier to strand separate. Once the origin has been located, these initiators recruit other proteins and form the pre-replication complex, which unzips the double-stranded DNA.

Elongation

DNA polymerase has 5'-3' activity. All known DNA replication systems require a free 3' hydroxyl group before synthesis can be initiated (note: the DNA template is read in 3' to 5' direction whereas a new strand is synthesized in the 5' to 3' direction—this is often confused). Four distinct mechanisms for DNA synthesis are recognized:

1. All cellular life forms and many DNA viruses, phages and plasmids use a primase to synthesize a short RNA primer with a free 3' OH group which is subsequently elongated by a DNA polymerase.

2. The retroelements (including retroviruses) employ a transfer RNA that primes DNA replication by providing a free 3′ OH that is used for elongation by the reverse transcriptase.

3. In the adenoviruses and the φ29 family of bacteriophages, the 3' OH group is provided by the side chain of an amino acid of the genome attached protein (the terminal protein) to which nucleotides are added by the DNA polymerase to form a new strand.

4. In the single stranded DNA viruses — a group that includes the circoviruses, the geminiviruses, the parvoviruses and others — and also the many phages and plasmids that use the rolling circle replication (RCR) mechanism, the RCR endonuclease creates a nick in the genome strand (single stranded viruses) or one of the DNA strands (plasmids). The 5′ end of the nicked strand is transferred to a tyrosine residue on the nuclease and the free 3′ OH group is then used by the DNA polymerase to synthesize the new strand.

The first is the best known of these mechanisms and is used by the cellular organisms. In this mechanism, once the two strands are separated, primase adds RNA primers to the template strands. The leading strand receives one RNA primer while the lagging strand receives several.

The leading strand is continuously extended from the primer by a DNA polymerase with high processivity, while the lagging strand is extended discontinuously from each primer forming Okazaki fragments. RNase removes the primer RNA fragments, and a low processivity DNA polymerase distinct from the replicative polymerase enters to fill the gaps. When this is complete, a single nick on the leading strand and several nicks on the lagging strand can be found. Ligase works to fill these nicks in, thus completing the newly replicated DNA molecule.

The primase used in this process differs significantly between bacteria and archaea/eukaryotes. Bacteria use a primase belonging to the DnaG protein superfamily which contains a catalytic domain of the TOPRIM fold type. The TOPRIM fold contains an α/β core with four conserved strands in a Rossmann-like topology. This structure is also found in the catalytic domains of topoisomerase Ia, topoisomerase II, the OLD-family nucleases and DNA repair proteins related to the RecR protein.

The primase used by archaea and eukaryotes, in contrast, contains a highly derived version of the RNA recognition motif (RRM). This primase is structurally similar to many viral RNA-dependent RNA polymerases, reverse transcriptases, cyclic nucleotide generating cyclases and DNA polymerases of the A/B/Y families that are involved in DNA replication and repair. In eukaryotic replication, the primase forms a complex with Pol α.

Multiple DNA polymerases take on different roles in the DNA replication process. In *E. coli*, DNA Pol III is the polymerase enzyme primarily responsible for DNA replication. It assembles into a replication complex at the replication fork that exhibits extremely high processivity, remaining intact for the entire replication cycle. In contrast, DNA Pol I is the enzyme responsible for replacing RNA primers with DNA. DNA Pol I has a 5' to 3' exonuclease activity in addition to its polymerase activity, and uses its exonuclease activity to degrade the RNA primers ahead of it as it extends the DNA strand behind it, in a process called nick translation. Pol I is much less processive than Pol III because its primary function in DNA replication is to create many short DNA regions rather than a few very long regions.

In eukaryotes, the low-processivity enzyme, Pol α, helps to initiate replication because it forms a complex with primase. In eukaryotes, leading strand synthesis is thought to be conducted by Pol ε; however, this view has recently been challenged, suggesting a role for Pol δ. Primer removal is completed Pol δ while repair of DNA during replication is completed by Pol ε.

As DNA synthesis continues, the original DNA strands continue to unwind on each side of the bubble, forming a replication fork with two prongs. In bacteria, which have a single origin of replication on their circular chromosome, this process creates a "theta structure" (resembling the Greek letter theta: θ). In contrast, eukaryotes have longer linear chromosomes and initiate replication at multiple origins within these.

Replication Fork

The replication fork is a structure that forms within the nucleus during DNA replication. It is created by helicases, which break the hydrogen bonds holding the two DNA strands together. The resulting structure has two branching "prongs", each one made up of a single strand of DNA. These two strands serve as the template for the leading and lagging strands, which will be created as DNA polymerase matches complementary nucleotides to the templates; the templates may be properly referred to as the leading strand template and the lagging strand template.

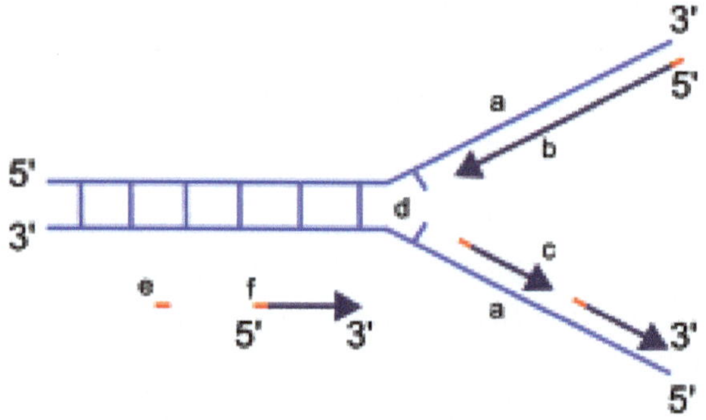

Scheme of the replication fork.
a: template, b: leading strand, c: lagging strand, d: replication fork, e: primer, f: Okazaki fragments

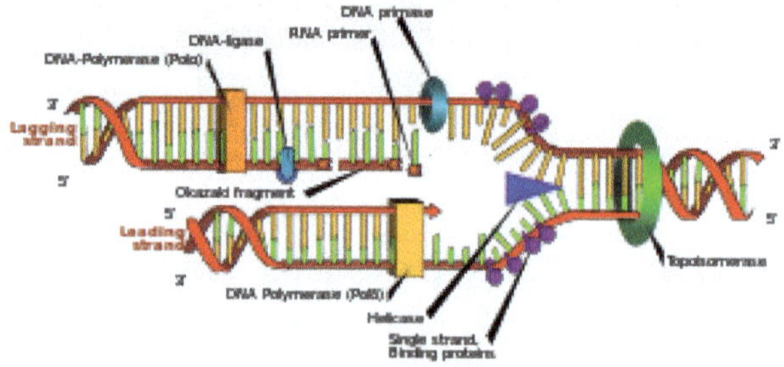

Many enzymes are involved in the DNA replication fork.

DNA is always synthesized in the 5' to 3' direction. Since the leading and lagging strand templates are oriented in opposite directions at the replication fork, a major issue is how to achieve synthesis of nascent (new) lagging strand DNA, whose direction of synthesis is opposite to the direction of the growing replication fork.

Leading Strand

The leading strand is the strand of nascent DNA which is being synthesized in the same direction as the growing replication fork. A polymerase "reads" the leading strand *template* and adds complementary nucleotides to the nascent leading strand on a continuous basis.

Lagging Strand

The lagging strand is the strand of nascent DNA whose direction of synthesis is opposite to the direction of the growing replication fork. Because of its orientation, replication of the lagging strand is more complicated as compared to that of the leading strand. As a consequence, the DNA polymerase on this strand is seen to "lag behind" the other strand.

The lagging strand is synthesized in short, separated segments. On the lagging strand *template*, a primase "reads" the template DNA and initiates synthesis of a short complementary RNA primer. A DNA polymerase extends the primed segments, forming Okazaki fragments. The RNA primers are then removed and replaced with DNA, and the fragments of DNA are joined together by DNA ligase.

Dynamics at the Replication Fork

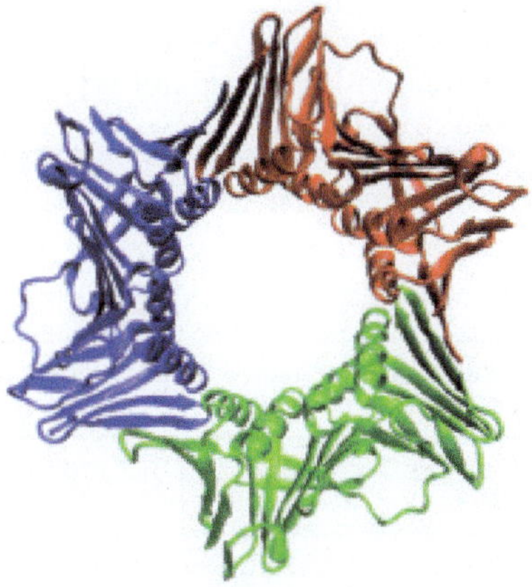

The assembled human DNA clamp, a trimer of the protein PCNA.

As helicase unwinds DNA at the replication fork, the DNA ahead is forced to rotate. This process results in a build-up of twists in the DNA ahead. This build-up forms a torsional resistance that would eventually halt the progress of the replication fork. Topoisomerases are enzymes that temporarily break the strands of DNA, relieving the tension caused by unwinding the two strands of the DNA helix; topoisomerases (including DNA gyrase) achieve this by adding negative supercoils to the DNA helix.

Bare single-stranded DNA tends to fold back on itself forming secondary structures; these structures can interfere with the movement of DNA polymerase. To prevent this, single-strand binding proteins bind to the DNA until a second strand is synthesized, preventing secondary structure formation.

Clamp proteins form a sliding clamp around DNA, helping the DNA polymerase maintain contact with its template, thereby assisting with processivity. The inner face of the clamp enables DNA to be threaded through it. Once the polymerase reaches the end of the template or detects double-stranded DNA, the sliding clamp undergoes a conformational change that releases the DNA polymerase. Clamp-loading proteins are used to initially load the clamp, recognizing the junction between template and RNA primers.

DNA Replication Proteins

At the replication fork, many replication enzymes assemble on the DNA into a complex molecular machine called the replisome. The following is a list of major DNA replication enzymes that participate in the replisome:

Enzyme	Function in DNA replication
DNA Helicase	Also known as helix destabilizing enzyme. Helicase separates the two strands of DNA at the Replication Fork behind the topoisomerase.
DNA Polymerase	The enzyme responsible for catalyzing the addition of nucleotide substrates to DNA in the 5' to 3' direction during DNA replication. Also performs proof-reading and error correction. There exist many different types of DNA Polymerase, (such as the DNA Polymerase III)each of which perform different functions in different types of cells.

DNA clamp	A protein which prevents elongating DNA polymerases from dissociating from the DNA parent strand.
Single-Strand Binding (SSB) Proteins	Bind to ssDNA and prevent the DNA double helix from re-annealing after DNA helicase unwinds it, thus maintaining the strand separation, and facilitating the synthesis of the nascent strand.
Topoisomerase	Relaxes the DNA from its super-coiled nature.
DNA Gyrase	Relieves strain of unwinding by DNA helicase; this is a specific type of topoisomerase
DNA Ligase	Re-anneals the semi-conservative strands and joins Okazaki Fragments of the lagging strand.
Primase	Provides a starting point of RNA (or DNA) for DNA polymerase to begin synthesis of the new DNA strand.
Telomerase	Lengthens telomeric DNA by adding repetitive nucleotide sequences to the ends of eukaryotic chromosomes. This allows germ cells and stem cells to avoid the Hayflick limit on cell division.

Replication Machinery

Replication machineries consist of factors involved in DNA replication and appearing on template ssDNAs. Replication machineries include primosotors are replication enzymes; DNA polymerase, DNA helicases, DNA clamps and DNA topoisomerases, and replication proteins; e.g. single-stranded DNA binding proteins (SSB). In the replication machineries these components coordinate. In most of the bacteria, all of the factors involved in DNA replication are located on replication forks and the complexes stay on the forks during DNA replication. These replication machineries are called replisomes or DNA replicase systems. These terms are generic terms for proteins located on replication forks. In eukaryotic and some bacterial cells the replisomes are not formed.

Since replication machineries do not move relatively to template DNAs such as factories, they are called a replication factory. In an alternative figure, DNA factories are similar to projectors and DNAs are like as cinematic films passing constantly into the projectors. In the replication factory model, after both DNA helicases for leading strands and lagging strands are loaded on the template DNAs, the helicases run along the DNAs into each other. The helicases remain associated for the remainder of replication process. Peter Meister et al. observed directly replication sites in budding yeast by monitoring green fluorescent protein(GFP)-tagged DNA polymerases α. They detected DNA replication of pairs of the tagged loci spaced apart symmetrically from a replication origin and found that the distance between the pairs decreased markedly by time. This finding suggests that the mechanism of DNA replication goes with DNA factories. That is, couples of replication factories are loaded on replication origins and the factories associated with each other. Also, template DNAs move into the factories, which bring extrusion of the template ssDNAs and nascent DNAs. Meister's finding is the first direct evidence of replication factory model. Subsequent research has shown that DNA helicases form dimers in many eukaryotic cells and bacterial replication machineries stay in single intranuclear location during DNA synthesis.

The replication factories perform disentanglement of sister chromatids. The disentanglement is essential for distributing the chromatids into daughter cells after DNA replication. Because sister chromatids after DNA replication hold each other by Cohesin rings, there is the only chance for the disentanglement in DNA replication. Fixing of replication machineries as replication factories can improve the success rate of DNA replication. If replication forks move freely in chromosomes, catenation of nuclei is aggravated and impedes mitotic segregation.

Termination

Eukaryotes initiate DNA replication at multiple points in the chromosome, so replication forks meet and terminate at many points in the chromosome; these are not known to be regulated in any particular way. Because eukaryotes have linear chromosomes, DNA replication is unable to reach the very end of the chromosomes, but ends at the telomere region of repetitive DNA close to the ends. This shortens the telomere of the daughter DNA strand. Shortening of the telomeres is a normal process in somatic cells. As a result, cells can only divide a certain number of times before the DNA loss prevents further division. (This is known as the Hayflick limit.) Within the germ cell line, which passes DNA to the next generation, telomerase extends the repetitive sequences of the telomere region to prevent degradation. Telomerase can become mistakenly active in somatic cells, sometimes leading to cancer formation. Increased telomerase activity is one of the hallmarks of cancer.

Termination requires that the progress of the DNA replication fork must stop or be blocked. Termination at a specific locus, when it occurs, involves the interaction between two components: (1) a termination site sequence in the DNA, and (2) a protein which binds to this sequence to physically stop DNA replication. In various bacterial species, this is named the DNA replication terminus site-binding protein, or Ter protein.

Because bacteria have circular chromosomes, termination of replication occurs when the two replication forks meet each other on the opposite end of the parental chromosome. *E. coli* regulates this process through the use of termination sequences that, when bound by the Tus protein, enable only one direction of replication fork to pass through. As a result, the replication forks are constrained to always meet within the termination region of the chromosome.

Regulation

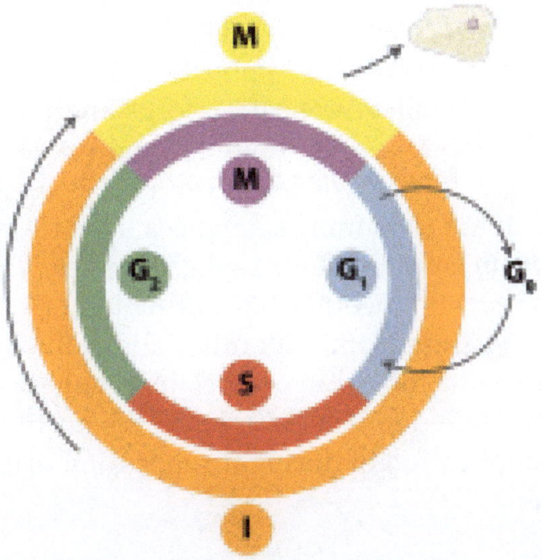

The cell cycle of eukaryotic cells.

Eukaryotes

Within eukaryotes, DNA replication is controlled within the context of the cell cycle. As the cell grows and divides, it progresses through stages in the cell cycle; DNA replication takes place during the S phase (synthesis phase). The progress of the eukaryotic cell through the cycle is controlled by cell cycle checkpoints. Progression through checkpoints is controlled through complex interactions between various proteins, including cyclins and cyclin-dependent kinases. Unlike bacteria, eukaryotic DNA replicates in the confines of the nucleus.

The G1/S checkpoint (or restriction checkpoint) regulates whether eukaryotic cells enter the process of DNA replication and subsequent division. Cells that do not proceed through this checkpoint remain in the G0 stage and do not replicate their DNA.

Replication of chloroplast and mitochondrial genomes occurs independently of the cell cycle, through the process of D-loop replication.

Replication Focus

In vertebrate cells, replication sites concentrate into positions called replication foci. Replication sites can be detected by immunostaining daughter strands and replication enzymes and monitoring GFP-tagged replication factors. By these methods it is found that replication foci of varying size and positions appear in S phase of cell division and their number per nucleus is far smaller than the number of genomic replication forks.

P. Heun et al.(2001) tracked GFP-tagged replication foci in budding yeast cells and revealed that replication origins move constantly in G1 and S phase and the dynamics decreased significantly in S phase. Traditionally, replication sites were fixed on spatial structure of chromosomes by nuclear matrix or lamins. The Heun's results denied the traditional concepts, budding yeasts don't have lamins, and support that replication origins self-assemble and form replication foci.

By firing of replication origins, controlled spatially and temporally, the formation of replication foci is regulated. D. A. Jackson et al.(1998) revealed that neighboring origins fire simultaneously in mammalian cells. Spatial juxtaposition of replication sites brings clustering of replication forks. The clustering do rescue of stalled replication forks and favors normal progress of replication forks. Progress of replication forks is inhibited by many factors; collision with proteins or with complexes binding strongly on DNA, deficiency of dNTPs, nicks on template DNAs and so on. If replication forks stall and the remaining sequences from the stalled forks are not replicated, the daughter strands have nick obtained un-replicated sites. The un-replicated sites on one parent's strand hold the other strand together but not daughter strands. Therefore, the resulting sister chromatids cannot separate from each other and cannot divide into 2 daughter cells. When neighboring origins fire and a fork from one origin is stalled, fork from other origin access on an opposite direction of the stalled fork and duplicate the un-replicated sites. As other mechanism of the rescue there is application of dormant replication origins that excess origins don't fire in normal DNA replication.

Bacteria

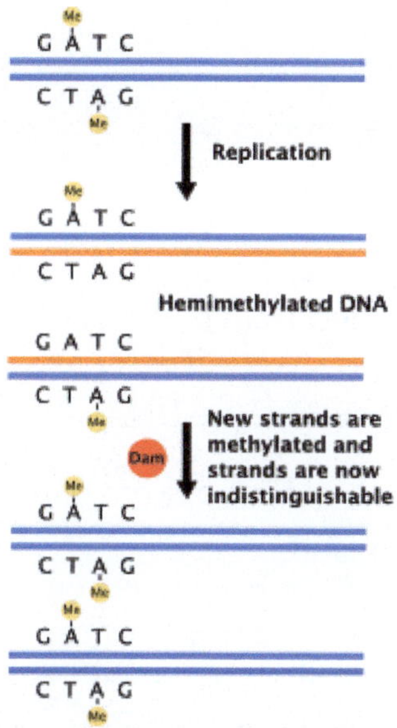

Dam methylates adenine of GATC sites after replication.

Most bacteria do not go through a well-defined cell cycle but instead continuously copy their DNA; during rapid growth, this can result in the concurrent occurrence of multiple rounds of replication. In *E. coli*, the best-characterized bacteria, DNA replication is regulated through several mechanisms, including: the hemimethylation and sequestering of the origin sequence, the ratio of adenosine triphosphate (ATP) to adenosine diphosphate (ADP), and the levels of protein DnaA. All these control the binding of initiator proteins to the origin sequences.

Because *E. coli* methylates GATC DNA sequences, DNA synthesis results in hemimethylated sequences. This hemimethylated DNA is recognized by the protein SeqA, which binds and sequesters the origin sequence; in addition, DnaA (required for initiation of replication) binds less well to hemimethylated DNA. As a result, newly replicated origins are prevented from immediately initiating another round of DNA replication.

ATP builds up when the cell is in a rich medium, triggering DNA replication once the cell has reached a specific size. ATP competes with ADP to bind to DnaA, and the DnaA-ATP complex is able to initiate replication. A certain number of DnaA proteins are also required for DNA replication – each time the origin is copied, the number of binding sites for DnaA doubles, requiring the synthesis of more DnaA to enable another initiation of replication.

Polymerase Chain Reaction

Researchers commonly replicate DNA *in vitro* using the polymerase chain reaction (PCR). PCR uses a pair of primers to span a target region in template DNA, and then polymerizes partner strands in each direction from these primers using a thermostable DNA polymerase. Repeating this process through multiple cycles amplifies the targeted DNA region. At the start of each cycle, the mixture of template and primers is heated, separating the newly synthesized molecule and template. Then, as the mixture cools, both of these become templates for annealing of new primers, and the polymerase extends from these. As a result, the number of copies of the target region doubles each round, increasing exponentially.

DNA Replication Process

Initiation

-> recognition of origin.

-> unwinding of dsDNA (double stranded DNA).

-> primer synthesis.

Elongation

-> addition of dNTPs to the existing strand,

-> formation of phosphodiester bonds.

-> proofreading.

-> extension of the DNA strand.

Termination

-> completion of the replication.

Events and process of Initiation

- The replication starts at a particular point called origin ori C.
- DNA A recognizes ori C.
- DNA B and DNA C join the DNA - DNA A complex, unwind the local AT-rich region, and move on the template downstream further to separate enough space.
- SSB protein binds the complex to stabilize ssDNA (single stranded DNA); Primase joins and forms a complex called primosome.
- Primase initiates synthesis of primers on the ssDNA template using NTP as substrates in the 5´- 3´ direction using ATP as source of energy.
- The short RNA fragments provide free 3´-OH groups for DNA elongation.
- Topoisomerase II binds to the dsDNA region just before the replication forks to release the supercoil constraints which are generated ahead of the replication forks.

Elongation

- The nascent DNA chain is continuously elongated by the addition of dNTPs . This proceeds via formation of a series of phosphodiester bonds; catalysed by DNA-pol III.
- The bases are added complementary to that of template following the Watson –Crick base pairing. eg A-T and G-C.

- The enzymes subunits (α, ε, and θ) catalyze the synthesis of leading and lagging strands, respectively.
- The leading strand is synthesized in the same direction as that of the replication fork.
- The latest Okazaki fragment is also synthesized in same direction as that of the replication fork.
- Primers on Okazaki fragments are digested by RNase.
- The gaps are filled by DNA-pol I in the 5´ --> 3´ direction.
- Ligase seals the nick between the 5´end of one fragment and the 3´end of the next fragment is sealed by.

Termination

- The replication of *E. coli* is bidirectional from one origin, and the two replication forks must meet at one point called ter.
- All the primers will be removed, and all the fragments will be connected by DNA-pol I and ligase.

Reverse Transcription

- In certain RNA viruses, the flow of genetic information is from single stranded RNA to DNA, opposite to the normal process. Reverse transcription is therefore the process in which ssRNA is used as the template to synthesize dsDNA.
- The following steps are involved in the process of Reverse Transcription :

  -> Synthesis of ssDNA complementary to ssRNA, forming a RNA-DNA hybrid.

  -> Hydrolysis of ssRNA in the RNA-DNA hybrid by RNase activity of reverse transcriptase, leaving ssDNA.

  -> Synthesis of the second ssDNA using the left ssDNA as the template, forming a DNA-DNA duplex.
- Reverse transcriptase is the enzyme for the reverse transcription. It has activity of three kinds of enzymes : RNA-dependent DNA polymerase, RNase, DNA-dependent DNA polymerase. It was discovered by David Baltimore in the year 1970.
- Human immunodeficiency virus and *Rous sarcoma* virus are the example of reverse transcription.

Transcription

Transcription is the Process of Synthesis of RNA

The synthesis of RNA molecules using DNA strands as the templates so that the genetic information can be transferred from DNA to RNA.

SIMILARITY	DIFFERENCE		
		REPLICATION	TRANSCRIPTION
•Both processes use DNA as the template •Phosphodiester bonds are formed in both cases •Both synthesis directions are from 5′ to 3′.	Template	double stranded	single stranded
	Substrate	dNTP	NTP
	Primer	yes	no
	Enzyme associated	DNA polymerase	RNA polymerase
	Product	dsDNA	ssRNA
	Base pair	A – T, G – C	A – U, T – A, G – C

- Only small portion of genome is transcribed depending on the development requirement, physiological need and environmental changes.
- The stretch of DNA transcribed into an RNA molecule is called a transcription unit and encodes for at least one gene.
- During transcription, RNA polymerase reads a DNA sequence and produces a complementary, antiparallel RNA strand. DNA is read from 3' --> 5' during transcription; the complementary RNA is created from the 5' --> 3' direction.
- The transcription direction on different strands is opposite. This feature is referred to as the asymmetric transcription.
- Transcription occurs in three phases : initiation, elongation, and termination.

Initiation : RNA-pol (enzyme responsible for RNA synthesis) recognizes the promoter and transcription begins.

Elongation : the RNA strand elongates.

Termination : synthesis stopped; nascent RNA is separated from the DNA template.

Template

The strand from which the RNA is gets transcribed is called the TEMPLATE strand or ANTISENSE strand. The CODING strand is the strand whose base sequence specifies the amino acid sequence of the encoded protein. Therefore, it is also called as SENSE strand.

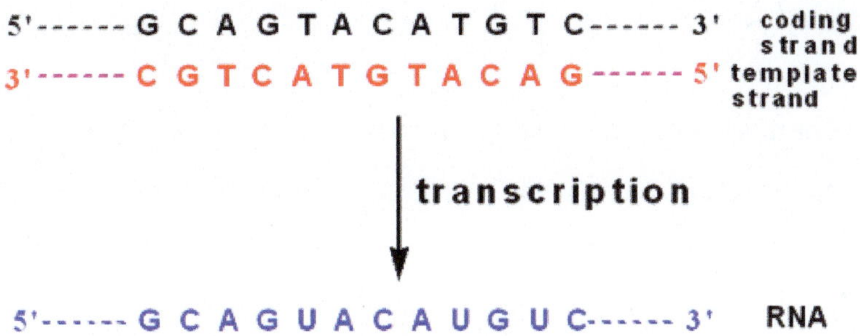

RNA Polymerase

DNA-dependent RNA polymerase is the enzyme responsible for the RNA synthesis.

- The prokaryotic RNA polymerase is a multiple-subunit protein of molecular mass 480kD.
- Eukaryotic systems have three kinds of RNA polymerases, RNA-pol I, II, III, each of which is a multiple-subunit protein and responsible for transcription of different RNAs.
- The prokaryotic RNA-polymerase can directly bind to the DNA template in the transcription process but the eukaryotic RNA-polymerase requires co-factors for binding to the DNA template together.
- Unlike DNA polymerases, RNA polymerases are deficient in 3'-5' exonuclease proofreading activity. Hence the error rate of transcription is higher than that in case of DNA replication.
- The direction of synthesis is 5 -----> 3'. No primer is needed for RNA synthesis

RNA-polymerase of E. Coli

The holoenzyme of RNA-polymerase in *E.coli* consists of 6 different subunits: $\alpha_2\beta\ \beta'\omega\sigma$. The σ subunit dissociates after initiation leaving behind the apoenzymes.

SUBUNIT	FUNCTION
α	Determination of the DNA to be transcribed
β	Role in catalysis of the polymerization process
β'	Binds and opens DNA template
ω	Restores denatured RNA polymerase to its functional form in vitro
σ	Recognition of the promoter region for synthesis initiation. The promoter is the DNA sequence that RNA – pol can bind.

Each transcriptable region is called operon. One operon includes several structural genes and upstream regulatory sequences.

Promoter Sequences are recognized by RNA polymerase for Initiating the Transcription

RNA synthesis begins when RNA polymerase binds to specific regions on the DNA called promoters.

In *E.coli*, the promoter region extends between 70 base pairs before (-70) and 30 bp (+30) beyond the transcription start region.

The sequences may vary from one promoter region to the other but comparison of the -10 and -35 regions have shown similarities. These regions are termed as consensus sequences. These sequences are important interaction sites for the σ subunit.

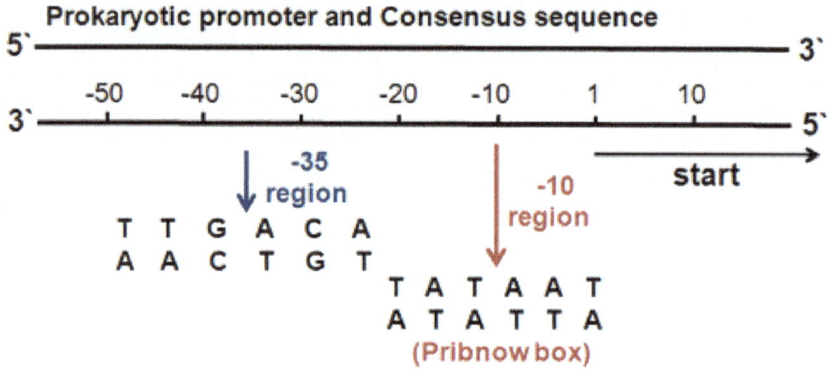

Escherichia Coli

Escherichia coli is a gram-negative, facultatively anaerobic, rod-shaped, coliform bacterium of the genus *Escherichia* that is commonly found in the lower intestine of warm-blooded organisms (endotherms). Most *E. coli* strains are harmless, but some serotypes can cause serious food poisoning in their hosts, and are occasionally responsible for product recalls due to food contamination. The harmless strains are part of the normal flora of the gut, and can benefit their hosts by producing vitamin K_2, and preventing colonization of the intestine with pathogenic bacteria. *E. coli* is expelled into the environment within fecal matter. The bacterium grows massively in fresh fecal matter under aerobic conditions for 3 days, but its numbers decline slowly afterwards.

E. coli and other facultative anaerobes constitute about 0.1% of gut flora, and fecal–oral transmission is the major route through which pathogenic strains of the bacterium cause disease. Cells are able to survive outside the body for a limited amount of time, which makes them potential indicator organisms to test environmental samples for fecal contamination. A growing body of research, though, has examined environmentally persistent *E. coli* which can survive for extended periods outside of a host.

The bacterium can be grown and cultured easily and inexpensively in a laboratory setting, and has been intensively investigated for over 60 years. *E. coli* is a chemoheterotroph whose chemically defined medium must include a source of carbon and energy. *E. coli* is the most widely studied prokaryotic model organism, and an important species in the fields of biotechnology and microbiology, where it has served as the host organism for the majority of work with recombinant DNA. Under favorable conditions, it takes only 20 minutes to reproduce.

Biology and Biochemistry

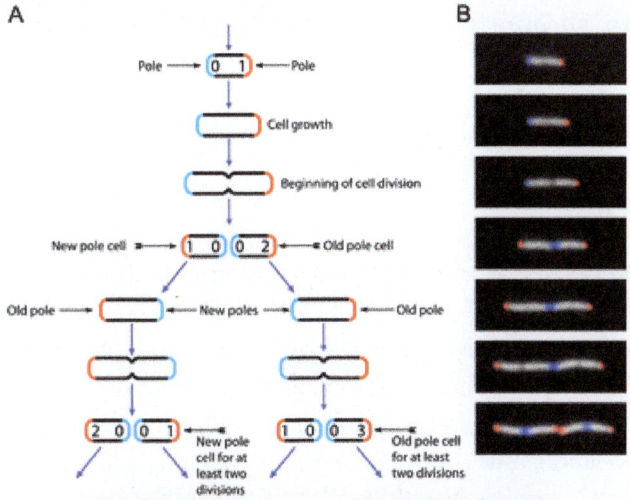

Model of successive binary fission in *E. coli*

A colony of *E. coli* growing

Type and Morphology

E. coli is a Gram-negative, facultative anaerobic (that makes ATP by aerobic respiration if oxygen is present, but is capable of switching to fermentation or anaerobic respiration if oxygen is absent) and nonsporulating bacterium. Cells are typically rod-shaped, and are about 2.0 μm long and 0.25–1.0 μm in diameter, with a cell volume of 0.6–0.7 μm^3.

E. coli stains Gram-negative because its cell wall is composed of a thin peptidoglycan layer and an outer membrane. During the staining process, *E. coli* picks up the color of the counterstain safranin and stains pink. The outer membrane surrounding the cell wall provides a barrier to certain antibiotics such that *E. coli* is not damaged by penicillin.

Strains that possess flagella are motile. The flagella have a peritrichous arrangement.

Metabolism

E. coli can live on a wide variety of substrates and uses mixed-acid fermentation in anaerobic conditions, producing lactate, succinate, ethanol, acetate, and carbon dioxide. Since many pathways in mixed-acid fermentation produce hydrogen gas, these pathways require the levels of hydrogen to be low, as is the case when *E. coli* lives together with hydrogen-consuming organisms, such as methanogens or sulphate-reducing bacteria.

Culture Growth

Optimum growth of *E. coli* occurs at 37 °C (98.6 °F), but some laboratory strains can multiply at temperatures up to 49 °C (120 °F). *E. coli* grows in a variety of defined laboratory media, such as lysogeny broth, or any medium that contains glucose, ammonium phosphate, monobasic, sodium chloride, magnesium sulfate, potassium phosphate, dibasic, and water. Growth can be driven by aerobic or anaerobic respiration, using a large variety of redox pairs, including the oxidation of pyruvic acid, formic acid, hydrogen, and amino acids, and the reduction of substrates such as oxygen, nitrate, fumarate, dimethyl sulfoxide, and trimethylamine N-oxide. *E. coli* is classified as a facultative anaerobe. It uses oxygen when it is present and available. It can, however, continue to grow in the absence of oxygen using fermentation or anaerobic respiration. The ability to continue growing in the absence of oxygen is an advantage to bacteria because their survival is increased in environments where water predominates.

Cell Cycle

The bacterial cell cycle is divided into three stages. The B period occurs between the completion of cell division and the beginning of DNA replication. The C period encompasses the time it takes to replicate the chromosomal DNA. The D period refers to the stage between the conclusion of DNA replication and the end of cell division. The doubling rate of *E. coli* is higher when more nutrients are available. However, the length of the C and D periods do not change, even when the doubling time becomes less than the sum of the C and D periods. At the fastest growth rates, replication begins before the previous round of replication has completed, resulting in multiple replication forks along the DNA and overlapping cell cycles.

Unlike eukaryotes, prokaryotes do not rely upon either changes in gene expression or changes in protein synthesis to control the cell cycle. This probably explains why they do not have similar proteins to those used by eukaryotes to control their cell cycle, such as cdk1. This has led to research on what the control mechanism is in prokaryotes. Recent evidence suggests that it may be membrane- or lipid-based.

Genetic Adaptation

E. coli and related bacteria possess the ability to transfer DNA via bacterial conjugation or transduction, which allows genetic material to spread horizontally through an existing population. The process of transduction, which uses the bacterial virus called a bacteriophage, is where the spread of the gene encoding for the Shiga toxin from the *Shigella* bacteria to *E. coli* helped produce *E. coli* O157:H7, the Shiga toxin-producing strain of *E. coli*.

Diversity

E. coli encompasses an enormous population of bacteria that exhibit a very high degree of both genetic and phenotypic diversity. Genome sequencing of a large number of isolates of *E. coli* and related bacteria shows that a taxonomic reclassification would be desirable. However, this has not been done, largely due to its medical importance, and *E. coli* remains one of the most diverse bacterial species: only 20% of the genes in a typical *E. coli* genome is shared among all strains.

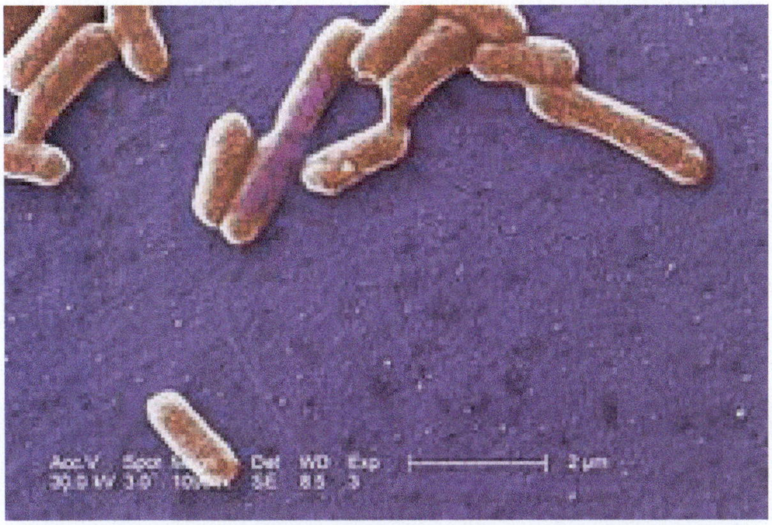

Scanning electron micrograph of an E. coli colony.

In fact, from the evolutionary point of view, the members of genus *Shigella* (*S. dysenteriae*, *S. flexneri*, *S. boydii*, and *S. sonnei*) should be classified as *E. coli* strains, a phenomenon termed taxa in disguise. Similarly, other strains of *E. coli* (e.g. the K-12 strain commonly used in recombinant DNA work) are sufficiently different that they would merit reclassification.

A strain is a subgroup within the species that has unique characteristics that distinguish it from other strains. These differences are often detectable only at the molecular level; however, they may result in changes to the physiology or lifecycle of the bacterium. For example, a strain may gain pathogenic capacity, the ability to use a unique carbon source, the ability to take upon a particular ecological niche, or the ability to resist antimicrobial agents. Different strains of *E. coli* are often host-specific, making it possible to determine the source of fecal contamination in environmental samples. For example, knowing which *E. coli* strains are present in a water sample allows researchers to make assumptions about whether the contamination originated from a human, another mammal, or a bird.

Serotypes

A common subdivision system of *E. coli*, but not based on evolutionary relatedness, is by serotype, which is based on major surface antigens (O antigen: part of lipopolysaccharide layer; H: flagellin; K antigen: capsule), e.g. O157:H7). It is, however, common to cite only the serogroup, i.e. the O-antigen. At present, about 190 serogroups are known. The common laboratory strain has a mutation that prevents the formation of an O-antigen and is thus not typeable.

Genome Plasticity and Evolution

Like all lifeforms, new strains of *E. coli* evolve through the natural biological processes of mutation, gene duplication, and horizontal gene transfer; in particular, 18% of the genome of the laboratory strain MG1655 was horizontally acquired since the divergence from *Salmonella*. *E. coli* K-12 and *E. coli* B strains are the most frequently used varieties for laboratory purposes. Some strains develop traits that can be harmful to a host animal. These virulent strains typically cause a bout of diarrhea that is often self-limiting in healthy adults but is frequently lethal to children in the developing world. More virulent strains, such as O157:H7, cause serious illness or death in the elderly, the very young, or the immunocompromised.

The genera *Escherichia* and *Salmonella* diverged around 102 million years ago (credibility interval: 57–176 mya) which coincides with the divergence of their hosts: the former being found in mammals and the latter in birds and reptiles. This was followed by a split of an *Escherichia* ancestor into five species (*E. albertii*, *E. coli*, *E. fergusonii*, *E. hermannii*, and *E. vulneris*). The last *E. coli* ancestor split between 20 and 30 million years ago.

The long-term evolution experiments using *E. coli*, begun by Richard Lenski in 1988, have allowed direct observation of major evolutionary shifts in the laboratory. In this experiment, one population of *E. coli* unexpectedly evolved the ability to aerobically metabolize citrate, which is extremely rare in *E. coli*. As the inability to grow aerobically is normally used as a diagnostic criterion with which to differentiate *E. coli* from other, closely related bacteria, such as *Salmonella*, this innovation may mark a speciation event observed in the laboratory.

Neotype Strain

E. coli is the type species of the genus (*Escherichia*) and in turn *Escherichia* is the type genus of the family Enterobacteriaceae, where the family name does not stem from the genus *Enterobacter* + "i" (sic.) + "aceae", but from "enterobacterium" + "aceae" (enterobacterium being not a genus, but an alternative trivial name to enteric bacterium).

The original strain described by Escherich is believed to be lost, consequently a new type strain (neotype) was chosen as a representative: the neotype strain is U5/41^T, also known under the deposit names DSM 30083, ATCC 11775, and NCTC 9001, which is pathogenic to chickens and has an O1:K1:H7 serotype. However, in most studies, either O157:H7, K-12 MG1655, or K-12 W3110 were used as a representative *E. coli*. The genome of the type strain has only lately been sequenced. Particularly the use of whole genome sequences yields highly supported phylogenies. Based on such data, five subspecies of *E. coli* were distinguished.

The link between phylogenetic distance ("relatedness") and pathology is small, *e.g.* the O157:H7 serotype strains, which form a clade ("an exclusive group")—group E below—are all enterohaemorragic strains (EHEC), but not all EHEC strains are closely related. In fact, four different species of *Shigella* are nested among *E. coli* strains (*vide supra*), while *E. albertii* and *E. fergusonii* are outside of this group. Indeed, all *Shigella* species were placed within a single subspecies of *E. coli* in a phylogenomic study that included the type strain, and for this reason an according reclassification is difficult. All commonly used research strains of *E. coli* belong to group A and are derived mainly from Clifton's K-12 strain (λ^+ F^+; O16) and to a lesser degree from d'Herelle's *Bacillus coli* strain (B strain)(O7).

Genomics

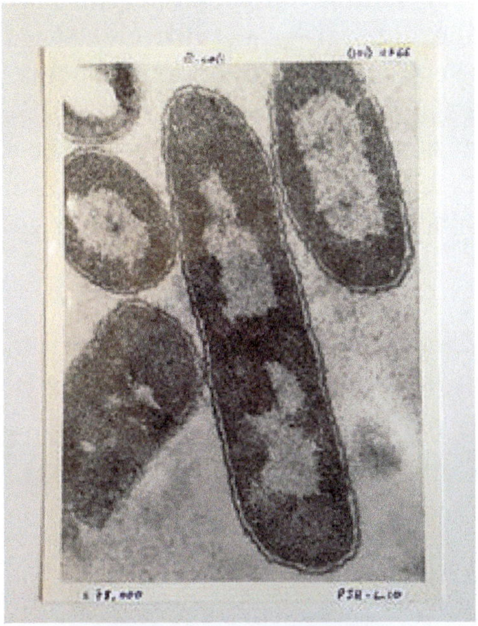

Early electron microscopy

The first complete DNA sequence of an *E. coli* genome (laboratory strain K-12 derivative MG1655) was published in 1997. It was found to be a circular DNA molecule 4.6 million base pairs in length, containing 4288 annotated protein-coding genes (organized into 2584 operons), seven ribosomal RNA (rRNA) operons, and 86 transfer RNA (tRNA) genes. Despite having been the subject of intensive genetic analysis for about 40 years, a large number of these genes were previously unknown. The coding density was found to be very high, with a mean distance between genes of only 118 base pairs. The genome was observed to contain a significant number of transposable genetic elements, repeat elements, cryptic prophages, and bacteriophage remnants.

Today, several hundred complete genomic sequences of *Escherichia* and *Shigella* species are available. The genome sequence of the type strain of *E. coli* has been added to this collection not before 2014. Comparison of these sequences shows a remarkable amount of diversity; only about 20% of each genome represents sequences present in every one of the isolates, while around 80% of each genome can vary among isolates. Each individual genome contains between 4,000 and 5,500 genes, but the total number of different genes among all of the sequenced *E. coli* strains (the pangenome) exceeds 16,000. This very large variety of component genes has been interpreted to mean that two-thirds of the *E. coli* pangenome originated in other species and arrived through the process of horizontal gene transfer.

Gene Nomenclature

Genes in *E. coli* are usually named by 4-letter acronyms that derive from their function (when known). For instance, recA is named after its role in homologous recombination plus the letter A. Functionally related genes are named recB, recC, recD etc. The proteins are named by uppercase acronyms, e.g. RecA, RecB, etc. When the genome of *E. coli* was sequenced, all genes were numbered (more or less) in their order on the genome and abbreviated by b numbers, such as b2819 (=recD) etc. The "b" names were created after Fred Blattner who led the genome sequence effort. Another numbering system was introduced with the sequence of another *E. coli* strain, W3110, which was sequenced in Japan and hence uses numbers starting by JW... (Japanese W3110), e.g. JW2787 (= recD). Hence, recD = b2819 = JW2787. Note, however, that most databases have their own numbering system, e.g. the EcoGene database uses EG10826 for recD. Finally, ECK numbers are specifically used for alleles in the MG1655 strain of *E. coli* K-12. Complete lists of genes and their synonyms can be obtained from databases such as EcoGene or Uniprot.

Proteomics

Proteome

Several studies have investigated the proteome of *E. coli*. By 2006, 1,627 (38%) of the 4,237 open reading frames (ORFs) had been identified experimentally.

Interactome

The interactome of *E. coli* has been studied by affinity purification and mass spectrometry (AP/MS) and by analyzing the binary interactions among its proteins.

Protein complexes. A study purified 4,339 proteins from cultures of strain K-12 and found interacting partners for 2,667 proteins, many of which had unknown functions at the time. A study found 5,993 interactions between proteins of the same *E. coli* strain, though these data showed little overlap with those of the publication.

Binary interactions. Rajagopala *et al.* have carried out systematic yeast two-hybrid screens with most *E. coli* proteins, and found a total of 2,234 protein-protein interactions. This study also integrated genetic interactions and protein structures and mapped 458 interactions within 227 protein complexes.

Normal Microbiota

E. coli belongs to a group of bacteria informally known as coliforms that are found in the gastrointestinal tract of warm-blooded animals. *E. coli* normally colonizes an infant's gastrointestinal tract within 40 hours of birth, arriving with food or water or from the individuals handling the child. In the bowel, *E. coli* adheres to the mucus of the large intestine. It is the primary facultative anaerobe of the human gastrointestinal tract. (Facultative anaerobes are organisms that can grow in either the presence or absence of oxygen.) As long as these bacteria do not acquire genetic elements encoding for virulence factors, they remain benign commensals.

Therapeutic use

Nonpathogenic *E. coli* strain Nissle 1917, also known as Mutaflor, and *E. coli* O83:K24:H31 (known as Colinfant) are used as probiotic agents in medicine, mainly for the treatment of various gastroenterological diseases, including inflammatory bowel disease.

Role in Disease

Most *E. coli* strains do not cause disease, but virulent strains can cause gastroenteritis, urinary tract infections, and neonatal meningitis. It can also be characterized by severe abdominal cramps, diarrhea that typically turns bloody within 24 hours, and sometimes fever. In rarer cases, virulent strains are also responsible for bowel necrosis (tissue death) and perforation without progressing to hemolytic-uremic syndrome, peritonitis, mastitis, septicemia, and gram-negative pneumonia.

There is one strain, *E.coli* #0157:H7, that produces the Shiga toxin (classified as a bioterrorism agent). This toxin causes premature destruction of the red blood cells, which then clog the body's filtering system, the kidneys, causing hemolytic-uremic syndrome (HUS). This in turn causes strokes due to small clots of blood which lodge in capillaries in the brain. This causes the body parts controlled by this region of the brain not to work properly. In addition, this strain causes the buildup of fluid (since the kidneys do not work), leading to edema around the lungs and legs and arms. This increase in fluid buildup especially around the lungs impedes the functioning of the heart, causing an increase in blood pressure.

Uropathogenic *E. coli* (UPEC) is one of the main causes of urinary tract infections. It is part of the normal flora in the gut and can be introduced in many ways. In particular for females, the direction of wiping after defecation (wiping back to front) can lead to fecal contamination of the urogenital orifices. Anal intercourse can also introduce this bacterium into the male urethra, and in switching from anal to vaginal intercourse, the male can also introduce UPEC to the female urogenital system.

In May 2011, one *E. coli* strain, O104:H4, was the subject of a bacterial outbreak that began in Germany. Certain strains of *E. coli* are a major cause of foodborne illness. The outbreak started when several people in Germany were infected with enterohemorrhagic *E. coli* (EHEC) bacteria, leading to hemolytic-uremic syndrome (HUS), a medical emergency that requires urgent treatment. The outbreak did not only concern Germany, but also 11 other countries, including regions in North America. On 30 June 2011, the German *Bundesinstitut für Risikobewertung (BfR)* (Federal Institute for Risk Assessment, a federal institute within the German Federal Ministry of Food, Agriculture and Consumer Protection) announced that seeds of fenugreek from Egypt were likely the cause of the EHEC outbreak.

Treatment

The mainstay of treatment is the assessment of dehydration and replacement of fluid and electrolytes. Administration of antibiotics has been shown to shorten the course of illness and duration of excretion of enterotoxigenic *E. coli* (ETEC) in adults in endemic areas and in traveller's diarrhoea, though the rate of resistance to commonly used antibiotics is increasing and they are generally not recommended. The antibiotic used depends upon susceptibility patterns in the particular geographical region. Currently, the antibiotics of choice are fluoroquinolones or azithromycin, with an emerging role for rifaximin. Oral rifaximin, a semisynthetic rifamycin derivative, is an effective and well-tolerated antibacterial for the management of adults with non-invasive traveller's diarrhoea. Rifaximin was significantly more effective than placebo and no less effective than ciprofloxacin in reducing the duration of diarrhoea. While rifaximin is effective in patients with *E. co-*

li-predominant traveller's diarrhoea, it appears ineffective in patients infected with inflammatory or invasive enteropathogens.

Prevention

ETEC is the type of *E. coli* that most vaccine development efforts are focused on. Antibodies against the LT and major CFs of ETEC provide protection against LT-producing ETEC expressing homologous CFs. Oral inactivated vaccines consisting of toxin antigen and whole cells, i.e. the licensed recombinant cholera B subunit (rCTB)-WC cholera vaccine Dukoral have been developed. There are currently no licensed vaccines for ETEC, though several are in various stages of development. In different trials, the rCTB-WC cholera vaccine provided high (85–100%) short-term protection. An oral ETEC vaccine candidate consisting of rCTB and formalin inactivated *E. coli* bacteria expressing major CFs has been shown in clinical trials to be safe, immunogenic, and effective against severe diarrhoea in American travelers but not against ETEC diarrhoea in young children in Egypt. A modified ETEC vaccine consisting of recombinant *E. coli* strains over expressing the major CFs and a more LT-like hybrid toxoid called LCTBA, are undergoing clinical testing.

Other proven prevention methods for *E. coli* transmission include handwashing and improved sanitation and drinking water, as transmission occurs through fecal contamination of food and water supplies.

Causes and Risk Factors

- Working around livestock
- Consuming unpasteurized dairy product
- Eating undercooked meat
- Drinking impure water

Model Organism in Life Science Research

Role in Biotechnology

Because of its long history of laboratory culture and ease of manipulation, *E. coli* plays an important role in modern biological engineering and industrial microbiology. The work of Stanley Norman Cohen and Herbert Boyer in *E. coli*, using plasmids and restriction enzymes to create recombinant DNA, became a foundation of biotechnology.

E. coli is a very versatile host for the production of heterologous proteins, and various protein expression systems have been developed which allow the production of recombinant proteins in *E. coli*. Researchers can introduce genes into the microbes using plasmids which permit high level expression of protein, and such protein may be mass-produced in industrial fermentation processes. One of the first useful applications of recombinant DNA technology was the manipulation of *E. coli* to produce human insulin.

Many proteins previously thought difficult or impossible to be expressed in *E. coli* in folded form have been successfully expressed in *E. coli*. For example, proteins with multiple disulphide bonds may be produced in the periplasmic space or in the cytoplasm of mutants rendered sufficiently oxidizing to allow disulphide-bonds to form, while proteins requiring post-translational modification such as glycosylation for stability or function have been expressed using the N-linked glycosylation system of *Campylobacter jejuni* engineered into *E. coli*.

Modified *E. coli* cells have been used in vaccine development, bioremediation, production of biofuels, lighting, and production of immobilised enzymes.

Model Organism

E. coli is frequently used as a model organism in microbiology studies. Cultivated strains (e.g. *E. coli* K12) are well-adapted to the laboratory environment, and, unlike wild-type strains, have lost their ability to thrive in the intestine. Many laboratory strains lose their ability to form biofilms. These features protect wild-type strains from antibodies and other chemical attacks, but require a large expenditure of energy and material resources.

In 1946, Joshua Lederberg and Edward Tatum first described the phenomenon known as bacterial conjugation using *E. coli* as a model bacterium, and it remains the primary model to study conjugation. *E. coli* was an integral part of the first experiments to understand phage genetics, and early researchers, such as Seymour Benzer, used *E. coli* and phage T4 to understand the topography of gene structure. Prior to Benzer's research, it was not known whether the gene was a linear structure, or if it had a branching pattern.

E. coli was one of the first organisms to have its genome sequenced; the complete genome of *E. coli* K12 was published by *Science* in 1997.

By evaluating the possible combination of nanotechnologies with landscape ecology, complex habitat landscapes can be generated with details at the nanoscale. On such synthetic ecosystems, evolutionary experiments with *E. coli* have been performed to study the spatial biophysics of adaptation in an island biogeography on-chip.

Studies are also being performed attempting to program *E. coli* to solve complicated mathematics problems, such as the Hamiltonian path problem.

History

In 1885, the German-Austrian pediatrician Theodor Escherich discovered this organism in the feces of healthy individuals. He called it *Bacterium coli commune* because it is found in the colon. Early classifications of prokaryotes placed these in a handful of genera based on their shape and motility (at that time Ernst Haeckel's classification of bacteria in the kingdom Monera was in place).

Bacterium coli was the type species of the now invalid genus *Bacterium* when it was revealed that the former type species ("*Bacterium triloculare*") was missing. Following a revision of *Bacterium*, it was reclassified as *Bacillus coli* by Migula in 1895 and later reclassified in the newly created genus *Escherichia*, named after its original discoverer.

Transcription Process

Sequence of events in Initiation

- RNA-polymerase recognizes the promoter region, binds there and forms closed complex. This leads to unwinding the DNA duplex.
- The unwound region is about 17 ± 1 bp (open complex).

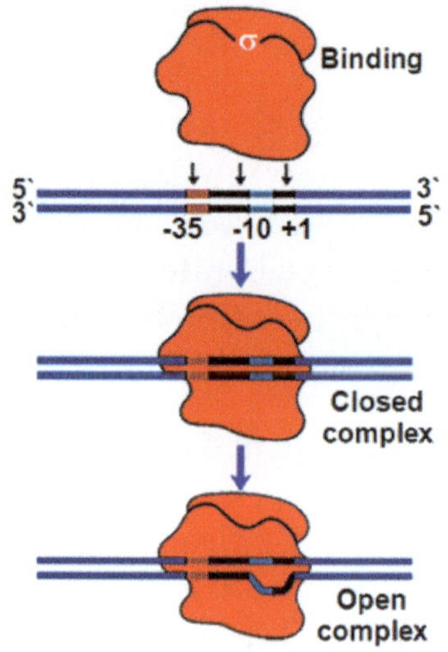

- The first nucleotide on RNA transcript is always purine triphosphate. GTP is more often than ATP.
- The pppGpN-OH structure remains on the RNA transcript until the RNA synthesis is completed.
- The σ subunit detaches from the RNA-pol once the first 3' - 5' phosphodiester bond gets synthesized.
- The nucleotide added in the newly synthesized RNA are complementary to that of template DNA.
- The core enzyme moves along the DNA template to enter the elongation phase.

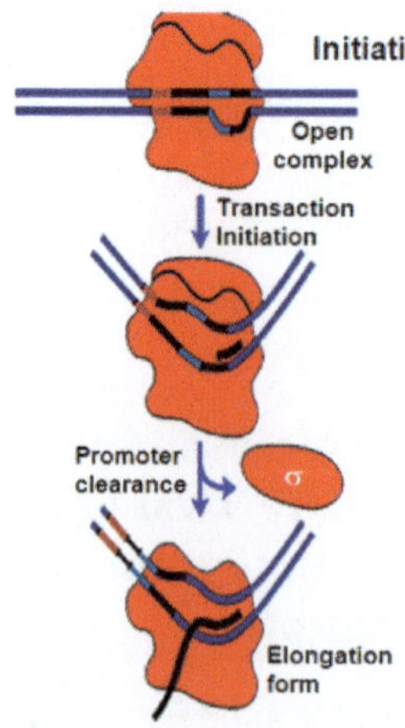

Elongation

- The release of the σ subunit brings about conformational change of the core enzyme.
- As RNA polymerase slides on the DNA template toward the 3' end, the free NTPs are sequentially added to the 3' -OH of the nascent RNA strand.

$$(NMP)_n + NTP \longrightarrow (NMP)_{n+1} + PPi$$

RNA strand — substrate — elongated RNA strand

- The bases are added complementary to that of template following the Watson –Crick base pairing. eg A-T and G-C.
- RNA-pol, DNA segment of ~40 nucleotides and the nascent RNA form a complex called the transcription bubble.
- The 3' segment of the nascent RNA hybridizes with the DNA template, and its 5' end extends out the transcription bubble as the synthesis is processing.

Transcription Bubble

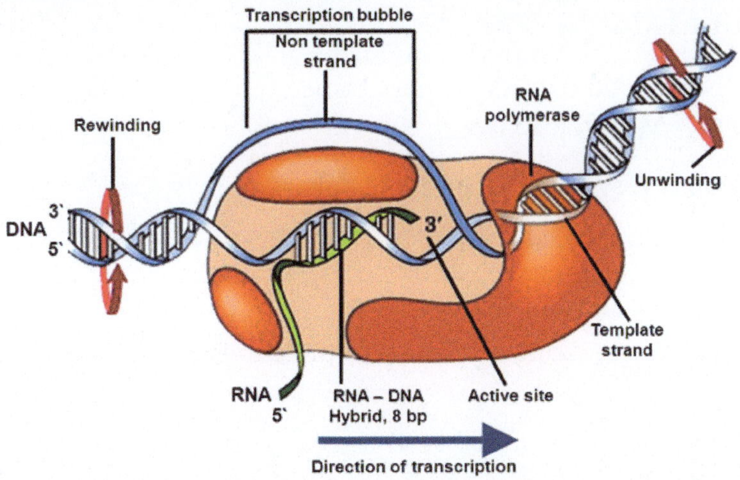

Termination

- The termination occurs in either ρ-dependent or ρ-independent manner.

ρ-dependent termination :

- The ρ factor, a hexamer, is a ATPase and a helicase. This protein associates itself with the RNA a specific sites and migrates in 5'-3'direction till it reaches termination site. The helicase activity of the ρ factor enables termination process using energy from ATP. The exact mechanism is not yet known.

ρ-independent termination :

- The termination signal is a stretch of 30-40 nucleotides on the RNA transcript, consisting of many GC followed by a series of U.
- The sequence specificity of this nascent RNA transcript is such that these are self complimentary seuences and will form particular stem-loop/ hair-pin like structures to terminate the transcription.
- When RNA pol arrives at a termination site resembling above structure, trancription pauses. The RNA-pol stops moving on the DNA template. The RNA transcript detaches from the transcription complex.

RNA Polymerase of Eukaryotes

- Eukaryotes have three RNA polymerases(I, II, III).

RNA polymerase	Function
RNA polymerase I	•Synthesis of pre-ribosomal RNA
RNA polymerase II	•Synthesis of m RNA and some specialized RNAs. •Ability to recognize a number of promoter regions varying greatly in sequence.
RNA polymerase III	•Synthesis of tRNA`s, 5S r RNA, small specialized RNAs.

Post-transcriptional Modifications

- The nascent RNA, also known as primary transcript, needs to be modified to become functional tRNAs, rRNAs, and mRNAs.
- The modification is very much essential to eukaryotic systems.
- Post transcriptional modification includes

  -> Capping at the 5'- end.

  -> Tailing at the 3'- end.

  -> mRNA splicing.

Capping at the 5'- End

- 7- methyl guanosine is added at the 5' end of almost all eukaryotic RNAs by a 5'-5' triphosphate linkage.
- The capping process occurs in nuclei, prior to the splicing. m^7GpppGp----
- The 5'cap protects the RNA from ribonucleases.
- The cap structure of mRNA will be recognized by the cap-binding protein required for translation.

- The cap binds to cap binding complex of proteins which promotes binding of the mRNA to the ribosome to initiate the process of translation.
- Generation of the 5' cap occurs in a series of three to four steps shown below.

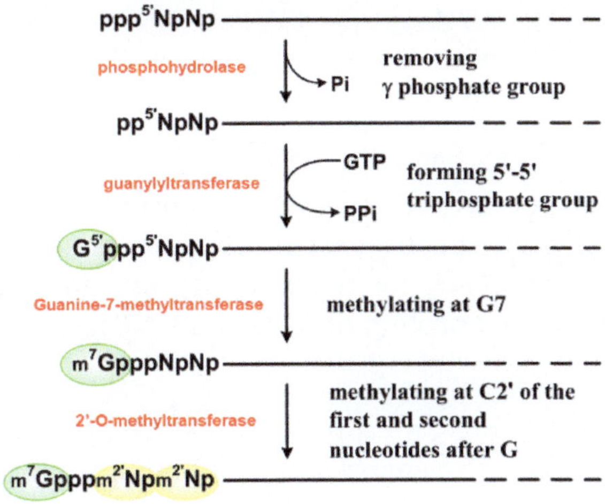

Poly-A Tail at 3' -END

- Eukaryotic RNAs posses a string of 80 to 250 A residues at the 3' end.
- This poly-A tail protects mRNA from enzymatic destruction, also acts as binding site for many proteins.
- There is no poly-T sequence on the DNA template from which the RNA is transcribed. The tailing process dose not depend on the template.
- The tailing process occurs in the nuclei, prior to the splicing.
- The tailing process takes place in multiple steps with the help of the enzyme polyadenylate polymerase.
- RNA + nATP -----> RNA-$(AMP)_n$ + nPP_i (n = 80 - 250).
- The enzyme uses the cleaved mRNA as a primer; does not require any template.

mRNA Splicing

- The matured mRNAs are much shorter than the DNA templates.
- The mRNA which has been transcribed contains sequences encompassing one gene. However, the structural genes are composed of coding and non-coding regions that are alternatively separated. Hence, these are also termed as "split genes".
- Exons are the coding sequences that appear on split genes and primary transcripts, and will be expressed to matured mRNA.
- Introns are the non-coding sequences that are transcripted into primary mRNAs.

- During splicing, the introns are selectively removed from the primary transcript while the exons are joined together to generate a continuous sequence that will code entirely for a functional protein.

Spilicing is Attained by Different Mechanisms

1. Out of the 4 classes of introns, group I and group II introns are self-splicing in nature wherein no enzymes, or energy source (ATP) is needed. Splicing occurs by two simple transesterification steps.
2. The next class of introns are not self-splicing in nature but require involvement of a large protein complex called spliceosome. They are hence called spliceosomal introns. The spliceosome is responsible for removing introns and splicing exons together. It is made up of small ribonucleoprotein particles (snRNPs) which recognize the intron-exon boundaries.
3. The fourth category of introns differ from the other three in the requirement of energy as ATP and an endonuclease. The endonuclease is functional in cleaving the phosphodiester bonds at the two ends of the intron while the two exons are joined by a reaction similar to that followed by DNA ligases.

Concept of Gene and Genome

Gene: Unit of Heredity

- The segment of DNA that carries genetic information are called genes.
- It is normally a stretch of DNA that codes for a protein or for an RNA chain that has a functional/structural role to play in the living cells.
- Genes hold the information to build and maintain an organism's cells and pass genetic traits to offspring.

Genome

- Genome is the sum total of an organism's hereditary information. It is encoded either in DNA or, for many types of virus, in RNA. The genome includes both the genes and the non-coding sequences of the DNA.

The Genetic Code

- The genetic code consists of the sequence of bases (A, G, C and U) found along the mRNA molecule.
- If triplets of bases represented each amino acid, this would be able to produce $4^3 = 64$ combinations, enough combinations to code for the 20 amino acids.
- The Genetic code therefore allows for correspondence between triplets of bases in DNA and the amino acid sequence of a polypeptide (protein).

Stop and Start Codons

- Francis Crick and Sydney Brenner determined that the DNA is read in sets of 3 nucleotides for each amino acid.
- Marshall Nirenberg and Hargovind Khorana identified the codons that specify each amino acid.
- Start codon = AUG (codes for methionine) - site where translation begins.
- Stop codons = UAA, UAG and UGA - sites where translation ends.
- Degeneracy : Most amino acids have more than one codon. For example, glycine is encoded by GGU GGC GGA and GGG.

First base	Second base U	Second base C	Second base A	Second base G	Third base
U	UUU Phe	UCU Ser	UAU Tyr	UGU Cys	U
U	UUC Phe	UCC Ser	UAC Tyr	UGC Cys	C
U	UUA Leu	UCA Ser	UAA Stop	UGA Stop	A
U	UUG Leu	UCG Ser	UAG Stop	UGG Trp	G
C	CUU Leu	CCU Pro	CAU His	CGU Arg	U
C	CUC Leu	CCC Pro	CAC His	CGC Arg	C
C	CUA Leu	CCA Pro	CAA Gln	CGA Arg	A
C	CUG Leu	CCG Pro	CAG Gln	CGG Arg	G
A	AUU Ile	ACU Thr	AAU Asn	AGU Ser	U
A	AUC Ile	ACC Thr	AAC Asn	AGC Ser	C
A	AUA Ile	ACA Thr	AAA Lys	AGA Arg	A
A	AUG Met	ACG Thr	AAG Lys	AGG Arg	G
G	GUU Val	GCU Ala	GAU Asp	GGU Gly	U
G	GUC Val	GCC Ala	GAC Asp	GGC Gly	C
G	GUA Val	GCA Ala	GAA Glu	GGA Gly	A
G	GUG Val	GCG Ala	GAG Glu	GGG Gly	G

The Wobble Hypothesis

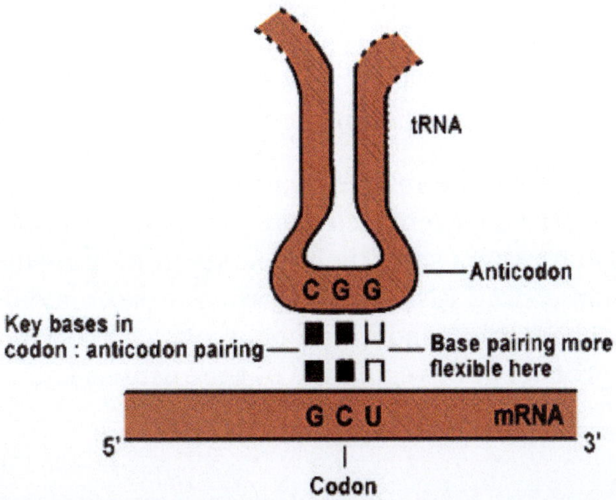

In 1966, Francis Crick proposed the Wobble hypothesis wherein the same amino acid may be coded by more than one number of codons. This is called the Degeneracy of the genetic code. According to the Wobble hypothesis, the third position of a codon is often interchangeable. For example : GCU codes for alanine and so does GCC, tRNA with anticodon CGG can bind to this codon and bring alanine.

Gene

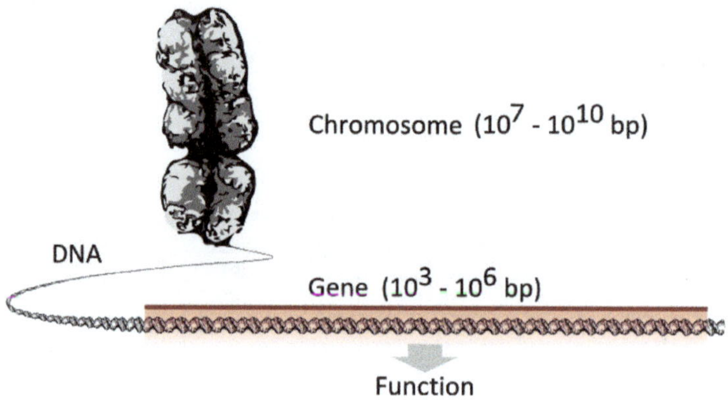

A gene is a region of DNA that encodes function. A chromosome consists of a long strand of DNA containing many genes. A human chromosome can have up to 500 million base pairs of DNA with thousands of genes.

A gene is a locus (or region) of DNA which is made up of nucleotides and is the molecular unit of heredity. The transmission of genes to an organism's offspring is the basis of the inheritance of phenotypic traits. These genes make up different DNA sequences called genotypes. Genotypes along with environmental and developmental factors determine what the phenotypes will be. Most biological traits are under the influence of polygenes (many different genes) as well as gene–environment interactions. Some genetic traits are instantly visible, such as eye color or number of limbs, and some are not, such as blood type, risk for specific diseases, or the thousands of basic biochemical processes that comprise life.

Genes can acquire mutations in their sequence, leading to different variants, known as alleles, in the population. These alleles encode slightly different versions of a protein, which cause different phenotypical traits. Usage of the term "having a gene" (e.g., "good genes," "hair colour gene") typically refers to containing a different allele of the same, shared gene. Genes evolve due to natural selection or survival of the fittest of the alleles.

The concept of a gene continues to be refined as new phenomena are discovered. For example, regulatory regions of a gene can be far removed from its coding regions, and coding regions can be split into several exons. Some viruses store their genome in RNA instead of DNA and some gene products are functional non-coding RNAs. Therefore, a broad, modern working definition of a gene is any discrete locus of heritable, genomic sequence which affect an organism's traits by being expressed as a functional product or by regulation of gene expression.

History

The existence of discrete inheritable units was first suggested by Gregor Mendel (1822–1884). From 1857 to 1864, in Brno (Czech Republic), he studied inheritance patterns in 8000 common edible pea plants, tracking distinct traits from parent to offspring. He described these mathematically as 2^n combinations where n is the number of differing characteristics in the original peas. Although he did not use the term *gene*, he explained his results in terms of discrete inherited units that give rise to observable physical characteristics. This description prefigured the distinction between genotype (the genetic material of an organism) and phenotype (the visible traits of that organism). Mendel was also the first to demonstrate independent assortment, the distinction between dominant and recessive traits, the distinction between a heterozygote and homozygote, and the phenomenon of discontinuous inheritance.

Gregor Mendel

Discovery of Discrete Inherited Units

Prior to Mendel's work, the dominant theory of heredity was one of blending inheritance, which suggested that each parent contributed fluids to the fertilisation process and that the traits of the parents blended and mixed to produce the offspring. Charles Darwin developed a theory of inheritance he termed pangenesis, from Greek pan ("all, whole") and genesis ("birth") / genos ("origin"). Darwin used the term *gemmule* to describe hypothetical particles that would mix during reproduction.

Mendel's work went largely unnoticed after its first publication, but was rediscovered in the late 19th century by Hugo de Vries, Carl Correns, and Erich von Tschermak, who (claimed to have) reached similar conclusions in their own research. Specifically, Hugo de Vries published his book *Intracellular Pangenesis*, in which he postulated that different characters have individual hereditary carriers and that inheritance of specific traits in organisms comes in particles. De Vries called these units "pangenes" (*Pangens* in German), after Darwin's 1868 pangenesis theory.

Sixteen years later, in 1905, the word *genetics* was first used by William Bateson, while Eduard Strasburger, amongst others, still used the term pangene for the fundamental physical and functional unit of heredity. The Danish botanist Wilhelm Johannsen shortened the name to "gene".

Discovery of DNA

Advances in understanding genes and inheritance continued throughout the 20th century. Deoxyribonucleic acid (DNA) was shown to be the molecular repository of genetic information by experiments in the 1940s to 1950s. The structure of DNA was studied by Rosalind Franklin and Maurice Wilkins using X-ray crystallography, which led James D. Watson and Francis Crick to publish a model of the double-stranded DNA molecule whose paired nucleotide bases indicated a compelling hypothesis for the mechanism of genetic replication.

In the early 1950s the prevailing view was that the genes in a chromosome acted like discrete entities, indivisible by recombination and arranged like beads on a string. The experiments of Benzer using mutants defective in the rII region of bacteriophage T4 (1955-1959) showed that individual genes have a simple linear structure and are likely to be equivalent to a linear section of DNA.

Collectively, this body of research established the central dogma of molecular biology, which states that proteins are translated from RNA, which is transcribed from DNA. This dogma has since been shown to have exceptions, such as reverse transcription in retroviruses. The modern study of genetics at the level of DNA is known as molecular genetics.

In 1972, Walter Fiers and his team at the University of Ghent were the first to determine the sequence of a gene: the gene for Bacteriophage MS2 coat protein. The subsequent development of chain-termination DNA sequencing in 1977 by Frederick Sanger improved the efficiency of sequencing and turned it into a routine laboratory tool. An automated version of the Sanger method was used in early phases of the Human Genome Project.

Modern Evolutionary Synthesis

The theories developed in the 1930s and 1940s to integrate molecular genetics with Darwinian evolution are called the modern evolutionary synthesis, a term introduced by Julian Huxley. Evolutionary biologists subsequently refined this concept, such as George C. Williams' gene-centric view of evolution. He proposed an evolutionary concept of the gene as a unit of natural selection with the definition: "that which segregates and recombines with appreciable frequency." In this view, the molecular gene *transcribes* as a unit, and the evolutionary gene *inherits* as a unit. Related ideas emphasizing the centrality of genes in evolution were popularized by Richard Dawkins.

Molecular Basis

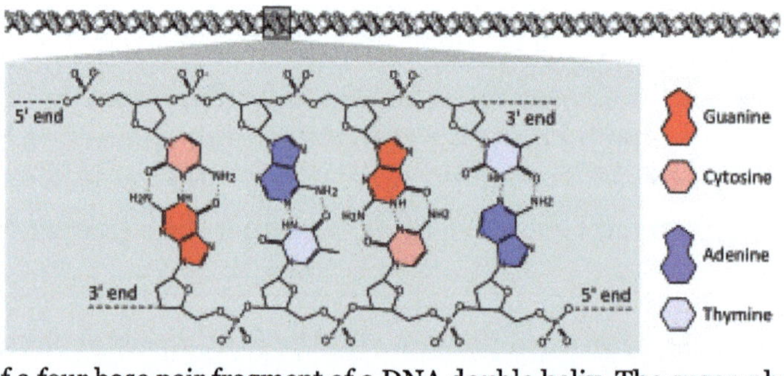

The chemical structure of a four base pair fragment of a DNA double helix. The sugar-phosphate backbone chains run in opposite directions with the bases pointing inwards, base-pairing A to T and C to G with hydrogen bonds.

DNA

The vast majority of living organisms encode their genes in long strands of DNA (deoxyribonucleic acid). DNA consists of a chain made from four types of nucleotide subunits, each composed of: a five-carbon sugar (2'-deoxyribose), a phosphate group, and one of the four bases adenine, cytosine, guanine, and thymine.

Two chains of DNA twist around each other to form a DNA double helix with the phosphate-sugar backbone spiralling around the outside, and the bases pointing inwards with adenine base pairing to thymine and guanine to cytosine. The specificity of base pairing occurs because adenine and thymine align to form two hydrogen bonds, whereas cytosine and guanine form three hydrogen bonds. The two strands in a double helix must therefore be complementary, with their sequence of bases matching such that the adenines of one strand are paired with the thymines of the other strand, and so on.

Due to the chemical composition of the pentose residues of the bases, DNA strands have directionality. One end of a DNA polymer contains an exposed hydroxyl group on the deoxyribose; this is known as the 3' end of the molecule. The other end contains an exposed phosphate group; this is the 5' end. The two strands of a double-helix run in opposite directions. Nucleic acid synthesis, including DNA replication and transcription occurs in the 5'→3' direction, because new nucleotides are added via a dehydration reaction that uses the exposed 3' hydroxyl as a nucleophile.

The expression of genes encoded in DNA begins by transcribing the gene into RNA, a second type of nucleic acid that is very similar to DNA, but whose monomers contain the sugar ribose rather than deoxyribose. RNA also contains the base uracil in place of thymine. RNA molecules are less stable than DNA and are typically single-stranded. Genes that encode proteins are composed of a series of three-nucleotide sequences called codons, which serve as the "words" in the genetic "language". The genetic code specifies the correspondence during protein translation between codons and amino acids. The genetic code is nearly the same for all known organisms.

Chromosomes

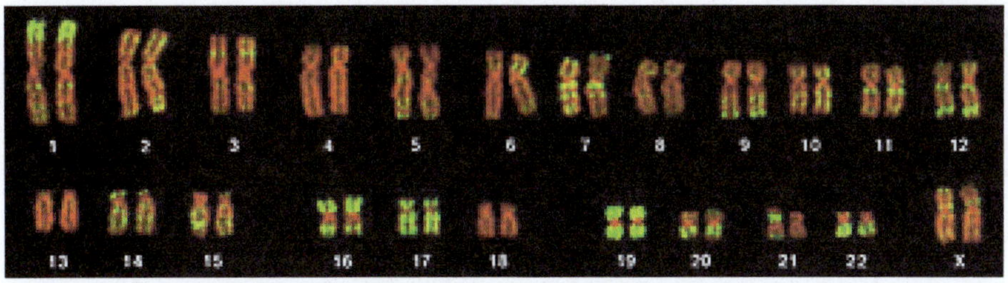

Fluorescent microscopy image of a human female karyotype, showing 23 pairs of chromosomes. The DNA is stained red, with regions rich in housekeeping genes further stained in green. The largest chromosomes are around 10 times the size of the smallest.

The total complement of genes in an organism or cell is known as its genome, which may be stored on one or more chromosomes. A chromosome consists of a single, very long DNA helix on which thousands of genes are encoded. The region of the chromosome at which a particular gene is located is called its locus. Each locus contains one allele of a gene; however, members of a population may have different alleles at the locus, each with a slightly different gene sequence.

The majority of eukaryotic genes are stored on a set of large, linear chromosomes. The chromosomes are packed within the nucleus in complex with storage proteins called histones to form a unit called a nucleosome. DNA packaged and condensed in this way is called chromatin. The manner in which DNA is stored on the histones, as well as chemical modifications of the histone itself, regulate whether a particular region of DNA is accessible for gene expression. In addition to genes, eukaryotic chromosomes contain sequences involved in ensuring that the DNA is copied without degradation of end regions and sorted into daughter cells during cell division: replication origins, telomeres and the centromere. Replication origins are the sequence regions where DNA replication is initiated to make two copies of the chromosome. Telomeres are long stretches of repetitive sequence that cap the ends of the linear chromosomes and prevent degradation of coding and regulatory regions during DNA replication. The length of the telomeres decreases each time the genome is replicated and has been implicated in the aging process. The centromere is required for binding spindle fibres to separate sister chromatids into daughter cells during cell division.

Prokaryotes (bacteria and archaea) typically store their genomes on a single large, circular chromosome. Similarly, some eukaryotic organelles contain a remnant circular chromosome with a small number of genes. Prokaryotes sometimes supplement their chromosome with additional small circles of DNA called plasmids, which usually encode only a few genes and are transferable between individuals. For example, the genes for antibiotic resistance are usually encoded on bacterial plasmids and can be passed between individual cells, even those of different species, via horizontal gene transfer.

Whereas the chromosomes of prokaryotes are relatively gene-dense, those of eukaryotes often contain regions of DNA that serve no obvious function. Simple single-celled eukaryotes have relatively small amounts of such DNA, whereas the genomes of complex multicellular organisms, including humans, contain an absolute majority of DNA without an identified function. This DNA has often been referred to as "junk DNA". However, more recent analyses suggest that, although protein-coding DNA makes up barely 2% of the human genome, about 80% of the bases in the genome may be expressed, so the term "junk DNA" may be a misnomer.

Structure and Function

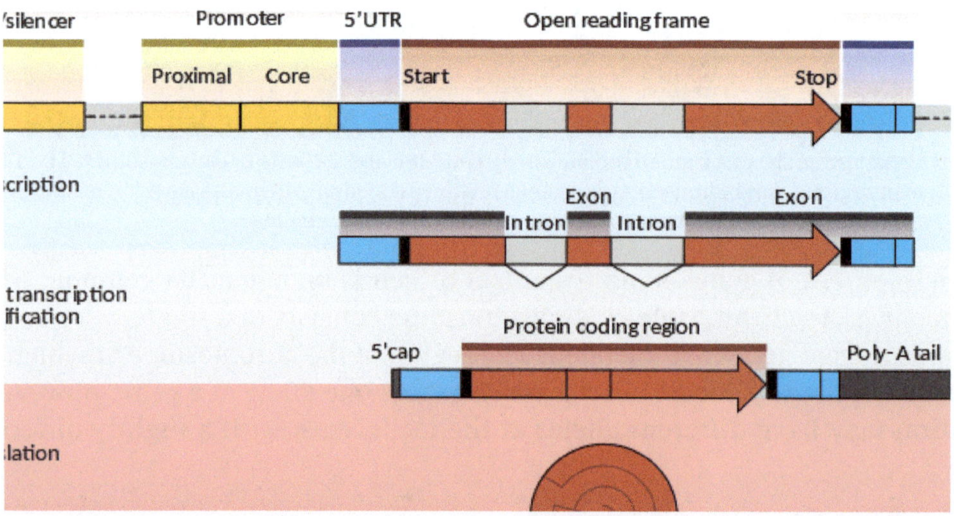

The structure of a eukaryotic protein-coding gene. Regulatory sequence controls when and where expression occurs for the protein coding region (red). Promoter and enhancer regions (yellow) regulate the transcription of the gene into a pre-mRNA which is modified to remove introns (light grey) and add a 5' cap and poly-A tail (dark grey). The mRNA 5' and 3' untranslated regions (blue) regulate translation into the final protein product.

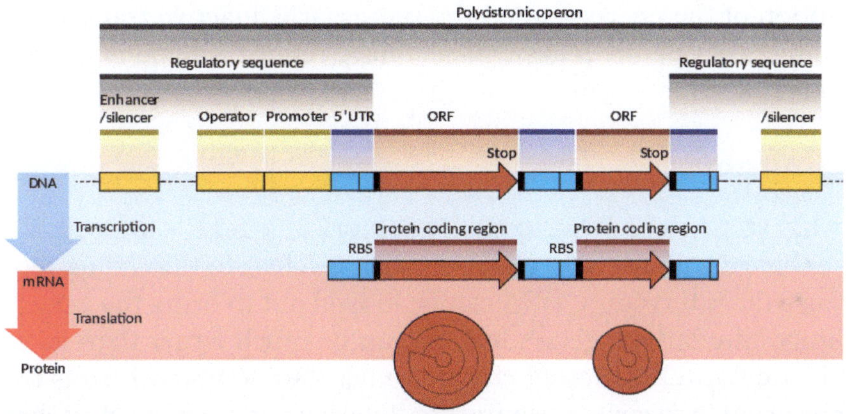

The structure of a prokaryotic operon of protein-coding genes. Regulatory sequence controls when expression occurs for the multiple protein coding regions (red). Promoter, operator and enhancer regions (yellow) regulate the transcription of the gene into an mRNA. The mRNA untranslated regions (blue) regulate translation into the final protein products.

The structure of a gene consists of many elements of which the actual protein coding sequence is often only a small part. These include DNA regions that are not transcribed as well as untranslated regions of the RNA.

Firstly, flanking the open reading frame, all genes contain a regulatory sequence that is required for their expression. In order to be expressed, genes require a promoter sequence. The promoter is recognized and bound by transcription factors and RNA polymerase to initiate transcription. A gene can have more than one promoter, resulting in messenger RNAs (mRNA) that differ in how far they extend in the 5' end. Promoter regions have a consensus sequence, however highly transcribed genes have "strong" promoter sequences that bind the transcription machinery well, whereas others have "weak" promoters that bind poorly and initiate transcription less frequently. Eukaryotic promoter regions are much more complex and difficult to identify than prokaryotic promoters.

Additionally, genes can have regulatory regions many kilobases upstream or downstream of the open reading frame. These act by binding to transcription factors which then cause the DNA to loop so that the regulatory sequence (and bound transcription factor) become close to the RNA polymerase binding site. For example, enhancers increase transcription by binding an activator protein which then helps to recruit the RNA polymerase to the promoter; conversely silencers bind repressor proteins and make the DNA less available for RNA polymerase.

The transcribed pre-mRNA contains untranslated regions at both ends which contain a ribosome binding site, terminator and start and stop codons. In addition, most eukaryotic open reading frames contain untranslated introns which are removed before the exons are translated. The sequences at the ends of the introns, dictate the splice sites to generate the final mature mRNA which encodes the protein or RNA product.

Many prokaryotic genes are organized into operons, with multiple protein-coding sequences that are transcribed as a unit. The genes in an operon are transcribed as a continuous messenger RNA, referred to as a polycistronic mRNA. The term cistron in this context is equivalent to gene. The transcription of an operon's mRNA is often controlled by a repressor that can occur in an active or inactive state depending on the presence of certain specific metabolites. When active, the repressor binds to a DNA sequence at the beginning of the operon, called the operator region, and represses transcription of the operon; when the repressor is inactive transcription of the operon can occur. The products of operon genes typically have related functions and are involved in the same regulatory network.

Functional Definitions

Defining exactly what section of a DNA sequence comprises a gene is difficult. Regulatory regions of a gene such as enhancers do not necessarily have to be close to the coding sequence on the linear molecule because the intervening DNA can be looped out to bring the gene and its regulatory region into proximity. Similarly, a gene's introns can be much larger than its exons. Regulatory regions can even be on entirely different chromosomes and operate *in trans* to allow regulatory regions on one chromosome to come in contact with target genes on another chromosome.

Early work in molecular genetics suggested the concept that one gene makes one protein. This concept (originally called the one gene-one enzyme hypothesis) emerged from an influential 1941 paper by George Beadle and Edward Tatum on experiments with mutants of the fungus Neurospora crassa. Norman Horowitz, an early colleague on the *Neurospora* research, reminisced in 2004 that "these experiments founded the science of what Beadle and Tatum called *biochemical genetics*. In actuality they proved to be the opening gun in what became molecular genetics and all the developments that have followed from that." The one gene-one protein concept has been refined since the discovery of genes that can encode multiple proteins by alternative splicing and coding sequences split in short section across the genome whose mRNAs are concatenated by trans-splicing.

A broad operational definition is sometimes used to encompass the complexity of these diverse phenomena, where a gene is defined as a union of genomic sequences encoding a coherent set of potentially overlapping functional products. This definition categorizes genes by their functional products (proteins or RNA) rather than their specific DNA loci, with regulatory elements classified as *gene-associated* regions.

Gene Expression

In all organisms, two steps are required to read the information encoded in a gene's DNA and produce the protein it specifies. First, the gene's DNA is *transcribed* to messenger RNA (mRNA).[:6.1] Second, that mRNA is *translated* to protein. RNA-coding genes must still go through the first step, but are not translated into protein. The process of producing a biologically functional molecule of either RNA or protein is called gene expression, and the resulting molecule is called a gene product.

Genetic Code

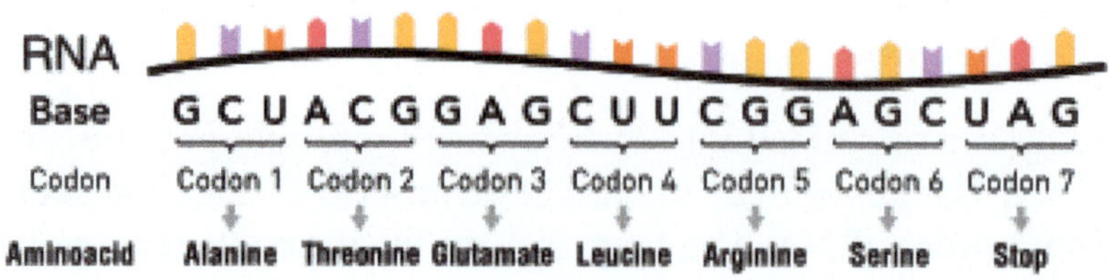

Schematic of a single-stranded RNA molecule illustrating a series of three-base codons. Each three-nucleotide codon corresponds to an amino acid when translated to protein

The nucleotide sequence of a gene's DNA specifies the amino acid sequence of a protein through the genetic code. Sets of three nucleotides, known as codons, each correspond to a specific amino acid. The principle that three sequential bases of DNA code for each amino acid was demonstrated in 1961 using frameshift mutations in the rIIB gene of bacteriophage T4.

Additionally, a "start codon", and three "stop codons" indicate the beginning and end of the protein coding region. There are 64 possible codons (four possible nucleotides at each of three positions, hence 4^3 possible codons) and only 20 standard amino acids; hence the code is redundant and multiple codons can specify the same amino acid. The correspondence between codons and amino acids is nearly universal among all known living organisms.

Transcription

Transcription produces a single-stranded RNA molecule known as messenger RNA, whose nucleotide sequence is complementary to the DNA from which it was transcribed. The mRNA acts as an intermediate between the DNA gene and its final protein product. The gene's DNA is used as a template to generate a complementary mRNA. The mRNA matches the sequence of the gene's DNA coding strand because it is synthesised as the complement of the template strand. Transcription is performed by an enzyme called an RNA polymerase, which reads the template strand in the 3' to 5' direction and synthesizes the RNA from 5' to 3'. To initiate transcription, the polymerase first recognizes and binds a promoter region of the gene. Thus, a major mechanism of gene regulation is the blocking or sequestering the promoter region, either by tight binding by repressor molecules that physically block the polymerase, or by organizing the DNA so that the promoter region is not accessible.

In prokaryotes, transcription occurs in the cytoplasm; for very long transcripts, translation may begin at the 5' end of the RNA while the 3' end is still being transcribed. In eukaryotes, transcription occurs in the nucleus, where the cell's DNA is stored. The RNA molecule produced by the polymerase is known as the primary transcript and undergoes post-transcriptional modifications before being exported to the cytoplasm for translation. One of the modifications performed is the splicing of introns which are sequences in the transcribed region that do not encode protein. Alternative splicing mechanisms can result in mature transcripts from the same gene having different sequences and thus coding for different proteins. This is a major form of regulation in eukaryotic cells and also occurs in some prokaryotes.

Translation

Translation is the process by which a mature mRNA molecule is used as a template for synthesizing a new protein.Translation is carried out by ribosomes, large complexes of RNA and protein responsible for carrying out the chemical reactions to add new amino acids to a growing polypeptide chain by the formation of peptide bonds. The genetic code is read three nucleotides at a time, in units called codons, via interactions with specialized RNA molecules called transfer RNA (tRNA). Each tRNA has three unpaired bases known as the anticodon that are complementary to the codon it reads on the mRNA. The tRNA is also covalently attached to the amino acid specified by the complementary codon. When the tRNA binds to its complementary codon in an mRNA strand, the ribosome attaches its amino acid cargo to the new polypeptide chain, which is synthesized from amino terminus to carboxyl terminus. During and after synthesis, most new proteins must fold to their active three-dimensional structure before they can carry out their cellular functions.

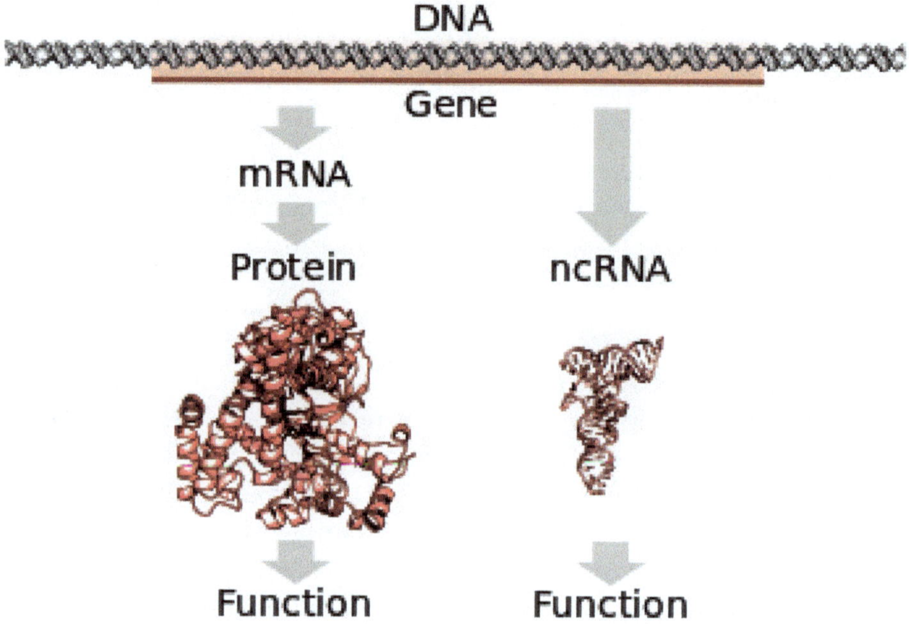

Protein coding genes are transcribed to an mRNA intermediate, then translated to a functional protein. RNA-coding genes are transcribed to a functional non-coding RNA. (PDB: 3BSE, 1OBB, 3TRA)

Regulation

Genes are regulated so that they are expressed only when the product is needed, since expression draws on limited resources. A cell regulates its gene expression depending on its external environment (e.g. available nutrients, temperature and other stresses), its internal environment (e.g. cell division cycle, metabolism, infection status), and its specific role if in a multicellular organism. Gene expression can be regulated at any step from transcriptional initiation, to RNA processing, to post-translational modification of the protein. The regulation of lactose metabolism genes in *E. coli* (*lac* operon) was the first such mechanism to be described in 1961.

RNA Genes

A typical protein-coding gene is first copied into RNA as an intermediate in the manufacture of the final protein product. In other cases, the RNA molecules are the actual functional products, as in the synthesis of ribosomal RNA and transfer RNA. Some RNAs known as ribozymes are capable of enzymatic function, and microRNA has a regulatory role. The DNA sequences from which such RNAs are transcribed are known as non-coding RNA genes.

Some viruses store their entire genomes in the form of RNA, and contain no DNA at all. Because they use RNA to store genes, their cellular hosts may synthesize their proteins as soon as they are infected and without the delay in waiting for transcription. On the other hand, RNA retroviruses, such as HIV, require the reverse transcription of their genome from RNA into DNA before their proteins can be synthesized. RNA-mediated epigenetic inheritance has also been observed in plants and very rarely in animals.

Inheritance

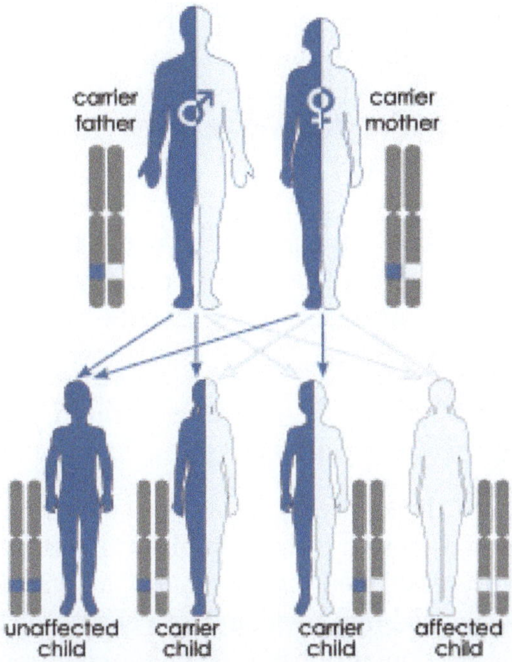

Inheritance of a gene that has two different alleles (blue and white). The gene is located on an autosomal chromosome. The blue allele is recessive to the white allele. The probability of each outcome in the children's generation is one quarter, or 25 percent.

Organisms inherit their genes from their parents. Asexual organisms simply inherit a complete copy of their parent's genome. Sexual organisms have two copies of each chromosome because they inherit one complete set from each parent.

Mendelian Inheritance

According to Mendelian inheritance, variations in an organism's phenotype (observable physical and behavioral characteristics) are due in part to variations in its genotype (particular set of genes). Each gene specifies a particular trait with different sequence of a gene (alleles) giving rise to different phenotypes. Most eukaryotic organisms (such as the pea plants Mendel worked on) have two alleles for each trait, one inherited from each parent

Alleles at a locus may be dominant or recessive; dominant alleles give rise to their corresponding phenotypes when paired with any other allele for the same trait, whereas recessive alleles give rise to their corresponding phenotype only when paired with another copy of the same allele. If you know the genotypes of the organisms, you can determine which alleles are dominant and which are recessive. For example, if the allele specifying tall stems in pea plants is dominant over the allele specifying short stems, then pea plants that inherit one tall allele from one parent and one short allele from the other parent will also have tall stems. Mendel's work demonstrated that alleles assort independently in the production of gametes, or germ cells, ensuring variation in the next generation. Although Mendelian inheritance remains a good model for many traits determined by single genes (including a number of well-known genetic disorders) it does not include the physical processes of DNA replication and cell division.

DNA Replication and Cell Division

The growth, development, and reproduction of organisms relies on cell division; the process by which a single cell divides into two usually identical daughter cells. This requires first making a duplicate copy of every gene in the genome in a process called DNA replication. The copies are made by specialized enzymes known as DNA polymerases, which "read" one strand of the double-helical DNA, known as the template strand, and synthesize a new complementary strand. Because the DNA double helix is held together by base pairing, the sequence of one strand completely specifies the sequence of its complement; hence only one strand needs to be read by the enzyme to produce a faithful copy. The process of DNA replication is semiconservative; that is, the copy of the genome inherited by each daughter cell contains one original and one newly synthesized strand of DNA.

The rate of DNA replication in living cells was first measured as the rate of phage T4 DNA elongation in phage-infected *E. coli* and found to be impressively rapid. During the period of exponential DNA increase at 37 °C, the rate of elongation was 749 nucleotides per second.

After DNA replication is complete, the cell must physically separate the two copies of the genome and divide into two distinct membrane-bound cells. In prokaryotes (bacteria and archaea) this usually occurs via a relatively simple process called binary fission, in which each circular genome attaches to the cell membrane and is separated into the daughter cells as the membrane invaginates to split the cytoplasm into two membrane-bound portions. Binary fission is extremely fast compared to the rates of cell division in eukaryotes. Eukaryotic cell division is a more complex process known as the cell cycle; DNA replication occurs during a phase of this cycle known as S phase, whereas the process of segregating chromosomes and splitting the cytoplasm occurs during M phase.

Molecular Inheritance

The duplication and transmission of genetic material from one generation of cells to the next is the basis for molecular inheritance, and the link between the classical and molecular pictures of genes. Organisms inherit the characteristics of their parents because the cells of the offspring contain copies of the genes in their parents' cells. In asexually reproducing organisms, the offspring will be a genetic copy or clone of the parent organism. In sexually reproducing organisms, a specialized form of cell division called meiosis produces cells called gametes or germ cells that are haploid, or contain only one copy of each gene. The gametes produced by females are called eggs or ova, and those produced by males are called sperm. Two gametes fuse to form a diploid fertilized egg, a single cell that has two sets of genes, with one copy of each gene from the mother and one from the father.

During the process of meiotic cell division, an event called genetic recombination or *crossing-over* can sometimes occur, in which a length of DNA on one chromatid is swapped with a length of DNA on the corresponding homologous non-sister chromatid. This can result in reassortment of otherwise linked alleles. The Mendelian principle of independent assortment asserts that each of a parent's two genes for each trait will sort independently into gametes; which allele an organism inherits for one trait is unrelated to which allele it inherits for another trait. This is in fact only true for genes that do not reside on the same chromosome, or are located very far from one another on the same chromosome. The closer two genes lie on the same chromosome, the more closely they will be associated in gametes and the more often they will appear together; genes that are very close are essentially never separated because it is extremely unlikely that a crossover point will occur between them. This is known as genetic linkage.

Molecular Evolution

Mutation

DNA replication is for the most part extremely accurate, however errors (mutations) do occur. The error rate in eukaryotic cells can be as low as 10^{-8} per nucleotide per replication, whereas for some RNA viruses it can be as high as 10^{-3}. This means that each generation, each human genome accumulates 1–2 new mutations. Small mutations can be caused by DNA replication and the aftermath of DNA damage and include point mutations in which a single base is altered and frameshift mutations in which a single base is inserted or deleted. Either of these mutations can change the gene by missense (change a codon to encode a different amino acid) or nonsense (a premature stop codon). Larger mutations can be caused by errors in recombination to cause chromosomal abnormalities including the duplication, deletion, rearrangement or inversion of large sections of a chromosome. Additionally, DNA repair mechanisms can introduce mutational errors when repairing physical damage to the molecule. The repair, even with mutation, is more important to survival than restoring an exact copy, for example when repairing double-strand breaks.

When multiple different alleles for a gene are present in a species's population it is called polymorphic. Most different alleles are functionally equivalent, however some alleles can give rise to different phenotypic traits. A gene's most common allele is called the wild type, and rare alleles are called mutants. The genetic variation in relative frequencies of different alleles in a population is due to both natural selection and genetic drift. The wild-type allele is not necessarily the ancestor of less common alleles, nor is it necessarily fitter.

Most mutations within genes are neutral, having no effect on the organism's phenotype (silent mutations). Some mutations do not change the amino acid sequence because multiple codons encode the same amino acid (synonymous mutations). Other mutations can be neutral if they lead to amino acid sequence changes, but the protein still functions similarly with the new amino acid (e.g. conservative mutations). Many mutations, however, are deleterious or even lethal, and are removed from populations by natural selection. Genetic disorders are the result of deleterious mutations and can be due to spontaneous mutation in the affected individual, or can be inherited. Finally, a small fraction of mutations are beneficial, improving the organism's fitness and are extremely important for evolution, since their directional selection leads to adaptive evolution.

Sequence Homology

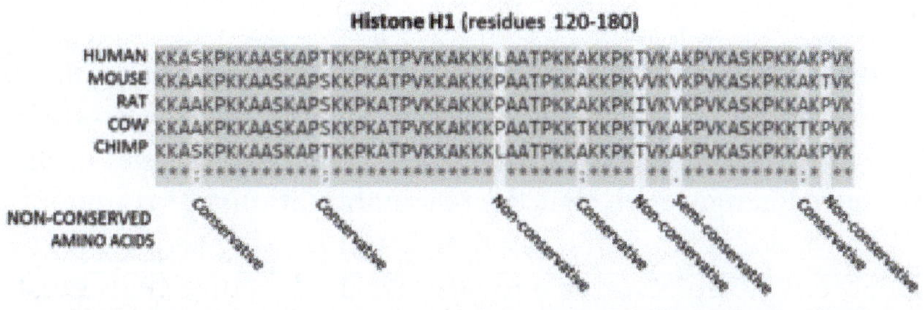

A sequence alignment, produced by ClustalO, of mammalian histone proteins

Genes with a most recent common ancestor, and thus a shared evolutionary ancestry, are known as homologs. These genes appear either from gene duplication within an organism's genome, where they are known as paralogous genes, or are the result of divergence of the genes after a speciation event, where they are known as orthologous genes, and often perform the same or similar functions in related organisms. It is often assumed that the functions of orthologous genes are more similar than those of paralogous genes, although the difference is minimal.

The relationship between genes can be measured by comparing the sequence alignment of their DNA. The degree of sequence similarity between homologous genes is called conserved sequence. Most changes to a gene's sequence do not affect its function and so genes accumulate mutations over time by neutral molecular evolution. Additionally, any selection on a gene will cause its sequence to diverge at a different rate. Genes under stabilizing selection are constrained and so change more slowly whereas genes under directional selection change sequence more rapidly. The sequence differences between genes can be used for phylogenetic analyses to study how those genes have evolved and how the organisms they come from are related.

Origins of New Genes

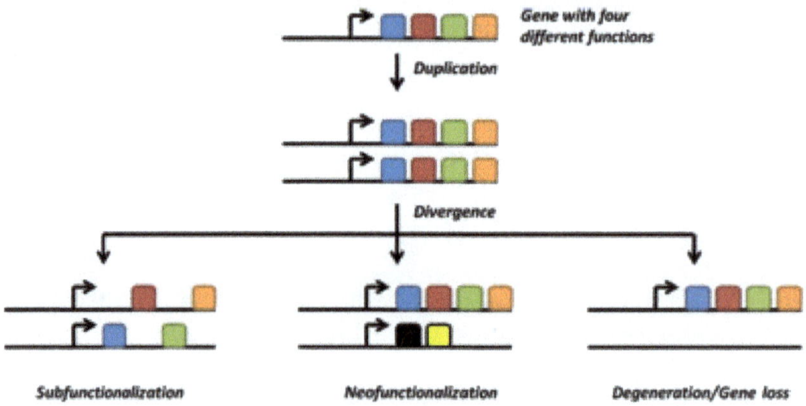

Evolutionary fate of duplicate genes

The most common source of new genes in eukaryotic lineages is gene duplication, which creates copy number variation of an existing gene in the genome. The resulting genes (paralogs) may then diverge in sequence and in function. Sets of genes formed in this way comprise a gene family. Gene duplications and losses within a family are common and represent a major source of evolutionary biodiversity. Sometimes, gene duplication may result in a nonfunctional copy of a gene, or a functional copy may be subject to mutations that result in loss of function; such nonfunctional genes are called pseudogenes.

"Orphan" genes, whose sequence shows no similarity to existing genes, are less common than gene duplicates. Estimates of the number of genes with no homologs outside humans range from 18 to 60. Two primary sources of orphan protein-coding genes are gene duplication followed by extremely rapid sequence change, such that the original relationship is undetectable by sequence comparisons, and de novo conversion of a previously non-coding sequence into a protein-coding gene. De novo genes are typically shorter and simpler in structure than most eukaryotic genes, with few if any introns. Over long evolutionary time periods, de novo gene birth may be responsible for a significant fraction of taxonomically-restricted gene families.

Horizontal gene transfer refers to the transfer of genetic material through a mechanism other than reproduction. This mechanism is a common source of new genes in prokaryotes, sometimes thought to contribute more to genetic variation than gene duplication. It is a common means of spreading antibiotic resistance, virulence, and adaptive metabolic functions. Although horizontal gene transfer is rare in eukaryotes, likely examples have been identified of protist and alga genomes containing genes of bacterial origin.

Genome

The genome is the total genetic material of an organism and includes both the genes and non-coding sequences.

Number of Genes

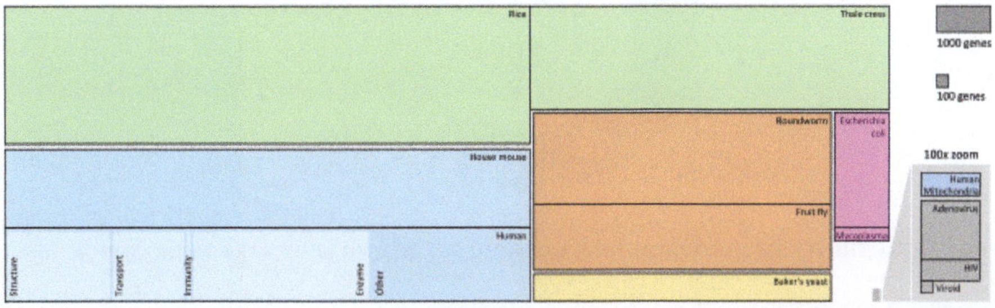

Representative genome sizes for plants (green), vertebrates (blue), invertebrates (red), fungus (yellow), bacteria (purple), and viruses (grey). An inset on the right shows the smaller genomes expanded 100-fold.

The genome size, and the number of genes it encodes varies widely between organisms. The smallest genomes occur in viruses (which can have as few as 2 protein-coding genes), and viroids (which act as a single non-coding RNA gene). Conversely, plants can have extremely large genomes, with rice containing >46,000 protein-coding genes. The total number of protein-coding genes (the Earth's proteome) is estimated to be 5 million sequences.

Although the number of base-pairs of DNA in the human genome has been known since the 1960s, the estimated number of genes has changed over time as definitions of genes, and methods of detecting them have been refined. Initial theoretical predictions of the number of human genes were as high as 2,000,000. Early experimental measures indicated there to be 50,000–100,000 *transcribed* genes (expressed sequence tags). Subsequently, the sequencing in the Human Genome Project indicated that many of these transcripts were alternative variants of the same genes, and the total number of protein-coding genes was revised down to ~20,000 with 13 genes encoded on the mitochondrial genome. Of the human genome, only 1–2% consists of protein-coding genes, with the remainder being 'noncoding' DNA such as introns, retrotransposons, and noncoding RNAs. Every multicellular organism has all its genes in each cell of its body but not every gene functions in every cell .

Essential Genes

Essential genes are the set of genes thought to be critical for an organism's survival. This definition assumes the abundant availability of all relevant nutrients and the absence of environmental stress. Only a small portion of an organism's genes are essential. In bacteria, an estimated 250–400 genes are essential for *Escherichia coli* and *Bacillus subtilis*, which is less than 10% of their genes. Half of these genes are orthologs in both organisms and are largely involved in protein synthesis. In the budding yeast *Saccharomyces cerevisiae* the number of essential genes is slightly higher, at 1000 genes (~20% of their genes). Although the number is more difficult to measure in higher eukaryotes, mice and humans are estimated to have around 2000 essential genes (~10% of their genes). The synthetic organism, *Syn 3*, has a minimal genome of 473 essential genes and quasi-essential genes (necessary for fast growth), although 149 have unknown function.

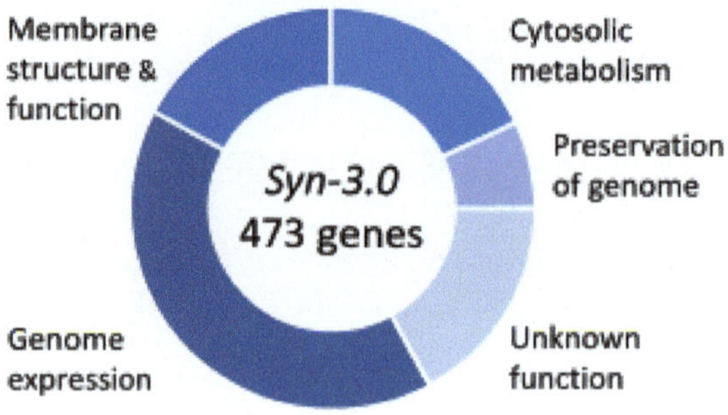

Gene functions in the minimal genome of the synthetic organism, *Syn 3*.

Essential genes include Housekeeping genes (critical for basic cell functions) as well as genes that are expressed at different times in the organisms development or life cycle. Housekeeping genes are used as experimental controls when analysing gene expression, since they are constitutively expressed at a relatively constant level.

Genetic and Genomic Nomenclature

Gene nomenclature has been established by the HUGO Gene Nomenclature Committee (HGNC) for each known human gene in the form of an approved gene name and symbol (short-form abbreviation), which can be accessed through a database maintained by HGNC. Symbols are chosen to be unique, and each gene has only one symbol (although approved symbols sometimes change). Symbols are preferably kept consistent with other members of a gene family and with homologs in other species, particularly the mouse due to its role as a common model organism.

Genetic Engineering

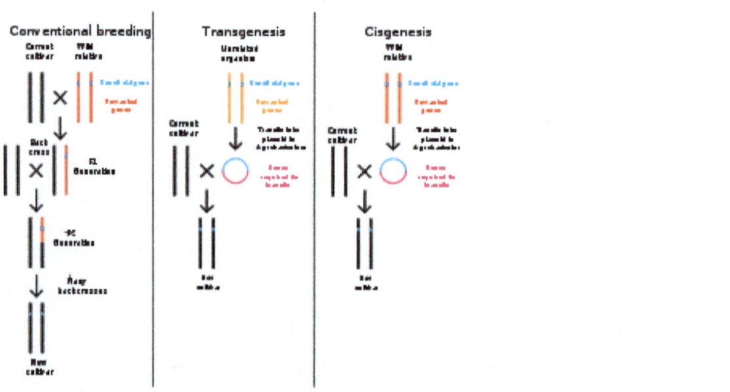

Comparison of conventional plant breeding with transgenic and cisgenic genetic modification.

Genetic engineering is the modification of an organism's genome through biotechnology. Since the 1970s, a variety of techniques have been developed to specifically add, remove and edit genes in an organism. Recently developed genome engineering techniques use engineered nuclease enzymes to create targeted DNA repair in a chromosome to either disrupt or edit a gene when the break is repaired. The related term synthetic biology is sometimes used to refer to extensive genetic engineering of an organism.

Genetic engineering is now a routine research tool with model organisms. For example, genes are easily added to bacteria and lineages of knockout mice with a specific gene's function disrupted are used to investigate that gene's function. Many organisms have been genetically modified for applications in agriculture, industrial biotechnology, and medicine.

For multicellular organisms, typically the embryo is engineered which grows into the adult genetically modified organism. However, the genomes of cells in an adult organism can be edited using gene therapy techniques to treat genetic diseases.

Genome

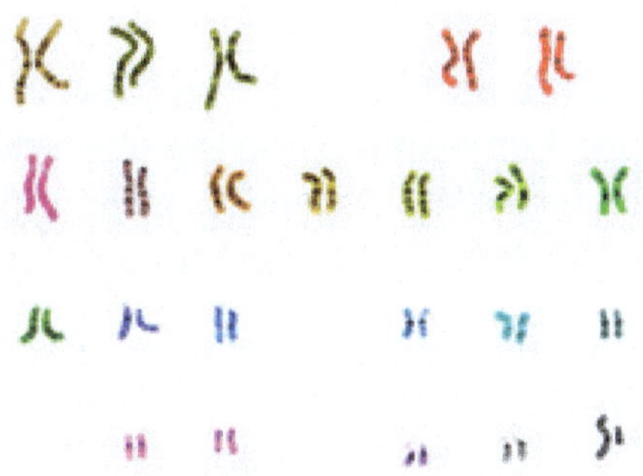

An image of the 46 chromosomes making up the diploid genome of a human male.
(The mitochondrial chromosome is not shown.)

In modern molecular biology and genetics, a genome is the genetic material of an organism. It consists of DNA (or RNA in RNA viruses). The genome includes both the genes (the coding regions), the noncoding DNA and the genetic material of the mitochondria and chloroplasts.

Origin of Term

The term *genome* was created in 1920 by Hans Winkler, professor of botany at the University of Hamburg, Germany. The Oxford Dictionary suggests the name is a blend of the words *gene* and *chromosome*. A few related *-ome* words already existed—such as *biome*, *rhizome*, forming a vocabulary into which *genome* fits systematically.

Overview

Some organisms have multiple copies of chromosomes: diploid, triploid, tetraploid and so on. In classical genetics, in a sexually reproducing organism (typically eukarya) the gamete has half the number of chromosomes of the somatic cell and the genome is a full set of chromosomes in a diploid cell. The halving of the genetic material in gametes is accomplished by the segregation of homologous chromosomes during meiosis. In haploid organisms, including cells of bacteria, archaea, and in organelles including mitochondria and chloroplasts, or viruses, that similarly contain genes, the single or set of circular or linear chains of DNA (or RNA for some viruses), likewise constitute the genome. The term *genome* can be applied specifically to mean what is stored on a complete set of nuclear DNA (i.e., the «nuclear genome") but can also be applied to what is stored within organelles that contain their own DNA, as with the "mitochondrial genome" or the "chloroplast genome". Additionally, the genome can comprise non-chromosomal genetic elements such as viruses, plasmids, and transposable elements.

Typically, when it is said that the genome of a sexually reproducing species has been "sequenced", it refers to a determination of the sequences of one set of autosomes and one of each type of sex chromosome, which together represent both of the possible sexes. Even in species that exist in only one sex, what is described as a "genome sequence" may be a composite read from the chromosomes of various individuals. Colloquially, the phrase "genetic makeup" is sometimes used to signify the genome of a particular individual or organism. The study of the global properties of genomes of related organisms is usually referred to as genomics, which distinguishes it from genetics which generally studies the properties of single genes or groups of genes.

Both the number of base pairs and the number of genes vary widely from one species to another, and there is only a rough correlation between the two (an observation is known as the C-value paradox). At present, the highest known number of genes is around 60,000, for the protozoan causing trichomoniasis, almost three times as many as in the human genome.

An analogy to the human genome stored on DNA is that of instructions stored in a book:

- The book (genome) would contain 23 chapters (chromosomes);
- Each chapter contains 48 to 250 million letters (A,C,G,T) without spaces;
- Hence, the book contains over 3.2 billion letters total;
- The book fits into a cell nucleus the size of a pinpoint;
- At least one copy of the book (all 23 chapters) is contained in most cells of our body. The only exception in humans is found in mature red blood cells which become enucleated during development and therefore lack a genome.

Sequencing and Mapping

In 1976, Walter Fiers at the University of Ghent (Belgium) was the first to establish the complete nucleotide sequence of a viral RNA-genome (Bacteriophage MS2). The next year Fred Sanger completed the first DNA-genome sequence: Phage Φ-X174, of 5386 base pairs. The first complete genome sequences among all three domains of life were released within a short period during the mid-1990s: The first bacterial genome to be sequenced was that of Haemophilus influenzae, completed by a team at The Institute for Genomic Research in 1995. A few months later, the first eukaryotic genome was completed, with sequences of the 16 chromosomes of budding yeast *Saccharomyces cerevisiae* published as the result of a European-led effort begun in the mid-1980s. The first genome sequence for an archaeon, *Methanococcus jannaschii*, was completed in 1996, again by The Institute for Genomic Research.

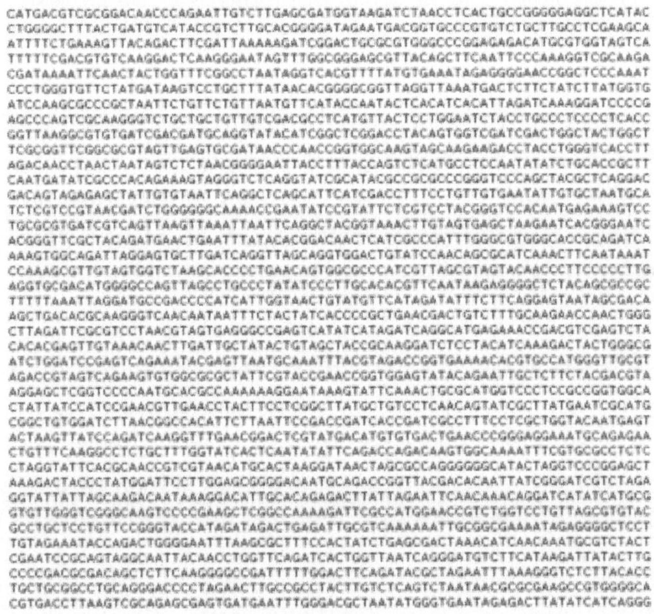

Part of DNA sequence - prototypification of complete genome of virus

The development of new technologies has made it dramatically easier and cheaper to do sequencing, and the number of complete genome sequences is growing rapidly. The US National Institutes of Health maintains one of several comprehensive databases of genomic information. Among the thousands of completed genome sequencing projects include those for rice, a mouse, the plant *Arabidopsis thaliana*, the puffer fish, and the bacteria E. coli. In December 2013, scientists first sequenced the entire *genome* of a Neanderthal, an extinct species of humans. The genome was extracted from the toe bone of a 130,000-year-old Neanderthal found in a Siberian cave.

New sequencing technologies, such as massive parallel sequencing have also opened up the prospect of personal genome sequencing as a diagnostic tool, as pioneered by Manteia Predictive Medicine. A major step toward that goal was the completion in 2007 of the full genome of James D. Watson, one of the co-discoverers of the structure of DNA.

Whereas a genome sequence lists the order of every DNA base in a genome, a genome map identifies the landmarks. A genome map is less detailed than a genome sequence and aids in navigating around the genome. The Human Genome Project was organized to map and to sequence the human genome. A fundamental step in the project was the release of a detailed genomic map by Jean Weissenbach and his team at the Genoscope in Paris.

Reference genome sequences and maps continue to be updated, removing errors and clarifying regions of high allelic complexity. The decreasing cost of genomic mapping has permitted genealogical sites to offer it as a service, to the extent that one may submit one's genome to crowd sourced scientific endeavours such as DNA.land at the New York Genome Center, an example both of the economies of scale and of citizen science.

Genome Compositions

Genome composition is used to describe the make up of contents of a haploid genome, which should include genome size, proportions of non-repetitive DNA and repetitive DNA in details. By comparing the genome compositions between genomes, scientists can better understand the evolutionary history of a given genome.

When talking about genome composition, one should distinguish between prokaryotes and eukaryotes as there are significant differences with contents structure. In prokaryotes, most of the genome (85–90%) is non-repetitive DNA, which means coding DNA mainly forms it, while non-coding regions only take a small part. On the contrary, eukaryotes have the feature of exon-intron organization of protein coding genes; the variation of repetitive DNA content in eukaryotes is also extremely high. In mammals and plants, the major part of the genome is composed of repetitive DNA.

Most biological entities that are more complex than a virus sometimes or always carry additional genetic material besides that which resides in their chromosomes. In some contexts, such as sequencing the genome of a pathogenic microbe, "genome" is meant to include information stored on this auxiliary material, which is carried in plasmids. In such circumstances then, "genome" describes all of the genes and information on non-coding DNA that have the potential to be present.

In eukaryotes such as plants, protozoa and animals, however, "genome" carries the typical connotation of only information on chromosomal DNA. So although these organisms contain chloroplasts or mitochondria that have their own DNA, the genetic information contained in DNA within these organelles is not considered part of the genome. In fact, mitochondria are sometimes said to have their own genome often referred to as the "mitochondrial genome". The DNA found within the chloroplast may be referred to as the "plastome".

Genome Size

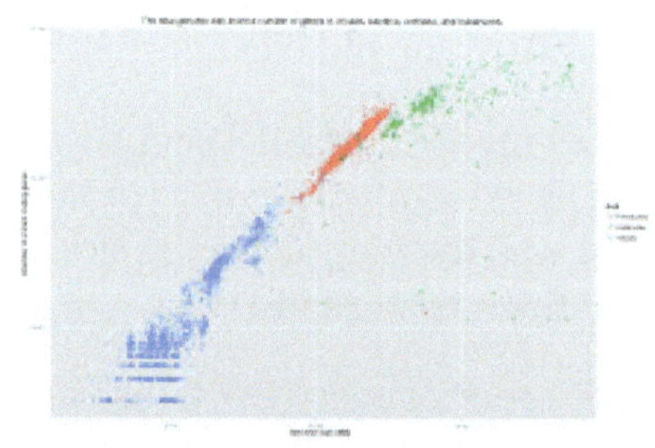

Log-log plot of the total number of annotated proteins in genomes submitted to GenBank as a function of genome size.

Genome size is the total number of DNA base pairs in one copy of a haploid genome. In humans, the nuclear genome comprises approximately 3.2 billion nucleotides of DNA, divided into 24 linear molecules, the shortest 50 000 000 nucleotides in length and the longest 260 000 000 nucleotides, each contained in a different chromosome. The genome size is positively correlated with the morphological complexity among prokaryotes and lower eukaryotes; however, after mollusks and all the other higher eukaryotes above, this correlation is no longer effective. This phenomenon also indicates the mighty influence coming from repetitive DNA act on the genomes.

Since genomes are very complex, one research strategy is to reduce the number of genes in a genome to the bare minimum and still have the organism in question survive. There is experimental work being done on minimal genomes for single cell organisms as well as minimal genomes for multi-cellular organisms. The work is both *in vivo* and *in silico*.

Here is a table of some significant or representative genomes.

Organism type	Organism	Genome size (base pairs)		Approx. no. of genes	Note
Virus	Porcine circovirus type 1	1,759	1.8kb		Smallest viruses replicating autonomously in eukaryotic cells.
Virus	Bacteriophage MS2	3,569	3.5kb		First sequenced RNA-genome
Virus	SV40	5,224	5.2kb		
Virus	Phage Φ-X174	5,386	5.4kb		First sequenced DNA-genome
Virus	HIV	9,749	9.7kb		
Virus	Phage λ	48,502	48kb		Often used as a vector for the cloning of recombinant DNA.
Virus	Megavirus	1,259,197	1.3Mb		Until 2013 the largest known viral genome.
Virus	*Pandoravirus salinus*	2,470,000	2.47Mb		Largest known viral genome.
Bacterium	*Nasuia deltocephalinicola* (strain NAS-ALF)	112,091	112kb		Smallest non-viral genome.
Bacterium	*Carsonella ruddii*	159,662	160kb		
Bacterium	*Buchnera aphidicola*	600,000	600kb		
Bacterium	*Wigglesworthia glossinidia*	700,000	700Kb		
Bacterium	*Haemophilus influenzae*	1,830,000	1.8Mb		First genome of a living organism sequenced, July 1995
Bacterium	*Escherichia coli*	4,600,000	4.6Mb	4288	
Bacterium	*Solibacter usitatus* (strain Ellin 6076)	9,970,000	10Mb		

Organism type	Organism	Genome size (base pairs)		Approx. no. of genes	Note
Bacterium – cyanobacterium	*Prochlorococcus* spp. (1.7 Mb)	1,700,000	1.7Mb	1884	Smallest known cyanobacterium genome
Bacterium – cyanobacterium	*Nostoc punctiforme*	9,000,000	9Mb	7432	7432 "open reading frames"
Amoeboid	*Polychaos dubium* (*"Amoeba" dubia*)	670,000,000,000	670Gb		Largest known genome. (Disputed)
Plant	*Genlisea tuberosa*	61,000,000	61Mb		Smallest recorded flowering plant genome, 2014.
Plant	*Arabidopsis thaliana*	135,000,000	135 Mb	27,655	First plant genome sequenced, December 2000.
Plant	*Populus trichocarpa*	480,000,000	480Mb	73013	First tree genome sequenced, September 2006
Plant	*Fritillaria assyrica*	130,000,000,000	130Gb		
Plant	*Paris japonica* (Japanese-native, pale-petal)	150,000,000,000	150Gb		Largest plant genome known
Plant – moss	*Physcomitrella patens*	480,000,000	480Mb		First genome of a bryophyte sequenced, January 2008.
Fungus – yeast	*Saccharomyces cerevisiae*	12,100,000	12.1Mb	6294	First eukaryotic genome sequenced, 1996
Fungus	*Aspergillus nidulans*	30,000,000	30Mb	9541	
Nematode	*Pratylenchus coffeae*	20,000,000	20Mb		Smallest animal genome known
Nematode	*Caenorhabditis elegans*	100,300,000	100Mb	19000	First multicellular animal genome sequenced, December 1998
Insect	*Drosophila melanogaster* (fruit fly)	175,000,000	175Mb	13600	Size variation based on strain (175-180Mb; standard *y w* strain is 175Mb)
Insect	*Apis mellifera* (honey bee)	236,000,000	236Mb	10157	)
Insect	*Bombyx mori* (silk moth)	432,000,000	432Mb	14623	14,623 predicted genes
Insect	*Solenopsis invicta* (fire ant)	480,000,000	480Mb	16569	

Organism type	Organism	Genome size (base pairs)		Approx. no. of genes	Note
Mammal	*Mus musculus*	2,700,000,000	2.7Gb	20210	
Mammal	*Homo sapiens*	3,289,000,000	3.3Gb	20000	*Homo sapiens* estimated genome size 3.2 billion bp Initial sequencing and analysis of the human genome
Mammal	*Bonobo*	3,286,640,000	3.3Gb	20000	*Pan paniscus* estimated genome size 3.29 billion bp
Fish	*Tetraodon nigroviridis* (type of puffer fish)	385,000,000	390Mb		Smallest vertebrate genome known estimated to be 340 Mb – 385 Mb.
Fish	*Protopterus aethiopicus* (marbled lungfish)	130,000,000,000	130Gb		Largest vertebrate genome known

Proportion of Non-repetitive DNA

The proportion of non-repetitive DNA is calculated by using the length of non-repetitive DNA divided by genome size. Protein-coding genes and RNA-coding genes are generally non-repetitive DNA. A bigger genome does not mean more genes, and the proportion of non-repetitive DNA decreases along with increasing genome size in higher eukaryotes.

It had been found that the proportion of non-repetitive DNA can vary a lot between species. Some *E. coli* as prokaryotes only have non-repetitive DNA, lower eukaryotes such as *C. elegans* and fruit fly, still possess more non-repetitive DNA than repetitive DNA. Higher eukaryotes tend to have more repetitive DNA than non-repetitive ones. In some plants and amphibians, the proportion of non-repetitive DNA is no more than 20%, becoming a minority component.

Proportion of Repetitive DNA

The proportion of repetitive DNA is calculated by using length of repetitive DNA divide by genome size. There are two categories of repetitive DNA in genome: tandem repeats and interspersed repeats.

Tandem Repeats

Tandem repeats are usually caused by slippage during replication, unequal crossing-over and gene conversion, satellite DNA and microsatellites are forms of tandem repeats in the genome. Although tandem repeats count for a significant proportion in genome, the largest proportion in mammalian is the other type, interspersed repeats.

Interspersed Repeats

Interspersed repeats mainly come from transposable elements (TEs), but they also include some protein coding gene families and pseudogenes. Transposable elements are able to integrate into the genome at another site within the cell. It is believed that TEs are an important driving force on genome evolution of higher eukaryotes. TEs can be classified into two categories, Class 1 (retrotransposons) and Class 2 (DNA transposons).

Retrotransposons

Retrotransposons can be transcribed into RNA, which are then duplicated at another site into the genome. Retrotransposons can be divided into Long terminal repeats (LTRs) and Non-Long Terminal Repeats (Non-LTR).

Long terminal repeats (LTRs)

similar to retroviruses, which have both gag and pol genes to make cDNA from RNA and proteins to insert into genome, but LTRs can only act within the cell as they lack the env gene in retroviruses. It has been reported that LTRs consist of the largest fraction in most plant genome and might account for the huge variation in genome size.

Non-long terminal repeats (Non-LTRs)

can be divided into long interspersed elements (LINEs), short interspersed elements (SINEs) and Penelope-like elements. In *Dictyostelium discoideum*, there is another DIRS-like elements belong to Non-LTRs. Non-LTRs are widely spread in eukaryotic genomes.

Long interspersed elements (LINEs)

are able to encode two Open Reading Frames (ORFs) to generate transcriptase and endonuclease, which are essential in retrotransposition. The human genome has around 500,000 LINEs, taking around 17% of the genome.

Short interspersed elements (SINEs)

are usually less than 500 base pairs and need to co-opt with the LINEs machinery to function as nonautonomous retrotransposons. The Alu element is the most common SINEs found in primates, it has a length of about 350 base pairs and takes about 11% of the human genome with around 1,500,000 copies.

DNA Transposons

DNA transposons generally move by "cut and paste" in the genome, but duplication has also been observed. Class 2 TEs do not use RNA as intermediate and are popular in bacteria, in metazoan it has also been found.

Genome Evolution

Genomes are more than the sum of an organism's genes and have traits that may be measured and studied without reference to the details of any particular genes and their products. Researchers compare traits such as karyotype (chromosome number), genome size, gene order, codon usage bias, and GC-content to determine what mechanisms could have produced the great variety of genomes that exist today.

Duplications play a major role in shaping the genome. Duplication may range from extension of short tandem repeats, to duplication of a cluster of genes, and all the way to duplication of entire chromosomes or even entire genomes. Such duplications are probably fundamental to the creation of genetic novelty.

Horizontal gene transfer is invoked to explain how there is often an extreme similarity between small portions of the genomes of two organisms that are otherwise very distantly related. Horizontal gene transfer seems to be common among many microbes. Also, eukaryotic cells seem to have experienced a transfer of some genetic material from their chloroplast and mitochondrial genomes to their nuclear chromosomes.

Translation

- Translation is the first stage of protein biosynthesis.
- During translation, the genetic code in mRNA is read and converted into protein by means of the protein synthesizing machinery comprising of ribosomes, tRNA, amino acids, and a number of enzymes/ proteins.
- Translation occurs in the cell cytoplasm, where the large and small subunits of the ribosome bind to the mRNA.
- The ribosome facilitates decoding by inducing the binding of tRNAs (anticodon) with complementary sequences of mRNA (codon).
- The t RNAs carry specific amino acids that are chained together into a polypeptide as the mRNA is "read" by the ribosome.

Transfer RNA (tRNA)

- tRNA is a small RNA chain, about 80 nucleotides long ,that transfers a specific amino acid to a growing polypeptide chain at the ribosomal site of protein synthesis during translation.
- It has sites for amino acid attachment and an anticodon region for codon recognition.
- One end of the molecule attached to a specific amino acid.
- The other end has an exposed sequence of 3-bases. These are called the anticodon.
- There are different tRNA molecule for each of the possible triplets meaning 64 anticodons. There is at least one t RNA for each aminoacids.
- The anticodons of the tRNAs each have a complimentary codon in the mRNA. For example the codon GCC would be the compliment of the anticodon CGG.

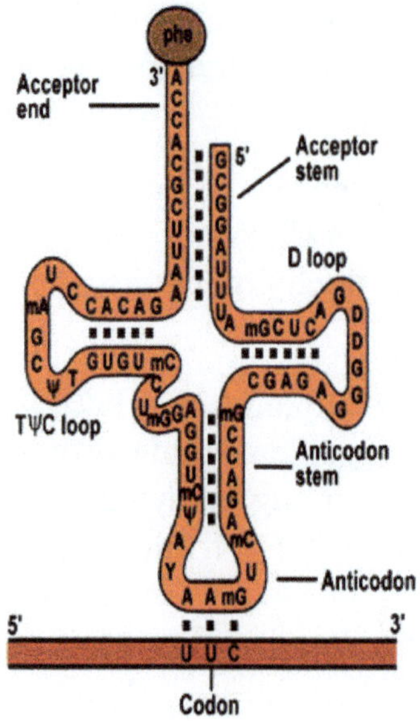

Ribosomal RNA(rRNA)

There are 4 kinds of ribosomal RNAs :

In eukaryotes, these are 18S rRNA. One of these molecules, along with different protein molecules, constitute the small subunit of the ribosome.

The other three are 28S, 5.8S, and 5S rRNA. One each o f these molecules, along with different proteins, form the large subunit of the ribosome.

rRNA (ribosomal RNA) - associates with proteins to form ribosomes.

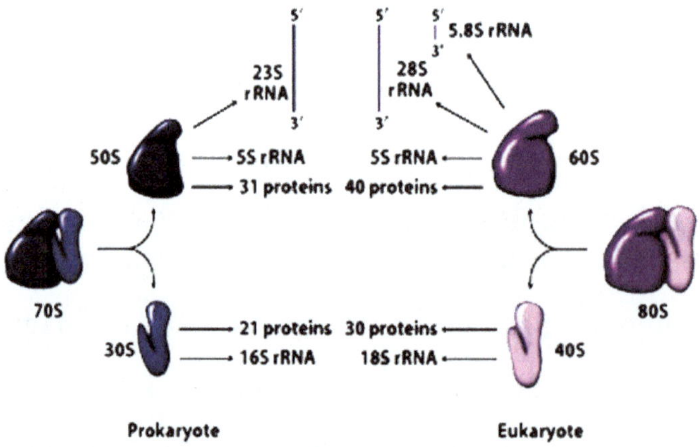

The Role of Ribosomes

- Ribosomes are the 'decoding' units of the cell.
- Each ribosome consists of two subunits, and is an assemblage of risosomal RNA(rRNA) and proteins. Subunits are separate in the cytoplasm, but join during protein synthesis (translation).

- Ribosomes have binding sites for both tRNA and mRNA molecules.
- The ribosome has two primary functions :

  -> decode the mRNA.

  -> form peptide bonds.

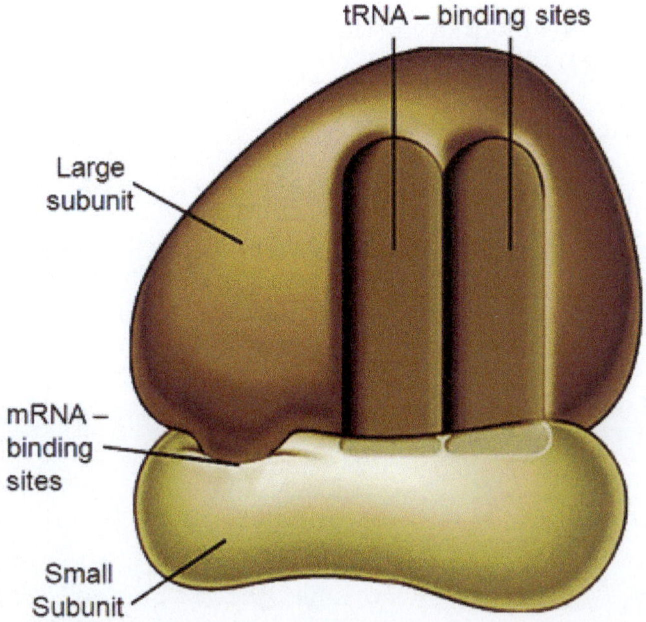

Translation, like replication and transcription also occurs in three steps of Initiation, Elongation and Termination

Initiation :- A ribosome binds to the mRNA and starts to code at the FMet codon (START codon: AUG, GUG or UUG).

Elongation :- tRNA brings the corresponding amino acid to each codon as the ribosome moves down the mRNA strand.

Termination :- Reading of the final mRNA codon (STOP codon: UAA, UAG, UGA), which ends the synthesis of the peptide chain and releases it.

Translation Process

Activation of the Amino acid

- Prior to initiation, attachment of the correct amino acid to the tRNA takes place, utilizing ATP as energy.
- This reaction occurs in the cytoplasm.
- The enzyme involved in this reaction is aminoacyl tRNA synthetases. It is a Mg^{2+} dependent enzyme.
- Upon attachment of the amino acid, the tRNA becomes activated or charged.

Initiation

- There is a specific sequence of bases on the mRNA, upstream of the start codon, that allows the ribosome to recognize and bind onto the start site to initiate the translation process.
- Such sequences are referred to as the Shine Dalgarno sequences.
- It begins with an initiation complex made up ribosomes, 3 initiation factors (IF-1, IF-2, IF-3), mRNA and formyl methionine tRNA. The formyl methionine is delivered to the initiation site on the mRNA, codon AUG.
- Role of initiation factors :

  -> IF-1 prevents premature binding of tRNAs onto the mRNA template strand.

  -> IF-2 facilitates binding of formyl methionine tRNA to 30S ribosomal unit.

  -> IF-3 prevents premature binding of the larger 50S ribosomal subunit with its 30S counterpart.

- Formation of the initiation complex requires hydrolysis of GTP for energy.

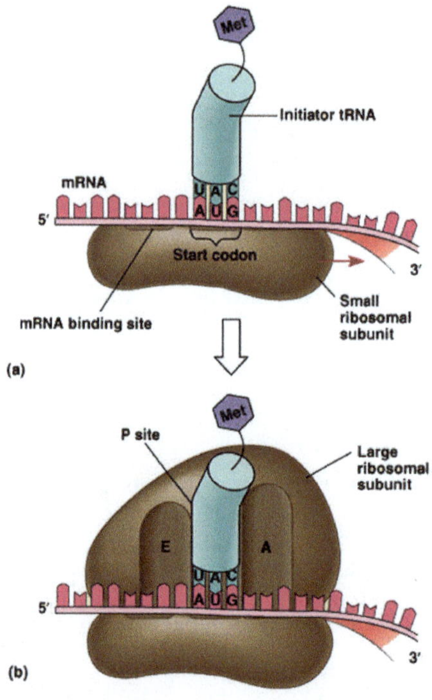

Elongation

- The ribosome moves along mRNA; polypeptide lengthens by covalent attachment or successive amino acids, each of which is carried to the ribosome by their respective charged/ activated tRNA and which base pairs correctly with its corresponding codon identified in the mRNA.
- Fmet-tRNA becomes P site (peptide) and new tRNA binds at A site.

- Elongation requires proteins called EF-Tu, EF-Ts, EF-G (elongation factors) found in the cytoplasm.
- Peptide bond forms at P site between amino acids; GTP hydrolysed for energy.
- Peptidyl transferase is the enzymatic component of the ribosome which forms peptide bonds between amino acids.

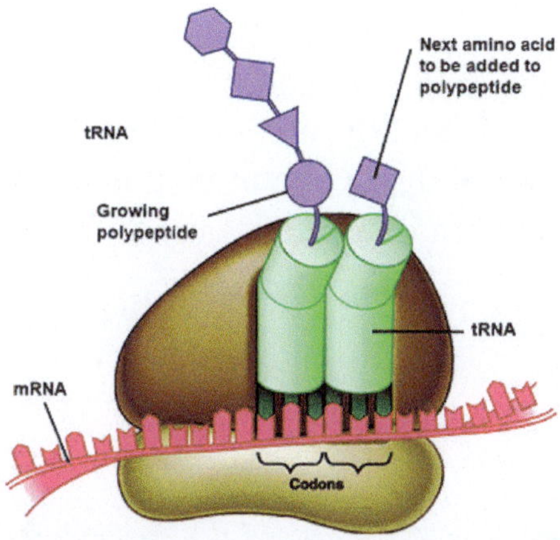

- Translocation (movement of ribosome in 5' --> 3' on mRNA).
- Initiation tRNA to E site, second tRNA to P site, and a 3rd tRNA binds to A site; assisted by elongation factor ET-G.
- "Empty" tRNA at E site leaves.
- GTP is hydrolysed for energy.
- Elongation continues in same manner.

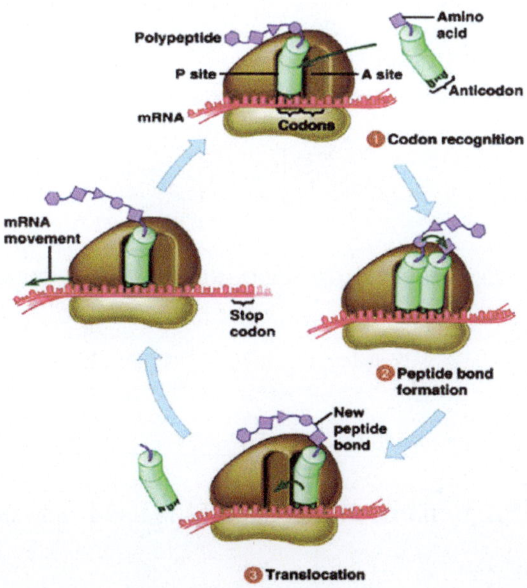

Termination and Release

- Stop codon : Termination or completion of the growing polypeptide chain occurs when a nonsense (stop) codon is encountered.
- tRNA does not bind to a stop codon. Proteins termed as termination or release factors (RF-1, RF-2 and RF-3) recognize this chain terminating signal and cleave the completed polypeptide from the terminal tRNA.
- Release factor binds to stop codon and adds H_2O to the peptide end (p site) and releases the newly synthesised peptide and the empty tRNA. This is followed by dissociation of the ribosome (70S) into its 30S and 50S subunits; a new cycle of polypeptide synthesis may now begin.

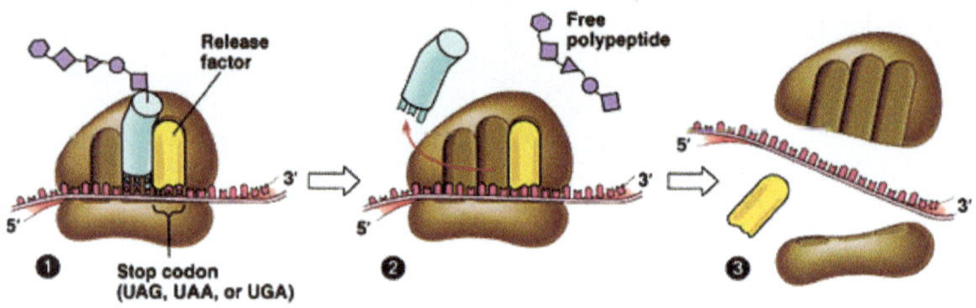

- Most of the structure of the protein that we see is usually the one that comes out of the ribosome after protein synthesis.

Post Translational Modifications

The newly synthesized polypeptide chain further undergoes modification to achieve its biologically active form. It involves :

- Folding of the polypeptide into correct three dimensional form. For this, they need chaperone proteins and ATP for assistance.
- Specific enzymes and cofactors for removal of sequences/ residues, addition of phosphate, carbohydrate, methyl, carboxyl or prosthetic groups, modification of certain terminal residues.

Transcription vs. Translation

Transcription

- Process by which genetic information encoded in DNA is copied onto messenger RNA.
- Occurs in the nucleus.
- DNA ----> mRNA.

Translation

- Process by which information encoded in mRNA is used to assemble a protein at a ribosome.
- Occurs on a ribosome.
- mRNA ----> protein.

References

- "What kinds of gene mutations are possible?". Genetics Home Reference. United States National Library of Medicine. 11 May 2015. Retrieved 19 May 2015
- Russell JB, Jarvis GN (2001). "Practical mechanisms for interrupting the oral-fecal lifecycle of Escherichia coli". Journal of Molecular Microbiology and Biotechnology. 3 (2): 265–72. PMID 11321582
- Williams, George C. (2001). Adaptation and Natural Selection a Critique of Some Current Evolutionary Thought. ([Online-Ausg.]. ed.). Princeton: Princeton University Press. ISBN 9781400820108
- Stanley N. Cohen; Annie C. Y. Chang (1 May 1973). "Recircularization and Autonomous Replication of a Sheared R-Factor DNA Segment in Escherichia coli Transformants". PNAS. Retrieved 17 July 2010
- Darnton NC, Turner L, Rojevsky S, Berg HC (Mar 2007). "On torque and tumbling in swimming Escherichia coli". Journal of Bacteriology. 189 (5): 1756–64. doi:10.1128/JB.01501-06. PMC 1855780. PMID 17189361
- Feng P; Weagant S; Grant, M (2002-09-01). "Enumeration of Escherichia coli and the Coliform Bacteria". Bacteriological Analytical Manual (8th ed.). FDA/Center for Food Safety & Applied Nutrition. Retrieved 2007-01-25
- Todar, K. "Pathogenic E. coli". Online Textbook of Bacteriology. University of Wisconsin–Madison Department of Bacteriology. Retrieved 2007-11-30
- Tortora, Gerard (2010). Microbiology: An Introduction. San Francisco, CA: Benjamin Cummings. pp. 85–87, 161, 165,. ISBN 0-321-55007-2
- Marande W, Burger G (October 2007). "Mitochondrial DNA as a genomic jigsaw puzzle". Science. AAAS. 318 (5849): 415. Bibcode:2007Sci...318..415M. doi:10.1126/science.1148033. PMID 17947575
- Evans Jr., Doyle J.; Dolores G. Evans. "Escherichia Coli". Medical Microbiology, 4th edition. The University of Texas Medical Branch at Galveston. Archived from the original on 2007-11-02. Retrieved 2007-12-02.
- Guerzoni, D; McLysaght, A (November 2011). "De novo origins of human genes.". PLOS Genetics. 7 (11): e1002381. doi:10.1371/journal.pgen.1002381. PMC 3213182. PMID 22102832.
- Deng C (2007). "In celebration of Dr. Mario R. Capecchi's Nobel Prize". International Journal of Biological Sciences. 3 (7): 417–419. doi:10.7150/ijbs.3.417. PMC 2043165. PMID 17998949
- Farrar J, Hotez P, Junghanss T, Kang G, Lalloo D, White NJ, eds. (2013). Manson's Tropical Diseases (23rd ed.). Oxford: Elsevier/Saunders. ISBN 9780702053061
- Pennacchio, L. A.; Bickmore, W.; Dean, A.; Nobrega, M. A.; Bejerano, G. (2013). "Enhancers: Five essential questions". Nature Reviews Genetics. 14 (4): 288–95. doi:10.1038/nrg3458. PMID 23503198
- Summers, Rebecca (24 April 2013) Bacteria churn out first ever petrol-like biofuel New Scientist, Retrieved 27 April 2013
- Frederick R. Blattner; Guy Plunkett III; et al. (1997). "The Complete Genome Sequence of Escherichia coli K-12". Science. 277 (5331): 1453–1462. doi:10.1126/science.277.5331.1453. PMID 9278503
- Judson, Horace (1979). The Eighth Day of Creation: Makers of the Revolution in Biology. Cold Spring Harbor Laboratory Press. pp. 51–169. ISBN 0-87969-477-7

- Stanley N. Cohen; Annie C. Y. Chang (1 May 1973). "Recircularization and Autonomous Replication of a Sheared R-Factor DNA Segment in Escherichia coli Transformants". PNAS. Retrieved 17 July 2010

Permissions

All chapters in this book are published with permission under the Creative Commons Attribution Share Alike License or equivalent. Every chapter published in this book has been scrutinized by our experts. Their significance has been extensively debated. The topics covered herein carry significant information for a comprehensive understanding. They may even be implemented as practical applications or may be referred to as a beginning point for further studies.

We would like to thank the editorial team for lending their expertise to make the book truly unique. They have played a crucial role in the development of this book. Without their invaluable contributions this book wouldn't have been possible. They have made vital efforts to compile up to date information on the varied aspects of this subject to make this book a valuable addition to the collection of many professionals and students.

This book was conceptualized with the vision of imparting up-to-date and integrated information in this field. To ensure the same, a matchless editorial board was set up. Every individual on the board went through rigorous rounds of assessment to prove their worth. After which they invested a large part of their time researching and compiling the most relevant data for our readers.

The editorial board has been involved in producing this book since its inception. They have spent rigorous hours researching and exploring the diverse topics which have resulted in the successful publishing of this book. They have passed on their knowledge of decades through this book. To expedite this challenging task, the publisher supported the team at every step. A small team of assistant editors was also appointed to further simplify the editing procedure and attain best results for the readers.

Apart from the editorial board, the designing team has also invested a significant amount of their time in understanding the subject and creating the most relevant covers. They scrutinized every image to scout for the most suitable representation of the subject and create an appropriate cover for the book.

The publishing team has been an ardent support to the editorial, designing and production team. Their endless efforts to recruit the best for this project, has resulted in the accomplishment of this book. They are a veteran in the field of academics and their pool of knowledge is as vast as their experience in printing. Their expertise and guidance has proved useful at every step. Their uncompromising quality standards have made this book an exceptional effort. Their encouragement from time to time has been an inspiration for everyone.

The publisher and the editorial board hope that this book will prove to be a valuable piece of knowledge for students, practitioners and scholars across the globe.

Index

M

N

O

P

Q

R

S

T

V

W

Z